BOCHNER–RIESZ MEANS ON EUCLIDEAN SPACES

BOCHNER–RIESZ MEANS ON EUCLIDEAN SPACES

Shanzhen Lu
Beijing Normal University, China

Dunyan Yan
University of Chinese Academy of Sciences, China

Published by

World Scientific Publishing Co. Pte. Ltd.

5 Toh Tuck Link, Singapore 596224

USA office: 27 Warren Street, Suite 401-402, Hackensack, NJ 07601

UK office: 57 Shelton Street, Covent Garden, London WC2H 9HE

British Library Cataloguing-in-Publication Data
A catalogue record for this book is available from the British Library.

BOCHNER–RIESZ MEANS ON EUCLIDEAN SPACES

Copyright © 2013 by World Scientific Publishing Co. Pte. Ltd.

All rights reserved. This book, or parts thereof, may not be reproduced in any form or by any means, electronic or mechanical, including photocopying, recording or any information storage and retrieval system now known or to be invented, without written permission from the Publisher.

For photocopying of material in this volume, please pay a copying fee through the Copyright Clearance Center, Inc., 222 Rosewood Drive, Danvers, MA 01923, USA. In this case permission to photocopy is not required from the publisher.

ISBN 978-981-4458-76-4

In-house Editor: Angeline Fong

Printed in Singapore

Contents

Preface   vii

1   AN INTRODUCTION TO MULTIPLE FOURIER SERIES   1
    1.1   Basic properties of multiple Fourier series   3
    1.2   Poisson summation formula   10
    1.3   Convergence and the opposite results   15
    1.4   Linear summation   34

2   BOCHNER-RIESZ MEANS OF MULTIPLE FOURIER INTEGRAL   41
    2.1   Localization principle and classic results on fixed-point convergence   41
    2.2 L^p-convergence   45
    2.3   Some basic facts on multipliers   48
    2.4   The disc conjecture and Fefferman theorem   51
    2.5   The L^p-boundedness of Bochner-Riesz operator T_α with $\alpha > 0$ 59
    2.6   Oscillatory integral and proof of Carleson-Sjölin theorem . . .   61
    2.7   Kakeya maximal function   78
    2.8   The restriction theorem of the Fourier transform   89
    2.9   The case of radial functions   97
    2.10   Almost everywhere convergence   105
    2.11   Commutator of Bochner-Riesz operator   129

3   BOCHNER-RIESZ MEANS OF MULTIPLE FOURIER SERIES   141
    3.1   The case of being over the critical index   141
    3.2   The case of the critical index (general discussion)   146
    3.3   The convergence at fixed point   167
    3.4 L^p approximation   177
    3.5   Almost everywhere convergence (the critical index)   194
    3.6   Spaces related to the a.e. convergence of the Fourier series . .   208

3.7	The uniform convergence and approximation	244
3.8	$(C,1)$ means	251
3.9	The saturation problem of the uniform approximation	259
3.10	Strong summation	280

4 THE CONJUGATE FOURIER INTEGRAL AND SERIES 293

4.1	The conjugate integral and the estimate of the kernel	293
4.2	Convergence of Bochner-Riesz means for conjugate Fourier integral	303
4.3	The conjugate Fourier series	309
4.4	Kernel of Bochner-Riesz means of conjugate Fourier series	316
4.5	The maximal operator of the conjugate partial sum	319
4.6	The relations between the conjugate series and integral	324
4.7	Convergence of Bochner-Riesz means of conjugate Fourier series	332
4.8	$(C,1)$ means in the conjugate case	334
4.9	The strong summation of the conjugate Fourier series	337
4.10	Approximation of continuous functions	347

Bibliography   **367**

Index   **375**

Preface

This book mainly concerns with the Bochner-Riesz means of multiple Fourier integral and series, which shall be simply termed the Bochner-Riesz means for short. Bochner-Riesz means is an important branch of multiple Fourier analysis initiated in the 1930s by Bochner. At that time, Bochner and his fellows restricted their studies to the cases when its degree is greater than the critical index. In the late of 1940s, Minde Cheng, Bochner's student, carried out a systematic research about the Bochner-Riesz means. In the late 50s and early 60s, Stein contributed significantly to the research where the degree is at and below the critical index. In the same period, Minde Cheng and his students also started the research on approximation of functions by the Bochner-Riesz means. Since the 1970s, inspired by Stein's work, many researchers in Europe, America, Soviet Union and China have joined the study. A number of important or valuable results are achieved in this field. Although the development in this field took nearly half of a century with many achievements, there still remain many fundamental problems waiting to be resolved. To introduce these results and some remaining problems to Chinese scholars, Shanzhen Lu and Kunyang Wang published a monograph "Bochner-Riesz Means" in Chinese in 1988. We note that there have been some important progresses in this field since then. To supplement the previous book with these important results, we wish to publish a new monograph in English with the present title.

The book is aimed at giving a systematical introduction to fundamental theories of the Bochner-Riesz means and important achievements attained through the latest 50 years. Numerous results illustrate that the Bochner-Riesz means offer the most natural way for the extension of Fourier series from the case of single variable to that of several variables. This book consists of four chapters. The first chapter is a brief introduction to the theory of multiple Fourier series. Chapter 2 and 3 contain main topics of this book which are closely related to each other. Chapter 2 introduces the Bochner-

Riesz means of multiple Fourier integral, including Fefferman theorem which negates the Disc multiplier's conjecture and the famous Carleson-Sjölin theorem. For almost everywhere convergence of the Bochner-Riesz means below the critical index, we introduce Carbery-Rubio de Francia-Vega's work. Some recent results on commutators of the Bochner-Riesz means are also included in the last section of this chapter. Chapter 3 concerns with the Bochner-Riesz means of multiple Fourier series, including the theory and application of a class of function space (block space), developed by Taibleson, Weiss and other mathematicians, which is closely related to almost everywhere convergence of the Bochner-Riesz means. In addition, a class of function space named spaces generated by smooth blocks is also introduced in this chapter. Such spaces are closely related to the rate of convergence of the Bochner-Riesz means. Chapter 4 discusses the Bochner-Riesz means of conjugate Fourier integral and conjugate Fourier series, where the concepts of conjugate integral and conjugate series are based on the theory of singular integral by Calderón-Zygmund.

We would like to thank Kunyang Wang whose research work contributes nicely to this book. We have to give our thanks to Bolin Ma who provides some useful materials related to new progress in this field since 1988. In addition, Shanzhen Lu wishes to express his gratitude to Guido Weiss for their collaboration in the early 1980s. This book is dedicated to Guido Weiss on the occasion of his 85th birthday. The book is also in memory of Minde Cheng, Yongsheng Sun and Mitchell H. Taibleson.

Shanzhen Lu (Beijing Normal University)

Dunyan Yan (University of Chinese Academy of Sciences)

December, 2012

Chapter 1

AN INTRODUCTION TO MULTIPLE FOURIER SERIES

Let $Q := [-\pi, \pi)^n$. Sometimes $[-\pi, \pi)^n$ is denoted by Q^n for emphasizing n-dimension. We denote by $L^p(Q)$ the set of all complex-valued measurable functions whose modulus to the p-th power is integrable on Q and 2π-periodic in each of its variables for $1 \leq p < \infty$. The space $L^p(Q)$ is a Banach space with the norm

$$\|f\|_p = \left(\int_Q |f(x)|^p dx\right)^{\frac{1}{p}}.$$

We denote a set of all continuous functions on Q having period 2π in each of its variables by $C(Q)$, which is a Banach space with the uniform norm

$$\|f\|_C = \max\{|f(x)| : x \in Q\}.$$

Let $f \in L(Q) := L^1(Q)$. The Fourier coefficients of f are

$$C_m := C_m(f) = \frac{1}{(2\pi)^n} \int_Q f(x) e^{-ix\cdot m} dx, \ m \in \mathbb{Z}^n. \quad (1.0.1)$$

We denote the Fourier series of f by $\sigma(f)$ as follow:

$$\sigma(f)(x) \sim \sum_{m \in \mathbb{Z}^n} C_m(f) e^{im\cdot x}. \quad (1.0.2)$$

According to different methods of summation, partial sum of Fourier series has various forms. The rectangular and spherical partial sums are the two most common forms. The rectangular partial sum is defined by

$$S_m(f;x) = \sum_{0\leq |l_j|\leq m_j, j=1,\ldots,n} C_l(f)e^{il\cdot x}, \ m\in \mathbb{Z}_+^n, \qquad (1.0.3)$$

which can be expressed in the integral form defined as

$$S_m(f;x) = \frac{1}{\pi^n}\int_Q f(x-t)D_{m_1}(t_1)\cdots D_{m_n}(t_n)dt, \ m\in \mathbb{Z}_+^n, \qquad (1.0.4)$$

where

$$D_v(u) = \frac{\sin\left(v+\frac{1}{2}\right)u}{2\sin\frac{1}{2}u}$$

is the Dirichlet kernel.

The spherical partial sum is defined by

$$S_R^0(f;x) = \sum_{|m|<R} C_m(f)e^{im\cdot x}, \quad R>0,$$

where $|m| = (m_1^2+\cdots+m_n^2)^{\frac{1}{2}}$.

Obviously, there exist various ways of only taking the limit when we discuss the convergence or summation of the multiple Fourier series. We usually discuss several kinds of limits.

The rectangular limit is defined by

$$\lim_{m_1,\ldots,m_n\to\infty} S_m(f;x).$$

The restricted rectangular limit is defined as the rectangular limit, but we need to restrict

$$\frac{m_j}{m_k} \in \left[\frac{1}{\lambda}, \lambda\right]$$

as $m_1,\ldots,m_n \to \infty$ for some fixed constant $\lambda \geq 1$.

The spherical limit which we will mainly discuss is defined as

$$\lim_{R\to\infty} S_R(f;x).$$

For the rectangular sum, we introduce the arithmetic means (also called by the Fejér means). The rectangular Fejér means of f is the trigonometric polynomial

$$\sigma_m(f;x) = \frac{1}{(m_1+1)\cdots(m_n+1)}\sum_{j_1=0}^{m_1}\cdots\sum_{j_n=0}^{m_n} S_{j_1,\ldots,j_n}(f;x). \qquad (1.0.5)$$

Substituting the equation (1.0.4) into the equation (1.0.5), we can obtain the integral form

$$\sigma_m(f;x) = \frac{1}{\pi^n}\int_Q f(x-t)K_{m_1}(t_1)\cdots K_{m_n}(t_n)dt, \qquad (1.0.6)$$

where

$$K_j(u) = \frac{1}{j+1}\sum_{l=0}^{j} D_l(u) = \frac{1}{2(j+1)}\left(\frac{\sin\frac{j+1}{2}u}{\sin\frac{u}{2}}\right)^2$$

is the Fejér kernel.

For the spherical summation, we will mainly discuss the Bochner-Riesz means, that is, the following spherical trigonometric polynomial of R-th order

$$S_R^\alpha(f;x) = \sum_{|m|<R}\left(1-\frac{|m|^2}{R^2}\right)^\alpha C_m(f)e^{im\cdot x}, \quad R>0, \qquad (1.0.7)$$

where the power α is any complex number whose real part is larger than -1.

We remark that, in the definition of S_R^α, we always restrict $R>0$. For the sake of convenience, the notation S_0^α denotes $\lim_{R\to 0_+} S_R^\alpha$, that is, $S_0^\alpha(f;x) = C_0(f)$.

In many aspects, it is more suitable for us to regard the spherical sum theory of multiple trigonometric series as the extension of the theory with one variable. For example, the uniqueness theory of the spherical summation of multiple trigonometric series, such as the foundational results, obtained by Cheng [Che1] illustrated these facts. As far as the convergence and approximation theory of the Fourier series is concerned, a series of facts we will introduce in the following also reflect this point. It is one of the reasons why we are more interested in the theory of spherical summation.

1.1 Basic properties of multiple Fourier series

In order to discuss the Fourier series, we begin with proving a basic property of $L^p(Q)$, that is, the density of trigonometric polynomials. For the sake of convenience, we denote $C(Q)$ by $L^\infty(Q)$.

Theorem 1.1.1 *The set of the trigonometric polynomials of n variables are dense in $L^p(Q)$ for all $1 \leq p \leq +\infty$.*

Proof. By the fundamental facts in real analysis, we all know that $C(Q)$ is a dense subset of $L^p(Q)$. Hence, we first prove that the set of trigonometric polynomials is dense in $C(Q)$ by using the Stone theorem, and then finish the proof by noticing that the distance in L^p is not larger than the distance in $C(Q)$.

Here we complete the proof of the theorem by proving

$$\lim_{m_1,\ldots,m_n \to \infty} \|f - \sigma_m(f)\|_p = 0. \tag{1.1.1}$$

For the sake of simplicity, let $n = 2$. By the equation (1.0.6), we have that

$$\|f - \sigma_{m_1,m_2}(f)\|_p$$
$$= \left\| \frac{1}{\pi^2} \int_Q \Big(f(x_1 + t_1, x_2 + t_2) - f(x_1, x_2)\Big) K_{m_1}(t_1) K_{m_2}(t_2) dt_1 dt_2 \right\|_p$$
$$\leq \frac{1}{\pi^2} \int_Q \|f(\cdot + t_1, \cdot + t_2) - f(\cdot, \cdot)\|_p K_{m_1}(t_1) K_{m_2}(t_2) dt_1 dt_2.$$

We denote the L^p modulus of continuity of f by $\omega(f; \mu, \nu)_p$, that is,

$$\omega(f; \mu, \nu)_p = \sup_{|h| \leq \mu, |\tau| \leq \nu} \|f(\cdot + h, \cdot + \tau) - f(\cdot, \cdot)\|_p.$$

Hence, it implies that

$$\|f - \sigma_{m_1,m_2}(f)\|_p \leq \frac{4}{\pi^2} \int_0^\pi \int_0^\pi \omega(f; t_1, t_2)_p K_{m_1}(t_1) K_{m_2}(t_2) dt_1 dt_2.$$

Set

$$\mu = \frac{\log(m_1 + 1)}{m_1 + 1}$$

and

$$\nu = \frac{\log(m_2 + 1)}{m_2 + 1}.$$

Let $m_1, m_2 > 0$. We conclude that

$$\|f - \sigma_{m_1,m_2}(f)\|_p \leq \frac{4}{\pi^2} \left(\int_0^\mu \int_0^\nu + \int_\mu^\pi \int_0^\nu + \int_0^\mu \int_\nu^\pi + \int_\mu^\pi \int_\nu^\pi \right)$$
$$\times \omega(f; t_1, t_2)_p K_{m_1}(t_1) K_{m_2}(t_2) dt_1 dt_2$$
$$= \frac{4}{\pi^2} (I_1 + I_2 + I_3 + I_4).$$

1.1 Basic properties of multiple Fourier series

$$I_1 \leq \omega(f;\mu,\nu)_p \int_0^\pi \int_0^\pi K_{m_1}(t_1)K_{m_2}(t_2)dt_1 dt_2 \leq \frac{\pi^2}{4}\omega(f;\mu,\nu)_p.$$

Using the property of modulus of continuity, we have

$$\omega(f;t_1,t_2)_p \leq \omega(f;\mu,t_2)_p \left(\frac{t_1}{\mu}+1\right).$$

Thus, it implies that

$$I_2 = \int_\mu^\pi dt_1 \int_0^\nu \omega(f;t_1,t_2)_p K_{m_1}(t_1) K_{m_2}(t_2) dt_2$$

$$\leq \omega(f;\mu,\nu)_p \int_\mu^\pi \left(1+\frac{t_1}{\mu}\right) K_{m_1}(t_1) dt_1 \int_0^\pi K_{m_2}(t_2) dt_2$$

$$\leq C\omega(f;\mu,\nu)_p \left(\int_\mu^\pi \frac{t_1}{\mu} \frac{dt_1}{(m_1+1)t_1^2}+1\right)$$

$$\leq C\omega(f;\mu,\nu)_p.$$

Similarly, we have

$$I_3 \leq C\omega(f;\mu,\nu)_p.$$

It follows that

$$I_4 = \int_\mu^\pi dt_1 \int_\nu^\pi dt_2 \omega(f;t_1,t_2)_p K_{m_1}(t_1) K_{m_2}(t_2)$$

$$\leq \omega(f;\mu,\nu)_p \int_\mu^\pi \left(1+\frac{t_1}{\mu}\right) K_{m_1}(t_1) dt_1 \int_\nu^\pi \left(1+\frac{t_2}{\nu}\right) K_{m_2}(t_2) dt_2$$

$$\leq C\omega(f;\mu,\nu)_p.$$

Consequently, we have that

$$\|f - \sigma_{m_1,m_2}(f)\|_p \leq C\omega\left(f; \frac{\log(m_1+1)}{m_1+1}, \frac{\log(m_2+1)}{m_2+1}\right)_p, \quad (1.1.2)$$

for $m_1, m_2 > 0$.

This completes the proof of Theorem 1.1.1. ∎

Remark 1.1.1 *From the proof of the above theorem, we actually transformed the multiple problem into the repeated estimate of unitary variable. That is to say, the proof is trivial in essence. In many cases, multiple rectangular problem can be simplified into repeated unitary variable problem.*

Consequently, many results in the unitary case can naturally be extended to the multiple rectangular case. In this case, we can not tell any essential difference between the unitary and the multiple. We have very little interest in this situation.

Theorem 1.1.2 *Let $f \in L(Q)$. If $C_m(f) = 0$, for any $m \in \mathbb{Z}^n$, then we have $f(x) = 0$, for a.e. $x \in Q$.*

Proof. From the equations

$$C_m(f) = \frac{1}{(2\pi)^n} \int_Q f(x) e^{-im \cdot x} dx = 0,$$

with every $m \in \mathbb{Z}^n$, we conclude that

$$\int_Q f(x) P(x) dx = 0, \qquad (1.1.3)$$

for any trigonometric polynomial $P(x)$. For any fixed $g \in C(Q)$ and any positive number ε, by Theorem 1.1.1, there exists a trigonometric polynomial P_ε such that

$$\|g - P_\varepsilon\|_C < \varepsilon. \qquad (1.1.4)$$

It follows from the equations (1.1.3) and (1.1.4) that

$$\left| \int_Q f(x) g(x) dx \right| = \left| \int_Q f(x)(g(x) - P_\varepsilon(x)) dx \right| \leq \varepsilon \int_Q |f(x)| dx.$$

Then this implies that

$$\int_Q f(x) g(x) dx = 0$$

for any $g \in C(Q)$ which implies $f = 0$. This completes the proof of the theorem. ∎

Corollary 1.1.1 *Let $f \in L(Q)$ and $\sum_m |C_m(f)| < +\infty$. Then there exists $g \in C(Q)$ such that $f(x) = g(x)$ for almost everywhere $x \in Q$.*

For the space $L^2(Q)$, the following basic theorem is the same as the one for the case $n = 1$.

1.1 Basic properties of multiple Fourier series

Theorem 1.1.3 Let $f \in L^2(Q)$. Then we have

$$\sum_{m \in \mathbb{Z}^n} |C_m(f)|^2 = \frac{1}{(2\pi)^n} \|f\|_2^2. \tag{1.1.5}$$

Proof. Since the proof is the same as the proof of one dimensional case, we briefly write down the proof just for the review purpose here.

$$(2\pi)^n \sum_{|m|<R} |C_m(f)|^2 = \int_Q f(x) \overline{S_R(f,x)} dx$$

$$\leq \|f\|_2 \|S_R(f)\|_2$$

$$= \|f\|_2 \left((2\pi)^n \sum_{|m|<R} |C_m(f)|^2 \right)^{\frac{1}{2}}.$$

This is equivalent to

$$\sum_{|m|<R} |C_m(f)|^2 \leq \frac{1}{(2\pi)^n} \|f\|_2^2.$$

for any $R > 0$.

Thus $S_R(f)$ is a Cauchy sequence of $L^2(Q)$. Consequently, there exists $g \in L^2(Q)$ such that $S_R(f) \to g$ in the sense of L^2 norm. So we have $\sigma(g) = \sigma(f)$. It follows from Theorem 1.1.2 that $g = f$.

On the other hand, since $S_R(f)$ converges to f in L^2 as R goes to ∞, we have

$$\|S_R(f)\|_2 \to \|f\|_2.$$

It follows that

$$\left((2\pi)^n \sum_{m \in \mathbb{Z}^n} |C_m(f)|^2 \right)^{\frac{1}{2}} = \|f\|_2.$$

This is our desired result. ∎

Corollary 1.1.2 Suppose $f \in C^{(k)}(Q)$ with $k > \frac{n}{2}$. Then we have

$$\sum |C_m(f)| < +\infty.$$

Proof. Let $\alpha = (\alpha_1, \ldots, \alpha_n) \in \mathbb{Z}_+^n$ with $|\alpha|_1 = \alpha_1 + \cdots + \alpha_n \leq k$. Then we have
$$D^\alpha(f)(x) = \frac{\partial^{\alpha_1 + \cdots + \alpha_n} f(x)}{\partial x_1^{\alpha_1} \cdots \partial x_n^{\alpha_n}} \in C(Q).$$

By the integration by parts, we obtain that
$$\int_Q D^\alpha(f)(x) e^{-im \cdot x} dx = (im)^\alpha \int_Q f(x) e^{-im \cdot x} dx$$
$$= (im)^\alpha (2\pi)^n C_m(f)$$

for $m \in \mathbb{Z}^n$.

According to Theorem 1.1.3, we have
$$\sum_{|\alpha|_1 = k} \left(\sum_{m \in \mathbb{Z}^n} |C_m(f)|^2 |(im)^\alpha|^2 \right) < +\infty.$$

That is,
$$\sum_{m \in \mathbb{Z}^n} |C_m(f)|^2 \left(\sum_{|\alpha|_1 = k} |m_1|^{2\alpha_1} \cdots |m_n|^{2\alpha_n} \right) < +\infty. \tag{1.1.6}$$

Obviously, there exists a constant $C = C_{k,n} > 0$ such that
$$|m|^{2k} = (|m_1|^2 + \cdots + |m_n|^2)^k \leq C \sum_{|\alpha| = k} |m_1|^{2\alpha_1} \cdots |m_n|^{2\alpha_n}. \tag{1.1.7}$$

Set
$$b_m = \left(\sum_{|\alpha|_1 = k} |m_1|^{2\alpha_1} \cdots |m_n|^{2\alpha_n} \right)^{\frac{1}{2}}.$$

Then, the inequalities (1.1.6) and (1.1.7) can be written as
$$\sum_{m \in \mathbb{Z}^n} |C_m(f)|^2 b_m^2 < +\infty, \tag{1.1.8}$$

and
$$|m|^k \leq C b_m. \tag{1.1.9}$$

1.1 Basic properties of multiple Fourier series

It implies from (1.1.9) that

$$\sum_{m \neq 0} |C_m(f)| \leq C \sum_{m \neq 0} |C_m(f)| \frac{b_m}{|m|^k}$$

$$\leq C \left(\sum_{m \neq 0} |C_m(f)|^2 b_m^2 \right)^{\frac{1}{2}} \left(\sum_{m \neq 0} \frac{1}{|m|^{2k}} \right)^{\frac{1}{2}}. \quad (1.1.10)$$

Since $k > \frac{n}{2}$, we have

$$\sum_{m \neq 0} \frac{1}{|m|^{2k}} < +\infty.$$

Thus from (1.1.8) and (1.1.10), we can deduce that

$$\sum |C_m(f)| < +\infty.$$

∎

It should be pointed out that Corollary 1.1.2 gives a sufficient condition for the Fourier series to be absolutely convergent. Of course, this result is rather sketchy.

Corollary 1.1.3 (Riemann-Lebesgue Theorem) *If $f \in L(Q)$, then we have*

$$\lim_{|m| \to \infty} |C_m(f)| = 0.$$

Proof. For any $\varepsilon > 0$, there exists $g \in C(Q)$ such that the function $h = f - g$ satisfies $\|h\|_1 < \varepsilon$. Since there hold

$$C_m(f) = C_m(g) + C_m(h)$$

and

$$|C_m(h)| \leq \|h\|_1,$$

we have that

$$\varlimsup_{|m| \to \infty} |C_m(f)| \leq \varlimsup_{|m| \to \infty} |C_m(g)| + \varepsilon.$$

Theorem 1.1.3 impies that

$$\varlimsup_{|m| \to \infty} |C_m(g)| = 0.$$

Hence, we obtain
$$\overline{\lim_{|m|\to\infty}} |C_m(f)| \leq \varepsilon.$$
This completes the proof. ∎

1.2 Poisson summation formula

Poisson summation formula is an important tool for investigating multiple Fourier series.

Let $f \in L(\mathbb{R}^n)$. The formal series
$$\sum_{m \in \mathbb{Z}^n} f(x + 2\pi m), \quad x \in \mathbb{R}^n \tag{1.2.1}$$

is called the periodization of the function f. If the series (1.2.1) converges to a function in some sense (some metric sense or some limit sense), then this sum function is undoubtedly 2π-periodic.

Theorem 1.2.1 *Let $f \in L(\mathbb{R}^n)$. The series (1.2.1) converges absolutely almost everywhere and the sum function belongs to $L(Q)$, with the Fourier series being*
$$\sum_{m \in \mathbb{Z}^n} \widehat{f}(m) e^{im \cdot x}. \tag{1.2.2}$$

That is to say, $\widehat{f}(m)$, $m \in \mathbb{Z}^n$, are the Fourier coefficients of
$$\sum_{m \in \mathbb{Z}^n} f(x + 2\pi m).$$

Proof. By Levi's theorem, we have
$$\int_Q \sum |f(x+2\pi m)|dx = \sum \int_Q |f(x+2\pi m)|dx$$
$$= \sum_m \int_{Q+2\pi m} |f(x)|dx$$
$$= \int_{\mathbb{R}^n} |f(x)|dx < +\infty,$$

1.2 Poisson summation formula

and

$$\frac{1}{(2\pi)^n}\int_Q \left(\sum_{m'\in\mathbb{Z}^n} f(x+2\pi m')\right) e^{-im\cdot x} dx$$
$$= \frac{1}{(2\pi)^n}\sum_{m'} \int_Q f(x+2\pi m') e^{-im\cdot x} dx$$
$$= \frac{1}{(2\pi)^n}\sum_{m'} \int_{Q+2\pi m'} f(x) e^{-im\cdot(x-2\pi m')} dx$$
$$= \frac{1}{(2\pi)^n}\sum_{m'} \int_{Q+2\pi m'} f(x) e^{-im\cdot x} dx$$
$$= \frac{1}{(2\pi)^n}\int_{\mathbb{R}^n} f(x) e^{-im\cdot x} dx$$
$$= \widehat{f}(m).$$

This completes the proof Theorem 1.2.1. ∎

Theorem 1.2.2 *Let $f \in C(\mathbb{R}^n)$. If there exist $\delta > 0$ and $A > 0$ such that*

$$|f(x)| \le A(1+|x|)^{-n-\delta}, \quad |\widehat{f}(y)| \le A(1+|y|)^{-n-\delta}, \quad (1.2.3)$$

then we have

$$\sum_{m\in\mathbb{Z}^n} f(x+2\pi m) = \sum_{m\in\mathbb{Z}^n} \widehat{f}(m) e^{im\cdot x} \quad (1.2.4)$$

for all $x \in \mathbb{R}^n$. The equality 1.2.4 is the so called Poisson summation formula.

Proof. By the assumption (1.2.3), we have that $f \in L(\mathbb{R}^n)$ and

$$\sum_{m\in\mathbb{Z}^n} |\widehat{f}(m)| < +\infty.$$

According to Theorem 1.2.1, we get

$$\sum_{m\in\mathbb{Z}^n} f(x+2\pi m) = \sum_{m\in\mathbb{Z}^n} \widehat{f}(m) e^{im\cdot x},$$

for a.e. $x \in \mathbb{R}^n$.

Since the functions on the both sides of (1.2.3) are continuous, the above equation holds everywhere, that is, (1.2.4) holds. ∎

We need the following formulas which need weaker assumptions.

Theorem 1.2.3 *Suppose that $f \in L(\mathbb{R}^n)$ is continuous except at $x = 0$. Let*

$$g(y) = \int_{\mathbb{R}^n} f(x)e^{ix\cdot y}dx. \tag{1.2.5}$$

If the following three conditions

(i)
$$\sum |f(m)| < +\infty; \tag{1.2.6}$$

(ii)
$$|g(y)| = o\left(\frac{1}{|y|^{n-1}}\right), \quad |y| \to \infty; \tag{1.2.7}$$

(iii) *the series*

$$\sum_{j=0}^{\infty} \sum_{m \in S_j} g(x + 2\pi m) \tag{1.2.8}$$

is locally uniformly convergent on $\mathbb{R}^n$, where

$$S_j := \{x \in \mathbb{Z}^n : j \leq |x| < j+1\},$$

are satisfied, then

$$\sum_{j=0}^{\infty} \sum_{m \in S_j} g(x + 2\pi m) = \sum f(m)e^{im\cdot x} + C_0, \tag{1.2.9}$$

holds for any $x \in \mathbb{R}^n$, where

$$C_0 = (2\pi)^{-n} \lim_{\rho \to \infty} \int_{|y|<\rho} g(y)dy.$$

Proof. By the definition (1.2.5), we have $g \in C(\mathbb{R}^n)$. According to the condition (iii), we obtain that the series (1.2.8) is a continuous 2π-periodic function, which we denote by $G(x)$. We now compute the Fourier coefficients of G

$$C_m(G) = (2\pi)^{-n} \int_Q G(x)e^{-im\cdot x}dx.$$

1.2 Poisson summation formula

It easily follows from the condition (iii) that

$$C_m(G) = \sum_{j=0}^{\infty} \sum_{\mu \in S_j} \frac{1}{(2\pi)^n} \int_{Q+2\pi\mu} g(x)e^{-im\cdot x} dx.$$

Let $\rho > 2n\pi + 1$ and $N = [\frac{\rho}{2\pi}]$. We obviously have

$$\left| \sum_{j=0}^{N} \sum_{\mu \in S_j} \int_{Q+2\pi\mu} g(x)e^{-im\cdot x} dx - \int_{|x|<\rho} g(x)e^{-im\cdot x} dx \right| \quad (1.2.10)$$

$$\leq \int_{\rho-2n\pi<|x|<\rho+2n\pi} |g(x)| dx.$$

The condition (ii) shows that the quantity on the right side of (1.2.10) is $o(1)$ as $\rho \to \infty$.

Hence, we have

$$C_m(G) = \lim_{\rho \to \infty} \frac{1}{(2\pi)^n} \int_{|x|<\rho} g(x)e^{-im\cdot x} dx$$

$$= \lim_{\rho \to \infty} \int_{|x|<\rho} \left(\frac{1}{(2\pi)^n} g(-x) \right) e^{im\cdot x} dx. \quad (1.2.11)$$

Specially, $C_0(G) = C_0$.

By (1.2.5), we get that

$$\frac{1}{(2\pi)^n} g(-x) = \widehat{f}(x).$$

Since f is continuous except at $x = \mathbf{0}$, we have for each $m \neq \mathbf{0}$,

$$\lim_{\varepsilon \to 0^+} \int_{\mathbb{R}^n} \widehat{f}(x) e^{im\cdot x} e^{-\varepsilon|x|} dx = f(m). \quad (1.2.12)$$

Let

$$\varphi(t) = \int_{|x|=t} f(x) e^{im\cdot x} d\sigma(x)$$

and

$$\Phi_\rho(t) = \int_\rho^t \varphi(\tau) d\tau,$$

for $0 < \rho \leq t < \infty$.

According to (1.2.11) and (1.2.12), we get respectively

$$\lim_{\rho\to\infty}\int_0^\rho \varphi(t)dt = C_m(G) \tag{1.2.13}$$

and

$$\lim_{\varepsilon\to 0}\int_0^\infty \varphi(t)e^{-\varepsilon t}dt = f(m), \tag{1.2.14}$$

for $m \neq \mathbf{0}$.

By (1.2.13), we obtain that for any $\eta > 0$, there exists $M > 0$ such that

$$|\Phi_\rho(t)| < \eta,$$

if $M \leq \rho \leq t < \infty$. Since there holds

$$\int_\rho^\infty \varphi(t)e^{-\varepsilon t}dt = \Phi_\rho(t)e^{-\varepsilon t}\Big|_\rho^\infty + \varepsilon\int_\rho^\infty \Phi_\rho(t)e^{-\varepsilon t}dt$$

$$= \int_\rho^\infty \Phi_\rho(t)\varepsilon e^{-\varepsilon t}dt,$$

we have that

$$\left|\int_\rho^\infty \varphi(t)e^{-\varepsilon t}dt\right| \leq \eta\int_\rho^\infty \varepsilon e^{-\varepsilon t}dt = \eta e^{-\varepsilon\rho},$$

which implies

$$\varlimsup_{\varepsilon\to 0}\left|\int_\rho^\infty \varphi(t)e^{-\varepsilon t}dt\right| < \eta, \quad M \leq \rho \leq t < \infty.$$

By (1.2.13), we can choose $\rho > M$ big enough, such that

$$\left|C_m(G) - \int_0^\rho \varphi(t)dt\right| < \eta.$$

By making use of (1.2.14), we have that for $m \neq \mathbf{0}$,

$$f(m) = \int_0^\rho \varphi(t)dt + \lim_{\varepsilon\to 0}\int_\rho^\infty \varphi(t)e^{-\varepsilon t}dt.$$

Hence, we have

$$|C_m(G) - f(m)| < 2\eta, \quad m \neq \mathbf{0},$$

which leads to

$$C_m(G) = f(m), \ m \neq \mathbf{0}.$$

Now, by the condition (i), we know that the function on the right side of (1.2.9) is a continuous periodic function, whose Fourier coefficient is $C_m(G)$. So this function is $G(x)$, that is to say, the equation (1.2.9) holds. This completes the proof. ∎

Remark 1.2.1 *From the proof, we can conclude that the continuity of f is not intrinsic, as long as f is continuous at non-zero grid points. Actually the conditions can be much weaker, as long as those points are Lebesgue points.*

Remark 1.2.2 *We can make an explanation about Theorem 1.2.3 in aspect of form: the function g defined by (1.2.5) can be regarded as the Fourier inverse transform of f, then $\hat{g} = f$. Now we can trace back to the equation (1.2.4) in Theorem 1.2.2, just executing the summation in spherical method for the left side.*

The three theorems in this section are all called the Poisson summation formulas.

1.3 Convergence and the opposite results

For the convergence of Fourier series, the most wonderful result in the unitary case has been established by Carleson-Hunt. They have confirmed the conjecture posed by Lusin, that is,

$$f \in L^p([-\pi,\pi)) \Rightarrow S_m(f,x) \to f(x)$$

holds for a.e. $x \in [-\pi, \pi)$ and $1 < p < \infty$, as $m \to \infty$.

The case of multi-dimension is much more complicated. The convergence of the multi-dimensional rectangular partial sum is not good. Fefferman [Fe2] proved that there exists a continuous periodic binary function $f(x, y)$, whose rectangular Fourier partial sum $S_{m,n}(f; x, y)$ diverges everywhere as $m \to \infty$ and $n \to \infty$.

However, it is much better when we take the square limit. At this time, the multi-dimensional analogy of the conjecture still holds. In the 1970s, many researchers have proven that

$$f \in L^p(Q) \Rightarrow S_{k,\ldots,k}(f;x) \to f(x)$$

holds for a.e. $x \in Q$ and $1 < p < \infty$, as $k \to \infty$.

Later, Sjölin [Sj1] moved forward with the above results and proved that

$$f \in L(\log^+ L)^n \log^+ \log^+ L(Q) \Rightarrow S_{k,\ldots,k}(f;x) \to f(x)$$

holds for a.e. $x \in Q$, as $k \to \infty$.

Here, we prove the following theorem about the convergence of the spherical sum.

Theorem 1.3.1 *If $n > 1$, then for every $p \in \left(1, \frac{2n}{n+1}\right)$, there exists $f \in L^p(Q)$ such that*

$$\limsup_{R \to \infty} \left|S_R^0(f;x)\right| = +\infty, \tag{1.3.1}$$

for a.e. $x \in Q$.

Definition 1.3.1 *We call a set A in $\mathbb{R}$ a linearly independent set if for any finite elements of A, $a_1, \ldots, a_m$, the linear combination with integer coefficients*

$$p_1 a_1 + \cdots + p_m a_m$$

equals to zero if and only if the coefficients

$$p_1 = p_2 = \cdots = p_m = 0.$$

Definition 1.3.2 *Let $x \in \mathbb{R}^n$. If $\{|x + 2\pi m| : m \in \mathbb{Z}^n\}$ is a linearly independent set, then we say $x \in S$ and denote by*

$$S^c = \mathbb{R}^n \backslash S.$$

Lemma 1.3.1 *Let $n > 1$. Then $|S^c| = 0$.*

Proof. Arbitrarily arrange the elements of $\mathbb{Z}^n$ as $\{m_j\}_{j=1}^\infty$. Let

$$\mathscr{A} = \{A = (a_1, \ldots, a_k) : a_j \in \mathbb{Z}^n, j = 1, \ldots, k,\ k \in \mathbb{N}\}.$$

Choose any $A \in \mathscr{A}$ such that $A \neq \mathbf{0} := (0, \ldots, 0)$ and let

$$\Phi_A(y) = \sum_{j=1}^k a_j |y + 2\pi m_j|.$$

It is obvious that $\Phi_A(y)$ is an analytic function defined on simply connected domain $\mathbb{R}^n \backslash 2\pi \mathbb{Z}^n$ and continuous on $\mathbb{R}^n$. For $A \neq 0$, then Φ_A is not zero function (notice that $\Phi_A'(y)$ is unbounded near some points). Denote

1.3 Convergence and the opposite results

the set of all the zero points of Φ_A by $\mathbb{Z}_A$. By the theory of analytic function, we have $|\mathbb{Z}_A| = 0$. Then we also have

$$\left| \bigcup_{A \in \mathscr{A}} \mathbb{Z}_A \right| = 0,$$

together with the fact

$$S^c = \bigcup_{A \in \mathscr{A}} \mathbb{Z}_A,$$

this leads to $|S^c| = 0$. ■

Lemma 1.3.2 *Suppose that $K(t)$ is a real bounded measurable function defined on $(1, \infty)$ and for every $\lambda \geq 0$, there always exists*

$$\lim_{T \to +\infty} \frac{1}{T} \int_1^T K(t) e^{i\lambda t} dt = b(\lambda). \tag{1.3.2}$$

If the set $\{\lambda \geq 0 : b(\lambda) \neq 0\}$ is countable, denoted by $\{\lambda_j\}_{j=1}^\infty$, and linearly independent, then we have

$$\sum_{j=1}^\infty |b(\lambda_j)| < \infty. \tag{1.3.3}$$

Proof. Let

$$b(\lambda_j) = |b(\lambda_j)| e^{i\mu_j}$$

and

$$A(t) = \prod_{j=1}^N (1 + \cos(\lambda_j t - \mu_j)).$$

It is obvious that $A(t) \geq 0$. By the linear independence of $\{\lambda_j\}_{j=1}^\infty$, we get

$$\lim_{T \to \infty} \frac{1}{T} \int_1^T A(t) dt = 1.$$

Since we have

$$\int_1^T A(t) K(t) dt = \int_1^T K(t) \big(1 + \cos(\lambda_1 t - \mu_1) + \cdots + \cos(\lambda_N t - \mu_N)$$
$$+ \cos(\lambda_1 t - \mu_1) \cdot \cos(\lambda_2 t - \mu_2) + \cdots$$
$$+ \cos(\lambda_1 t - \mu_1) \cdots \cos(\lambda_N t - \mu_N)\big) dt,$$

by definition of $\{\lambda_j\}_{j=1}^\infty$, it follows that

$$\lim_{T\to\infty}\frac{1}{T}\int_1^T K(t)A(t)dt = \lim_{T\to\infty}\frac{1}{T}\int_1^T K(t)\sum_{j=1}^N \cos(\lambda_j t - \mu_j)dt$$

$$= \sum_{j=1}^N \Big(\mathrm{Re}b(\lambda_j)\cdot\cos\mu_j + \mathrm{Im}b(\lambda_j)\cdot\sin\mu_j\Big)$$

$$= \sum_{j=1}^N |b(\lambda_j)|.$$

Consequently, we have

$$\sum_{j=1}^N |b(\lambda_j)| \leq \overline{\lim_{T\to\infty}}\frac{1}{T}\int_1^T |K(t)|A(t)dt$$

$$\leq \sup\{|K(t)| : t > 0\}$$

$$< +\infty.$$

Thus (1.3.3) holds and we complete the proof of Lemma 1.3.2. ∎

We define

$$D_R^\alpha(x) = \sum_{|m|<R}\left(1 - \frac{|m|^2}{R^2}\right)^\alpha e^{im\cdot x}, \qquad (1.3.4)$$

for $\mathrm{Re}\,\alpha > -1$.

Lemma 1.3.3 *If* $\mathrm{Re}\,\alpha > \frac{n-1}{2}$, *then we have*

$$D_R^\alpha(x) = 2^\alpha \Gamma(\alpha+1)\left(2\pi^{\frac{n}{2}} R^{\frac{n}{2}-\alpha}\sum_{m\in\mathbb{Z}^n}\frac{J_{\frac{n}{2}+\alpha}(Rr_m)}{r_m^{\frac{n}{2}+\alpha}}\right), \qquad (1.3.5)$$

where $r_m = |x + 2\pi m|$.

Proof. In Section 1.2, we introduce

$$\Phi^\alpha(x) = \begin{cases} (1-|x|^2)^\alpha, & |x| < 1, \\ 0, & |x| \geq 1, \end{cases}$$

1.3 Convergence and the opposite results

and obtain that

$$B^\alpha(x) := \widehat{\Phi^\alpha}(x) = \frac{2^\alpha \Gamma(\alpha+1)}{(2\pi)^{\frac{n}{2}}} \frac{J_{\frac{n}{2}+\alpha}(|x|)}{|x|^{\frac{n}{2}+\alpha}}.$$

Due to the fact that $\text{Re}\,\alpha > \frac{n-1}{2}$, we have $B^\alpha \in L(\mathbb{R}^n)$ which leads to

$$\Phi^\alpha(x) = \mathscr{F}\left((2\pi)^n B^\alpha(-y)\right)(x) = (2\pi)^n \widehat{B^\alpha}(x).$$

According to Theorem 1.2.2, we have that

$$\sum_{m \in \mathbb{Z}^n} \mathscr{F}\left((2\pi)^n B^\alpha(Ry) R^n\right)(m) e^{im \cdot x} = \sum_{m \in \mathbb{Z}^n} (2\pi)^n B^\alpha(R(x+2\pi m)) R^n,$$

that is,

$$\sum_{m \in \mathbb{Z}^n} \Phi^\alpha\left(\frac{m}{R}\right) e^{im \cdot x} = 2^\alpha \Gamma(\alpha+1)(2\pi)^{\frac{n}{2}} R^n \sum_{m \in \mathbb{Z}^n} \frac{J_{\frac{n}{2}+\alpha}(R|x+2\pi m|)}{(R|x+2\pi m|)^{\frac{n}{2}+\alpha}},$$

which is just the equation (1.3.5). This completes the proof. ∎

Lemma 1.3.4 *Let $x \in \mathbb{R}^n$ with $x \neq 2\pi m$, for any $m \in \mathbb{Z}^n$. Then, for $\lambda \geq 0$, we have*

$$\lim_{T \to \infty} \frac{1}{T} \int_1^T D_R^{\frac{n-1}{2}}(x) e^{i\lambda R} dR = \begin{cases} \frac{C}{\lambda^n}, & \text{if } \lambda = r_m, \text{ for } m \in \mathbb{Z}^n, \\ 0, & \text{if } \lambda \neq r_m, \text{ for } m \in \mathbb{Z}^n, \end{cases} \quad (1.3.6)$$

where

$$C = 2^{n-1} \pi^{\frac{n-1}{2}} \Gamma\left(\frac{n+1}{2}\right) e^{i\frac{\pi n}{2}}.$$

Proof. Let $\alpha \in \left(\frac{n-1}{2}, \frac{n+1}{2}\right]$. By Lemma 1.3.3, we get

$$D_R^\alpha(x) = 2^\alpha \Gamma(\alpha+1)(2\pi)^{n/2} R^{\frac{n}{2}-\alpha} \sum \frac{J_{\frac{n}{2}+\alpha}(Rr_m)}{r_m^{\frac{n}{2}+\alpha}}$$

$$= C_{\alpha,n} R^{\frac{n-1}{2}-\alpha} \sum \frac{\cos(Rr_m + \theta_n)}{r_m^{\frac{n+1}{2}+\alpha}} + R^{\frac{n-3}{2}-\alpha} O\left(\sum \frac{1}{r_m^{\frac{n+3}{2}+\alpha}}\right)$$

$$= C_{\alpha,n} R^{\frac{n-1}{2}-\alpha} \sum \frac{\cos(Rr_m + \theta_n)}{r_m^{\frac{n+1}{2}+\alpha}} + O\left(\frac{1}{R}\right),$$

(1.3.7)

as $R \to \infty$, where
$$C_{\alpha,n} = 2^{\frac{n+1}{2}+\alpha}\pi^{\frac{n-1}{2}}\Gamma(\alpha+1), \quad \theta = -\frac{\pi}{2}(\frac{n}{2}+\alpha) - \frac{\pi}{4},$$
and 'O' is uniformly valid for $\alpha > \frac{n-1}{2}$. Thus, we have
$$\frac{1}{T}\int_1^T D_R^\alpha(x)e^{i\lambda R}dR = C_{\alpha,n}\sum\left(\frac{1}{T}\int_1^T \frac{\cos(Rr_m+\theta)}{R^{\alpha-\frac{n-1}{2}}r_m^{\frac{n+1}{2}+\alpha}}e^{i\lambda R}dR\right)$$
$$+ O\left(\frac{1}{T}\log T\right). \qquad (1.3.8)$$

Let
$$\varphi_T^\alpha(m,\lambda) = \frac{1}{T}\int_1^T \frac{\cos(Rr_m+\theta)}{R^{\alpha-\frac{n-1}{2}}}e^{i\lambda R}dR$$
$$= \frac{1}{2T}\int_1^T \frac{1}{R^{\alpha-\frac{n-1}{2}}}\left(e^{i[(\lambda+r_m)R+\theta]} + e^{i[(\lambda-r_m)R-\theta]}\right)dR.$$

By the integration by parts, we have that
$$|\varphi_T^\alpha(m,\lambda)| = \frac{1}{|\lambda-r_m|}O\left(\frac{\log T}{T}\right) \qquad (1.3.9)$$
holds uniformly for $\alpha \in \left(\frac{n-1}{2}, \frac{n+1}{2}\right]$ and $\lambda \neq r_m$, as $T \to \infty$.

Therefore, if $\lambda \in \mathbb{R}^+\backslash\{r_m : m \in \mathbb{Z}^n\}$, then we have
$$\frac{1}{T}\int_1^T D_R^\alpha(x)e^{i\lambda R}dR = \left(C_{\alpha,n}\sum_{m\in\mathbb{Z}^n}\frac{1}{|\lambda-r_m|r_m^{\frac{n+1}{2}+\alpha}} + 1\right)O\left(\frac{\log T}{T}\right).$$

By taking $\alpha \to \frac{n-1}{2}$, we get
$$\frac{1}{T}\int_1^T D_R^{\frac{n-1}{2}}(x)e^{i\lambda R}dR = O\left(\frac{\log T}{T}\right). \qquad (1.3.10)$$

Let $\lambda = r_m$. Then we have for $\alpha \in \left(\frac{n-1}{2}, \frac{n+1}{2}\right)$,
$$\varphi_T^\alpha(m,r_m) = \frac{1}{2T}\int_1^T R^{\frac{n-1}{2}-\alpha}e^{-i\theta}dR + \frac{1}{2T}\int_1^T R^{\frac{n-1}{2}-\alpha}e^{i(2r_mR+\theta)}dR$$
$$= \frac{1}{2}e^{-i\theta}\frac{1}{\frac{n+1}{2}-\alpha}\frac{T^{\frac{n+1}{2}-\alpha}-1}{T} + O\left(\frac{\log T}{T}\right)\frac{1}{r_m} \qquad (1.3.11)$$

1.3 Convergence and the opposite results

holds uniformly.

Substituting (1.3.9) and (1.3.11) into (1.3.8), we obtain

$$\frac{1}{T}\int_1^T D_R^\alpha(x)e^{ir_m R}dR = \frac{1}{2}e^{-i\theta}\frac{1}{\frac{n+1}{2}-\alpha}\frac{T^{\frac{n+1}{2}-\alpha}-1}{T}\frac{C_{\alpha,n}}{r_m^{\frac{n+1}{2}+\alpha}}$$

$$+O\left(\frac{\log T}{T}\right)\left(C_{\alpha,n}\sum_{l\in\mathbb{Z}^n, l\neq m}\frac{1}{|r_l-r_m|r_l^{\frac{n+1}{2}+\alpha}}+1\right).$$

By taking $\alpha \to \frac{n-1}{2}$, we have

$$\frac{1}{T}\int_1^T D_R^{\frac{n-1}{2}}(x)e^{ir_m R}dR = \frac{C}{r_m^n}\frac{T-1}{T}+O\left(\frac{\log T}{T}\right). \quad (1.3.12)$$

Combining (1.3.10) with (1.3.12), we obtain (1.3.6).
This completes the proof of Lemma 1.3.4. ∎

Lemma 1.3.5 *Let $x^0 \in S$. Then we have*

$$\limsup_{R\to\infty}\left|D_R^{\frac{n-1}{2}}(x^0)\right| = +\infty. \quad (1.3.13)$$

Proof. Suppose that (1.3.13) is not valid, that is, $D_R^{\frac{n-1}{2}}(x^0)$ is bounded about R on $(0, \infty)$. In fact, $D_R^{\frac{n-1}{2}}(x^0)$ is a real measurable function about R. By Lemma 1.3.4, we get that

$$\lim_{T\to\infty}\frac{1}{T}\int_1^T D_R^{\frac{n-1}{2}}(x^0)e^{i\lambda R}dR$$

exists everywhere about $\lambda \geq 0$ and the set of non-zero points is

$$\{r_m = |x^0 + 2\pi m| : m \in \mathbb{Z}^n\}.$$

For $x^0 \in S$, the set $\{r_m : m \in \mathbb{Z}^n\}$ is linearly independent. According to Lemma 1.3.2, we have

$$\sum_{m\in\mathbb{Z}^n}|b_m| < \infty,$$

where

$$b_m = \lim_{T\to\infty}\frac{1}{T}\int_1^T D_R^{\frac{n-1}{2}}(x^0)e^{ir_m R}dR = \frac{C}{r_m^n},$$

for $C \neq 0$.

However, it is evident that
$$\sum_{m \in \mathbb{Z}^n} \frac{1}{r_m^n} = \sum_{m \in \mathbb{Z}^n} \frac{1}{|x^0 + 2\pi m|^n} = +\infty.$$

This contradiction shows that (1.3.13) holds. ∎

Lemma 1.3.6 (M. Riesz convexity theorem) *Let $x^0 \in \mathbb{R}^n$ and $t > 0$. We define*
$$A^\alpha(t) = t^\alpha D_{\sqrt{t}}^\alpha(x^0) = \sum_{|m| < \sqrt{t}} (t - |m|^2)^\alpha e^{im \cdot x^0}.$$

Suppose that $V(t)$, $W(t)$ are both positive increasing functions on $(0, +\infty)$ and let
$$U_\theta(t) = (V(t))^{1-\theta}(W(t))^\theta,$$

for $0 \leq \theta \leq 1$.

If
$$|A^0(t)| \leq V(t), \ |A^\alpha(t)| \leq W(t),$$

for $\alpha > \beta > 0$, then we have
$$|A^\beta(t)| \leq C U_{\frac{\beta}{\alpha}}(t),$$

where $C = C_{\alpha,\beta}$ is a constant.

Proof. Because of $|A^0(t)| \leq V(t)$, $V(t)$ is a non-negative increasing function and
$$A^\beta(t) = \beta \int_0^t (t-s)^{\beta-1} A^0(s) ds. \qquad (1.3.14)$$

Then we get that
$$|A^\beta(t)| \leq t^\beta V(t).$$

Therefore, if
$$t^\beta \leq \left(\frac{W(t)}{V(t)}\right)^{\frac{\beta}{\alpha}},$$

or that is,
$$t \leq \left(\frac{W(t)}{V(t)}\right)^{\frac{1}{\alpha}},$$

1.3 Convergence and the opposite results

then the theorem turns to be trivial. Hence, we can assume that there exists $\xi(t) > 0$ such that

$$t - \xi = \left(\frac{W(t)}{V(t)}\right)^{\frac{1}{\alpha}}.$$

On one hand, for the case $0 < \alpha \leq 1$, we have

$$A^\beta(t) = \beta \int_0^t (t-s)^{\beta-1} A^0(s) ds$$
$$= \beta \left(\int_0^\xi + \int_\xi^t\right)(t-s)^{\beta-1} A^0(s) ds$$
$$= J_1 + J_2.$$

For J_2, we have the estimate

$$|J_2| = \left|\beta \int_\xi^t (t-s)^{\beta-1} A^0(s) ds\right| \leq (t-\xi)^\beta V(t) = U_{\frac{\beta}{\alpha}}(t).$$

Since we have

$$J_1 = \beta \int_0^\xi (t-s)^{\beta-\alpha}(t-s)^{\alpha-1} A^0(s) ds$$
$$= \beta(t-\xi)^{\beta-\alpha} \int_u^\xi (t-s)^{\alpha-1} A^0(s) ds, \ 0 \leq u \leq \xi.$$
$$= \frac{\beta}{\alpha}(t-\xi)^{\beta-\alpha}(A^\alpha(\xi) - A^\alpha(u)),$$

we have the estimate of J_1 as

$$|J_1| \leq 2\frac{\beta}{\alpha}(t-\xi)^{\beta-\alpha} W(t) = 2\frac{\beta}{\alpha} U_{\frac{\beta}{\alpha}}(t).$$

Consequently, we get the conclusion

$$|A^\beta(t)| \leq 3 U_{\frac{\beta}{\alpha}}(t).$$

On the other hand, for the case $\alpha > 1$ and $h = [\alpha]$, let

$$\zeta = \frac{t-\xi}{h+1}, \ \alpha - h = \delta \in [0,1).$$

We first assume that $\beta \in \mathbb{Z}_+$ and use induction to prove the statement. Clearly the conclusion holds for $\beta = 0$.

We assume that $\beta \in \mathbb{N}$ and
$$A^{\beta-1}(t) = O\left(U_{\frac{\beta-1}{\alpha}}(t)\right).$$

Then, for $\xi \le t' < t$, it follows that

$$\begin{aligned}
A^\beta(t) - A^\beta(t') &= \beta \int_{t'}^{t} A^{\beta-1}(s) ds \\
&= O\left((t-t') \cdot U_{\frac{\beta-1}{\alpha}}(t)\right) \\
&= O\left\{\left(\frac{W(t)}{V(t)}\right)^{\frac{1}{\alpha}} \cdot W(t)^{\frac{\beta-1}{\alpha}} V(t)^{t-\frac{\beta-1}{\alpha}}\right\} \\
&= O\left(U_{\frac{\beta}{\alpha}}(t)\right). \quad (1.3.15)
\end{aligned}$$

In the following, we define the finite difference. For $m \in \mathbb{Z}_+$, $m \le h$,

$$\triangle_{-\zeta}^m A^h(t) = \sum_{\nu=0}^{m} (-1)^\nu C_m^\nu A^h(t - \nu\zeta).$$

For $\delta \in (0,1)$, we define

$$\triangle_{-\zeta}^\delta f(t) = \delta \int_{t-\zeta}^{t} (t-s)^{\delta-1} f(s) ds.$$

It is obvious that

$$\triangle_{-\zeta}^\delta \cdot \triangle_{-\zeta}^m = \triangle_{-\zeta}^m \cdot \triangle_{-\zeta}^\delta.$$

We denote the operation by $\triangle_{-\zeta}^{m+\delta}$. Since we have

$$\begin{aligned}
A^{h-1}(t) &= \sum_{|m|<\sqrt{t}} (t - |m|^2)^{h-1} e^{im \cdot x^0} \\
&= \sum_{m \in \mathbb{Z}^n} \chi_{[0,\sqrt{t})}(|m|)(t - |m|^2)^{h-1} e^{im \cdot x^0},
\end{aligned}$$

1.3 Convergence and the opposite results

we conclude that

$$\int_{t-\zeta}^{t} A^{h-1}(s)ds$$

$$= \left(\sum_{|m|<\sqrt{t-\zeta}} + \sum_{\sqrt{t-\zeta}\leq|m|} \right) \int_{t-\zeta}^{t} \chi_{[0,\sqrt{s}]}(|m|)(s-|m|^2)^{h-1} e^{im\cdot x^0} ds$$

$$= \sum_{|m|<\sqrt{t-\zeta}} \int_{t-\zeta}^{t} (s-|m|^2)^{h-1} ds\, e^{im\cdot x^0}$$

$$+ \sum_{\sqrt{t-\zeta}\leq|m|<\sqrt{t}} \int_{|m|^2}^{t} (s-|m|^2)^{h-1} ds\, e^{im\cdot x^0}$$

$$= \frac{1}{h}(A^h(t) - A^h(t-\zeta)) = \frac{1}{h}\Delta_{-\zeta} A^h(t).$$

Then we obtain by induction that

$$\Delta_{-\zeta}^m A^h(t) = \frac{\Gamma(h+1)}{\Gamma(h-m+1)} \int_{t-\zeta}^{t} dt_1 \int_{t_1-\zeta}^{t_1} dt_2 \cdots \int_{t_{m-1}-\zeta}^{t_{m-1}} A^{h-m}(t_m) dt_m$$

$$= \frac{\Gamma(h+1)}{\Gamma(h-m+1)} \left(\zeta^m A^{h-m}(t) \right.$$

$$\left. + \int_{t-\zeta}^{t} dt_1 \cdots \int_{t_{m-1}-\zeta}^{t_{m-1}} \left(A^{h-m}(t_m) - A^{h-m}(t) \right) dt_m \right),$$

and

$$\Delta_{-\zeta}^{m+\delta} A^h(t) = \frac{\Gamma(h+1)}{\Gamma(h-m+1)} \zeta^m \Delta_{-\zeta}^{\delta} A^{h-m}(t) + \frac{\Gamma(h+1)}{\Gamma(h-m+1)} \Delta_{-\zeta}^{\delta}$$

$$\times \left(\int_{t-\zeta}^{t} dt_1 \cdots \int_{t_{m-1}-\zeta}^{t_{m-1}} \left(A^{h-m}(t_m) - A^{h-m}(t) \right) dt_m \right).$$

In the above equation, the symbol $A^{h-m}(t)$ denotes the value of A^{h-m} at t, which does not participate in the operation of finite difference. We hence get

$$\Delta_{-\zeta}^{\delta}(A^{h-m}(t)) = A^{h-m}(t)\zeta^{\delta}.$$

Substituting this into the above equation and transposing the terms, we have

$$\zeta^{m+\delta}A^{h-m}(t) = \frac{\Gamma(h-m+1)}{\Gamma(h+1)}\triangle_{-\zeta}^{m+\delta}A^h(t)$$
$$- \triangle_{-\zeta}^{\delta}\left(\int_{t-\zeta}^{t}dt_1 \cdots \int_{t_{m-1}-\zeta}^{t_{m-1}}\left(A^{h-m}(t_m) - A^{h-m}(t)\right)dt_m\right). \tag{1.3.16}$$

Substitute $m = h - \beta$ into (1.3.16) and note $\beta \in \mathbb{Z}_+$.

If $\delta = 0$, then $h = [\alpha] = \alpha$. By the condition $|A^\alpha(f)| \leq W(t)$, we have that
$$\triangle_{-\zeta}^{m+\delta}A^h(t) = O(W(t)).$$

If $\delta \in (0,1)$, then we first consider $\triangle_{-\zeta}^{\delta}A^h(t)$.

Let $0 \leq s < u < t$ and
$$\theta(t,u;v) = \frac{(t-u)^\delta}{\Gamma(\delta)\Gamma(1-\delta)(t-v)(u-v)^\delta}, \quad v \in (-\infty, u).$$

Then we get
$$\int_s^u \theta(t,u;v)(v-s)^{\delta-1}dv = \frac{(t-u)^\delta}{\Gamma(\delta)\Gamma(1-\delta)}\int_s^u \frac{(v-s)^{\delta-1}}{(t-v)(u-v)^\delta}dv$$
$$= \frac{(t-u)^\delta}{\Gamma(\delta)\Gamma(1-\delta)(u-s)}\int_0^1 \frac{v^{\delta-1}}{\left(\frac{t-s}{u-s} - v\right)(1-v)^\delta}dv. \tag{1.3.17}$$

Set $a = \frac{u-s}{t-s} \in (0,1)$. By the expansion
$$\frac{1}{1-av} = \sum_{k=0}^{\infty}(av)^k,$$

we obtain that
$$\int_0^1 \frac{v^{\delta-1}}{\left(\frac{t-s}{u-s} - v\right)(1-v)^\delta}dv = a\sum_{k=0}^{\infty}\int_0^1 v^{k+\delta-1}(1-v)^{-\delta}dv\, a^k$$
$$= a\sum_{k=0}^{\infty}\frac{\Gamma(k+\delta)\Gamma(1-\delta)}{\Gamma(k+1)}a^k$$
$$= a\Gamma(\delta)\Gamma(\delta)(1-a)^{-\delta}$$
$$= \Gamma(\delta)\Gamma(1-\delta)\frac{(u-s)(t-s)^{\delta-1}}{(t-u)^\delta}.$$

1.3 Convergence and the opposite results

Substituting this into (1.3.17), we have

$$(t-s)^{\delta-1} = \int_s^u \theta(t,u;v)(v-s)^{\delta-1}dv. \tag{1.3.18}$$

For $0 < u < t$, by making use of (1.3.18), we get

$$\int_0^u (t-s)^{\delta-1} A^h(s)ds = \int_0^u \left(\int_s^u \theta(t,u;v)(v-s)^{\delta-1}dv A^h(s)\right)ds$$
$$= \int_0^u \theta(t,u;v)\left(\int_0^v (v-s)^{\delta-1} A^h(s)\right)dv. \tag{1.3.19}$$

It is easy to deduce the following formula from the equation (1.3.14),

$$A^{h+\delta}(t) = \frac{\Gamma(h+\delta+1)}{\Gamma(h+1)\Gamma(\delta)} \int_0^t (t-s)^{\delta-1} A^h(s)ds, \tag{1.3.20}$$

where h, δ and t are arbitrary positive numbers, which yields

$$\int_0^v (v-s)^{\delta-1} A^h(s)ds = \frac{\Gamma(h+1)\Gamma(\delta)}{\Gamma(h+\delta+1)} A^\alpha(v), \quad \alpha = h+\delta. \tag{1.3.21}$$

Substituting this into (1.3.19), then we get

$$\int_0^u (t-s)^{\delta-1} A^h(s)ds = \frac{\Gamma(h+1)\Gamma(\delta)}{\Gamma(h+\delta+1)} \int_0^u \theta(t,u;v) A^\alpha(v)dv. \tag{1.3.22}$$

Therefore, we get the conclusion that

$$\Delta^\delta_{-\zeta} A^h(t) = \delta \int_{t-\zeta}^t (t-s)^{\delta-1} A^h(s)ds$$
$$= \delta \int_0^t (t-s)^{\delta-1} A^h(s)ds - \delta \int_0^{t-\zeta} (t-s)^{\delta-1} A^h(s)ds.$$

By (1.3.20) and (1.3.21), we derive from the above equation that

$$\Delta^\delta_{-\zeta} A^h(t) = \frac{\Gamma(h+1)\Gamma(\delta+1)}{\Gamma(h+\delta+1)} \left(A^\alpha(t) - \int_0^{t-\zeta} \theta(t,t-\zeta;v) A^\alpha(v)dv\right). \tag{1.3.23}$$

Due to the condition

$$|A^\alpha(v)| \leq W(v) \leq W(t), \quad 0 \leq v \leq t,$$

we have
$$\left|\int_0^u \theta(t,u;v)A^\alpha(v)dv\right| \leq W(t)\int_0^u \theta(t,u;v)dv$$
$$\leq W(t)\int_{-\infty}^u \theta(t,u;v)dv \qquad (1.3.24)$$
$$= W(t),$$

where we use the fact $\theta(t,u;v) > 0$ and
$$\int_{-\infty}^u \frac{(t-u)^\delta}{\Gamma(\delta)\Gamma(1-\delta)(t-v)(u-v)^\delta}dv = \int_0^\infty \frac{1}{\Gamma(\delta)\Gamma(1-\delta)(1+s)s^\delta}ds = 1.$$

Combining (1.3.23) and (1.3.24), we have
$$\triangle_{-\zeta}^\delta A^h(t) = O(W(t)).$$

Consequently, taking account of
$$m = h - \beta, \quad \alpha = h + \delta,$$

we have
$$\triangle_{-\zeta}^{\alpha-\beta} A^h(t) = \triangle_{-\zeta}^{h-\beta}\left(\triangle_{-\zeta}^\delta A^h(t)\right) = O(W(t)).$$

By the above discussion, the first term on the right side of (1.3.16) is $O(W(t))$. In the integral of the second term on the right side of (1.3.16), it is obvious that
$$t \geq t_m \geq t - m\zeta.$$

Since
$$\zeta = \frac{t-\xi}{h+1}$$

and
$$m\tau = \frac{h-\beta}{h+1}(t-\xi) < t - \xi,$$

we have
$$\xi < t_m < t.$$

Therefore, by (1.3.15), we have
$$\left|A^{h-m}(t_m) - A^{h-m}(t)\right| = O\left(U_{\frac{\beta}{\alpha}}(t)\right)$$

1.3 Convergence and the opposite results

and
$$\int_{t-\tau}^{t} dt_1 \cdots \int_{t_{m-1}-\tau}^{t_{m-1}} \left(A^\beta(t_m) - A^\beta(t) \right) dt_m = O\left(U_{\frac{\beta}{\alpha}}(t) \right) \tau^m,$$

where $\beta = h - m$.

The second term on the right side of (1.3.16) is $O\left(U_{\frac{\beta}{\alpha}}(t) \right) \tau^{\alpha-\beta}$, then we have
$$\tau^{\alpha-\beta} A^\beta(t) = O(W(t)) + \tau^{\alpha-\beta} O\left(U_{\frac{\beta}{\alpha}}(t) \right).$$

By noticing
$$\tau = \frac{1}{h+1}(t - \xi)$$

and
$$t - \xi = \left(\frac{W(t)}{V(t)} \right)^{\frac{1}{\alpha}},$$

we get
$$A^\beta(t) = O\left(U_{\frac{\beta}{\alpha}}(t) \right).$$

This is included in case (ii) when $\beta \in \mathbb{Z}_+$.

Now assume that β is not an integer and let
$$\beta = [\beta] + \gamma, \ 0 < \gamma < 1.$$

If $[\beta] < [\alpha]$, then by the proven result, we have that
$$A^{[\beta]}(t) = O\left(U_{\frac{[\beta]}{\alpha}}(t) \right)$$

and
$$A^{[\beta]+1}(t) = O\left(U_{\frac{[\beta]+1}{\alpha}}(t) \right).$$

Paying attention to (1.3.20), we apply the interpolation result between $A^{[\beta]}(t)$ and $A^{[\beta]+1}(t)$ as in case (i), and then have
$$A^\beta(t) = O\left(\left(U_{\frac{[\beta]}{\alpha}}^{1-\gamma} U_{\frac{[\beta]+1}{\alpha}}^{\gamma} \right)(t) \right) = O\left(U_{\frac{\beta}{\alpha}}(t) \right).$$

If $[\beta] = [\alpha]$, then we directly interpolate between $A^{[\beta]}$ and A^α and get
$$A^\beta(t) = O\left((U_{\frac{[\beta]}{\alpha}}(t))^{1-\frac{\gamma}{\delta}} (W(t))^{\frac{\gamma}{\delta}} \right),$$

where $\delta = \alpha - [\alpha] \geq \gamma > 0$.

Consequently, we conclude that

$$A^\beta(t) = O\left\{V(t)^{\left(1-\frac{|\alpha|}{\alpha}\right)\left(1-\frac{\gamma}{\delta}\right)}W(t)^{\frac{|\alpha|}{\alpha}\left(1-\frac{\gamma}{\delta}\right)+\frac{\gamma}{\delta}}\right\}$$
$$= O\left(V(t)^{1-\frac{\beta}{\alpha}}W(t)^{\frac{\beta}{\alpha}}\right)$$
$$= O\left(U_{\frac{\beta}{\alpha}}(t)\right).$$

This completes the proof of Lemma 1.3.6. ∎

Remark 1.3.1 *Lemma 1.3.6 is actually a part of Theorem 1.71 by Chandrasekharan and Minakshisundaram [CM1].*

Lemma 1.3.7 *Let $x^0 \in S$. Then we have*

$$\sup_{R>1} \frac{1}{R^{\frac{n-1}{2}}} \left|D_R^0(x^0)\right| = +\infty. \tag{1.3.25}$$

Proof. Suppose that (1.3.25) is false, that is, there exists $M > 0$ such that

$$\left|D_R^0(x^0)\right| \leq MR^{\frac{n-1}{2}}, \quad R \geq 1.$$

We use the notations in Lemma 1.3.6, that is,

$$\left|A^0(t)\right| = \left|D_{\sqrt{t}}^0(x^0)\right| \leq Mt^{\frac{n-1}{4}} \quad (t \geq 1).$$

On the other hand, by (1.3.7), we get that

$$\left|A^{\frac{n+1}{2}}(t)\right| = \left|t^{\frac{n+1}{2}}D_{\sqrt{t}}^{\frac{n+1}{2}}(x^0)\right| \leq Ct^{\frac{n}{2}}.$$

Then according to Lemma 1.3.6 and interpolating at $\beta = \frac{n-1}{2}$, we obtain that

$$\left|A^{\frac{n-1}{2}}(t)\right| \leq Ct^{\frac{n-1}{4}\left(1-\frac{n-1}{n+1}\right)}t^{\frac{n}{2}\cdot\frac{n-1}{n+1}} = Ct^{\frac{n-1}{2}},$$

for $t > 1$, which contradicts with the conclusion in Lemma 1.3.5. And thus (1.3.25) holds. ∎

1.3 Convergence and the opposite results

Lemma 1.3.8 *Let $x^0 \in S$. Then there holds*

$$\sup_{R>0} \left| \sum_{0<|m|<R} \frac{1}{|m|^{\frac{n-1}{2}}} e^{im \cdot x_0} \right| = +\infty. \tag{1.3.26}$$

Proof. We suppose that the item on the left side of (1.3.26) is a finite number M. Let

$$\sigma_t = \sum_{0<|m|<t} \frac{e^{im \cdot x_0}}{|m|^{\frac{n-1}{2}}},$$

for $t > 0$, which is left-continuous step function on $(0, +\infty)$. It obviously follows that

$$\sum_{1 \leq |m| < R} e^{im \cdot x_0} = \int_1^R t^{\frac{n-1}{2}} d\sigma_t$$

$$= R^{\frac{n-1}{2}} \sigma_R - \sigma_1 - \frac{n-1}{2} \int_1^R t^{\frac{n-3}{2}} \sigma_t dt$$

$$= R^{\frac{n-1}{2}} \sigma_R - \frac{n-1}{2} \int_1^R t^{\frac{n-3}{2}} \sigma_t dt.$$

Then we have

$$\left| \sum_{1 \leq |m| < R} e^{im \cdot x_0} \right| \leq M R^{\frac{n-1}{2}} + \frac{n-1}{2} \int_1^R M t^{\frac{n-3}{2}} dt \leq 2 M R^{\frac{n-1}{2}}.$$

Consequently, we have

$$\frac{1}{R^{\frac{n-1}{2}}} |D_R^0(x^0)| \leq 2M,$$

for $R \geq 1$, which contradicts with Lemma 1.3.7 and thus (1.3.26) is valid. ∎

Corollary 1.3.1 *The equality (1.3.26) is valid almost everywhere on $\mathbb{R}^n$.*

Lemma 1.3.9 *If $0 < \alpha < n$, then the series*

$$\sum_{m \neq 0} \frac{1}{|m|^\alpha} e^{im \cdot x} \tag{1.3.27}$$

is the Fourier series of the function $F = F_\alpha$, where F_α has the same singular properties at $x = 0$ as $\frac{1}{|x|^{n-\alpha}}$ and $F_\alpha \in C(Q \backslash \{0\})$.

Proof. Let $\eta \in C^\infty(\mathbb{R}^n)$ satisfying
$$\eta(x) = \begin{cases} 1, & \text{if } |x| \geq 1, \\ 0, & \text{if } |x| < \frac{1}{2}. \end{cases}$$

Define
$$G(x) = \frac{\eta(x)}{|x|^\alpha}.$$

If $x \neq 0$, we have
$$G(x) = \frac{1}{|x|^\alpha} + (\eta(x) - 1)\frac{1}{|x|^\alpha}.$$

By a theorem on the the Fourier transforms of radial functions (see [SW1]), we get that as a generalized slowly increasing function, the Fourier transform of $\frac{1}{|x|^\alpha}$ is

$$\mathscr{F}\left(\frac{1}{|x|^\alpha}\right)(y) = \gamma_\alpha \frac{1}{|y|^{n-\alpha}}, \qquad (1.3.28)$$

where
$$\gamma_\alpha = \frac{1}{2^\alpha \pi^{n/2}} \frac{\Gamma(\frac{n-\alpha}{2})}{\Gamma(\alpha/2)}.$$

On the other hand,
$$(\eta(x) - 1)\frac{1}{|x|^\alpha}$$

is an integrable function supported on $\{x : |x| < 1\}$, whose normal Fourier transform denoted by b_1, and $b_1 \in C_0^\infty(\mathbb{R}^n)$, where $C_0^\infty(\mathbb{R}^n)$ refers to a class of infinitely differentiable functions whose derivatives of any order all vanish at infinity. We define

$$f(y) = (2\pi)^n \left(\frac{\gamma_\alpha}{|y|^{n-\alpha}} + b_1(y)\right). \qquad (1.3.29)$$

For any $\beta = (\beta_1, \ldots, \beta_n) \in \mathbb{Z}_+^n$ and $\gamma = (\gamma_1, \ldots, \gamma_n) \in \mathbb{Z}_+^n$, the Fourier transform is as follows:

$$\mathscr{F}\left[(-ix)^\gamma (D^\beta f)(x)\right](y) = D^\gamma \left[\mathscr{F}(D^\beta f)\right](y)$$
$$= D^\gamma \left[(iy)^\beta \mathscr{F}(f)\right](y)$$
$$= D^\gamma \left[(iy)^\beta G(y)\right].$$

Here, we utilize a theorem on the operation law of the Fourier transform (see [SW1]).

1.3 Convergence and the opposite results

Since $G(y)(iy)^\beta$ vanishes when $|y| < 1/2$ and equals to $(iy)^\beta$ when $|y| > 1$, for any $\beta \in \mathbb{Z}_+^n$, as long as $\gamma \in \mathbb{Z}_+^n$ big enough, we can make sure that

$$D^\gamma \left((iy)^\beta G(y)\right) \in L(\mathbb{R}^n).$$

Then we have

$$(2\pi)^n \mathscr{F}\left\{D^\gamma\left[(iy)^\beta G(y)\right]\right\}(-x) = (-ix)^\gamma \left(D^\beta f\right)(x).$$

The left side of the above equation is the normal the Fourier transform of the function in $L(\mathbb{R}^n)$ and thus it is bounded. Taking $\beta = 0$, we get that

$$|f(x)| = O\left(|x|^{-N}\right), \qquad (1.3.30)$$

as $|x| \to \infty$, for any $N > 0$. So we obtain that $f \in L(\mathbb{R}^n)$ and hence it follows from $\hat{f} = G$ in normal sense that

$$\hat{f}(0) = G(0) = 0$$

and

$$\hat{f}(m) = \frac{1}{|m|^\alpha} \quad |m| \geq 1.$$

From (1.3.30),

$$\sum_{m \neq 0} f(x + 2\pi m)$$

is uniformly convergent on Q. Set

$$b(x) = (2\pi)^n b_1(x) + \sum_{m \neq 0} f(x + 2\pi m),$$

which leads to $b \in C(\mathbb{R}^n)$. Moreover, by Theorem 1.2.1, the Fourier series of the function in $L(Q)$

$$F_\alpha(x) := \sum f(x + 2\pi m)$$

is

$$\sigma(F_\alpha)(x) = \sum \hat{f}(m) e^{im \cdot x} = \sum_{m \neq 0} \frac{1}{|m|^\alpha} e^{im \cdot x}.$$

Obviously, if $x \neq 0$, we have

$$F(x) = F_\alpha(x) = \frac{(2\pi)^n \gamma_\alpha}{|x|^{n-\alpha}} + b(x). \qquad (1.3.31)$$

This completes the proof. ∎

The proof of Theorem 1.3.1. Suppose that $n > 1$ and denote the value of $(2\pi)^n \gamma_\alpha$ at $\alpha = \frac{n-1}{2}$ by γ. Taking $\alpha = \frac{n-1}{2}$, we have

$$F(x) = \frac{\gamma}{|x|^{\frac{n+1}{2}}} + b(x),$$

where $b \in C(\mathbb{R}^n))$. Then for $p \in \left(1, \frac{2n}{n+1}\right)$, we have $F \in L^p(Q)$. Since we have

$$S_R^0(F; x) = \sum_{0 < |m| < R} \frac{1}{|m|^{\frac{n-1}{2}}} e^{im \cdot x},$$

by the above equation and Corollary 1.3.1.

This finishes the proof of Theorem 1.3.1. ∎

At present, it is still an unresolved problem whether the Fourier series of $L^p(Q)$, $p \geq 2$, $n > 1$ is convergent almost everywhere.

1.4 Linear summation

Similar to the unitary case, in order to investigate functions by the method of Fourier series which has bad convergence property, we introduce various kinds of linear means. We only discuss some special linear means here.

Suppose that $\Phi \in C(\mathbb{R}^n)$ satisfies

(i) $|\Phi(x)| \leq A(1 + |x|)^{-n-\delta}$, $\delta > 0$;

(ii) $\Phi(0) = 1$.

For $f \in L(Q)$ and $\varepsilon > 0$, we define the series by

$$\sigma_\varepsilon^\Phi(f)(x) = \sum_{m \in \mathbb{Z}^n} \Phi(\varepsilon m) C_m(f) e^{im \cdot x}, \quad (1.4.1)$$

which is an absolutely convergent series by the condition (i), which is called the Φ means of $\sigma(f)$. We consider the relation between $\sigma_\varepsilon^\Phi(f)(x)$ and $f(x)$ as $\varepsilon \to 0^+$.

1.4 Linear summation

Theorem 1.4.1 *Suppose $f \in L^p(Q)$ ($1 \le p \le \infty$). If Φ also satisfies the following condition:*

(iii) $|\widehat{\Phi}(y)| \le A(1+|y|)^{-n-\delta}, \delta > 0$,

then (1.4.1) converges to f in the norm of $L^p(Q)$ and converges to f at the Lebesgue points of f.

Proof. By the condition (iii), we have that the function

$$\varphi(y) := (2\pi)^n \widehat{\Phi}(-y) \quad (\in C(\mathbb{R}^n)) \tag{1.4.2}$$

belongs to $L(\mathbb{R}^n)$. It follows that $\widehat{\varphi} = \Phi$. By letting

$$\varphi_\varepsilon(y) = \varepsilon^{-n} \varphi(\varepsilon^{-1} y),$$

we get $\widehat{\varphi_\varepsilon}(x) = \Phi(\varepsilon x)$. Since φ_ε satisfies the conditions of Theorem 1.2.2, we have

$$K_\varepsilon(x) := \sum \varphi_\varepsilon^0(x+2\pi m) = \sum \Phi(\varepsilon m) e^{im\cdot x} \in C(Q). \tag{1.4.3}$$

Taking convolution on Q, we get that

$$\begin{aligned} f * K_\varepsilon(x) &:= \frac{1}{(2\pi)^n} \int_Q f(y) K_\varepsilon(x-y) dy \\ &= \sum_{m \in \mathbb{Z}^n} \left(\frac{1}{(2\pi)^n} \int_Q f(y) e^{-im\cdot y} dy \, \Phi(\varepsilon m) e^{im\cdot x} \right) \\ &= \sum_{m \in \mathbb{Z}^n} C_m(f) \Phi(\varepsilon m) e^{im\cdot x} \\ &= \sigma_\varepsilon^\Phi(f)(x). \end{aligned} \tag{1.4.4}$$

By the generalized Minkowski's inequality, we have

$$\left\| \sigma_\varepsilon^\Phi(f) \right\|_p \le \frac{1}{(2\pi)^n} \|f\|_p \|K_\varepsilon\|_1.$$

It easily follows that σ_ε^Φ is a bounded linear operator from $L^p(Q)$ to $L^p(Q)$ and its operator norm satisfies that

$$\begin{aligned} \|\sigma_\varepsilon^\Phi\| &\le \frac{1}{(2\pi)^n} \|K_\varepsilon\|_1 \\ &\le \frac{1}{(2\pi)^n} \sum_{m \in \mathbb{Z}^n} \int_Q |\varphi_\varepsilon(x+2\pi m)| dx \\ &= \frac{1}{(2\pi)^n} \int_{\mathbb{R}^n} |\varphi_\varepsilon(x)| dx \\ &= \frac{1}{(2\pi)^n} \int_{\mathbb{R}^n} |\varphi(x)| dx. \end{aligned} \tag{1.4.5}$$

Furthermore, for any triangle polynomial $T(x)$, by the condition (ii), we have
$$\lim_{\varepsilon \to 0^+} \sigma_\varepsilon^\Phi(T)(x) = \sum_{m \in \mathbb{Z}^n} \lim_{\varepsilon \to 0^+} \Phi(\varepsilon m) C_m(T) e^{im \cdot x}$$
$$= T(x)$$
holds uniformly for $x \in Q$. Therefore, σ_ε^Φ converges to the identity operator in the $L^p(Q)$ norm. Then we obtain
$$\sigma_\varepsilon^\Phi(f) \to f,$$
in the sense of L^p norm, as $\varepsilon \to 0^+$, for any $f \in L^p(Q)$.

Now we turn to proving the convergence of σ_ε^Φ at the Lebesgue points. We might as well take the origin point into consideration. By (1.4.3) and (1.4.4), we obtain

$$\sigma_\varepsilon^\Phi(f)(0) = \frac{1}{(2\pi)^n} \int_Q f(y) K_\varepsilon(-y) dy$$
$$= \frac{1}{(2\pi)^n} \int_Q f(y) \varphi_\varepsilon(-y) dy$$
$$+ \frac{1}{(2\pi)^n} \int_Q f(y) \sum_{m \neq 0} \varphi_\varepsilon(-y + 2\pi m) dy.$$

For $m \neq 0$, $y \in Q$, we have
$$|\varphi_\varepsilon(-y + 2\pi m)| = \left| \varepsilon^{-n} \varphi\left(\frac{-y + 2\pi m}{\varepsilon}\right) \right|$$
$$\leq A \varepsilon^{-n} \left(1 + \left|\frac{-y + 2\pi m}{\varepsilon}\right|\right)^{-n-\delta}$$
$$= A \varepsilon^\delta (\varepsilon + |-y + 2\pi m|)^{-n-\delta}$$
$$\leq A \frac{\varepsilon^\delta}{|m|^{n+\delta}},$$

which yields
$$\left| \sum_{m \neq 0} \varphi_\varepsilon(-y + 2\pi m) \right| \leq A \varepsilon^\delta \sum_{m \neq 0} \frac{1}{|m|^{n+\delta}} \leq C \varepsilon^\delta.$$

It follows from the above estimate that
$$\sigma_\varepsilon^\Phi(f)(0) = \frac{1}{|Q|} \int_Q f(y) \varphi_\varepsilon(-y) dy + O(\varepsilon^\delta).$$

1.4 Linear summation

By the condition (ii), that is,

$$\frac{1}{(2\pi)^n}\int_{\mathbb{R}^n}\varphi(y)dy = 1,$$

we get

$$\frac{1}{|Q|}\int_{\mathbb{R}^n}\varphi_\varepsilon(-y)dy = 1.$$

Hence, we have

$$\sigma_\varepsilon^\Phi(f)(0) - f(0) = \frac{1}{|Q|}\int_{\mathbb{R}^n}(f(y)x_Q(y) - f(0))\varphi_\varepsilon(-y)dy + \mathrm{O}(\varepsilon^\delta).$$

For the kernel φ_ε, we have the following estimate

$$|\varphi_\varepsilon(-y)| \le C\varepsilon^{-n}\left(1+\frac{|y|}{\varepsilon}\right)^{-n-\delta} \le \begin{cases} C\varepsilon^{-n}, & \text{if } |y| < \varepsilon, \\ C\dfrac{\varepsilon^\delta}{|y|^{n+\delta}}, & \text{if } |y| \ge \varepsilon. \end{cases}$$

Since $x = 0$ is the Lebesgue point of f, for any $\eta > 0$, there exists $0 < \sigma < 1$ such that, for $r \in (0, \sigma)$, we have

$$\frac{1}{r^n}\int_{|y|<r}|f(y) - f(0)|dy < \eta.$$

Letting $\varepsilon > \sigma$, we have

$$\int_{\mathbb{R}^n}(f(y)x_Q(y) - f(0))\,\varphi_\varepsilon(-y)dy$$

$$= \left(\int_{|y|<\varepsilon} + \int_{\varepsilon\le|y|<\sigma} + \int_{|y|\ge\sigma}\right)(f(y)x_Q(y) - f(0))\,\varphi_\varepsilon(-y)dy$$

$$:= I_1 + I_2 + I_3.$$

We have the estimates of I_1, I_2 and I_3 as

$$|I_1| \le \int_{|y|<\varepsilon}|f(y) - f(0)|\frac{C}{\varepsilon^n}dy < C\eta,$$

$$|I_2| \leq \int_{\varepsilon \leq |y| < \sigma} |f(y) - f(0)| \frac{C\varepsilon^\delta}{|y|^{n+\delta}} dy$$

$$= C\varepsilon^\delta \int_\varepsilon^\sigma \left(\frac{1}{r^{n+\delta}} \int_{|y|=r} |f(y) - f(0)| d\sigma(y) \right) dr$$

$$= C\varepsilon^\delta \left(\frac{1}{r^{n+\delta}} \int_0^r \int_{|y|=\tau} |f(y) - f(0)| d\sigma(y) d\tau \right) \Big|_{r=\varepsilon}^{r=\sigma}$$

$$+ (n+\delta) \int_\varepsilon^\sigma \left(\int_{|y|<r} |f(y) - f(0)| dy \right) \frac{dr}{r^{n+\delta+1}}$$

$$\leq C\eta + C\varepsilon^\delta \int_\varepsilon^\sigma \frac{\eta}{\gamma^{\delta+1}} d\gamma$$

$$\leq C\eta,$$

and

$$|I_3| \leq \int_{|y| \geq \sigma,\, y \in Q} |f(y)| \frac{C\varepsilon^\delta}{|y|^{n+\delta}} dy + \int_{|y|>\sigma} |f(0)| \frac{C\varepsilon^\delta}{|y|^{n+\delta}} dy$$

$$\leq C\|f\|_1 \frac{1}{\sigma^{n+\delta}} \varepsilon^\delta + |f(0)| C\varepsilon^\delta \int_{|y|>\sigma} \frac{dy}{|y|^{n+\delta}}$$

$$= O(\varepsilon^\delta).$$

Due to the above discussion, we have

$$\limsup_{\varepsilon \to 0^+} \left| \sigma_\varepsilon^\Phi(f)(0) - f(0) \right| \leq C\eta,$$

for any $\eta > 0$. This implies

$$\lim_{\varepsilon \to 0^+} \sigma_\varepsilon^\Phi(f)(0) = f(0).$$

This completes the proof. ∎

Definition 1.4.1 *For $f \in L(Q)$,*

$$A_\varepsilon(f)(x) = \sum_{m \in \mathbb{Z}^n} e^{-\varepsilon|m|} C_m(f) e^{im \cdot x} \tag{1.4.6}$$

is called the Abel-Poisson means of f.

We denote by $e^{-\varepsilon} = r \in (0,1)$, then (1.4.6) can be rewritten as

$$P(f)(r; x) = \sum_{m \in \mathbb{Z}^n} r^{|m|} C_m(f) e^{im \cdot x}. \tag{1.4.7}$$

1.4 Linear summation

Definition 1.4.2 *For $f \in L(Q)$,*

$$W_\varepsilon(f;x) = \sum_{m\in\mathbb{Z}^n} e^{-|\varepsilon m|^2} C_m(f) e^{im\cdot x} \tag{1.4.8}$$

is called the Gauss-Weierstrass means of f.

Definition 1.4.3 *Suppose that α is a complex number whose real part is bigger than -1 and $f \in L(Q)$. The triangle polynomial*

$$S_R^\alpha(f;x) = \sum_{|m|<R} \left(1 - \frac{|m|^2}{R^2}\right)^\alpha C_m(f) e^{im\cdot x} \tag{1.4.9}$$

is called the Bochner-Riesz means of f and α is called its degree (or index).

Substituting the integral expression of Fourier coefficients $C_m(f)$ into (1.4.9), we get the integral form of $S_R^\alpha(f;x)$ as

$$S_R^\alpha(f;x) = \frac{1}{(2\pi)^n} \int_Q f(y) D_R^\alpha(x-y) dy, \tag{1.4.10}$$

where the kernel $D_R^\alpha(x)$ is defined as in (1.3.4).

We all know that the kernel of Abel-Poisson means $\Phi(x) = e^{-|x|}$ is the Fourier transform of the function

$$\varphi(y) = C_n \left(1 + |y|^2\right)^{-\frac{n+1}{2}},$$

satisfying the conditions in Theorem 1.4.1. Therefore, $A_\varepsilon(f)(x)$ is convergent in L^p norm and also convergent at Lebesgue points. The Gauss-Weierstrass kernel $e^{-|x|^2}$ is the Fourier transform of the function

$$\varphi(y) = \pi^{\frac{n}{2}} e^{-\frac{1}{4}|y|^2},$$

also satisfying the conditions in Theorem 1.4.1. So $W_\varepsilon(f)$ is also convergent in L^p norm and convergent at Lebesgue points at the same time.

As far as Bochner-Riesz means is concerned, the Fourier transform of the kernel

$$\Phi^\alpha(x) = \begin{cases} (1-|x|^2)^\alpha, & |x| < 1, \\ 0, & |x| \geq 1, \end{cases}$$

is

$$B^\alpha(x) = \frac{2^\alpha \Gamma(\alpha+1)}{(2\pi)^{\frac{n}{2}}} \frac{J_{\frac{n}{2}+\alpha}(x)}{|x|^{\frac{n}{2}+\alpha}}.$$

If $\alpha > \alpha_0 = \frac{n-1}{2}$, or more generally, $\operatorname{Re}\alpha > \alpha_0$, then we have $B^\alpha \in L(\mathbb{R}^n)$. Consequently, Φ^α is the Fourier transform of $(2\pi)^n B^\alpha$ and satisfies the conditions in Theorem 1.4.1. However, for the case of $\operatorname{Re}\alpha \leq \alpha_0$, Φ^α is no longer the Fourier transform of the integrable functions. The index (or order) $\alpha_0 = \frac{n-1}{2}$ is called the critical index both in Bochner-Riesz summation of the Fourier series and the Fourier integral. The index whose real part is bigger than α_0 is called over the critical index.

For the indivisible intrinsic relations between the Fourier integral and the Fourier series, we will discuss the Fourier integral firstly in the next chapter and focus on the discussion below the critical index. Chapter 3 will get back to the Bochner-Riesz means of the series and focus on the discussion for the critical index situations.

Chapter 2

BOCHNER-RIESZ MEANS OF MULTIPLE FOURIER INTEGRAL

2.1 Localization principle and classic results on fixed-point convergence

Suppose that $f \in L(\mathbb{R}^n)$ and $\hat{f}$ is the the Fourier transform of f. Then the Bochner-Riesz means of the Fourier integral of f is defined by

$$B_R^\alpha(f;x) = \int_{\mathbb{R}^n} f(x+y) B_R^\alpha(y) dy, \qquad (2.1.1)$$

for $\text{Re}\alpha > -1$, where the kernel $B_R^\alpha(y)$ is

$$B_R^\alpha(y) = \frac{2^\alpha \Gamma(\alpha+1)}{(2\pi)^{\frac{n}{2}}} \frac{J_{\frac{n}{2}+\alpha}(R|y|)}{(R|y|)^{\frac{n}{2}+\alpha}} R^n, \qquad (2.1.2)$$

and the index $\alpha_0 = \frac{n-1}{2}$ is called the critical index.

In this section, we mainly consider the case of

$$\frac{n-3}{2} < \alpha \leq \frac{n-1}{2}.$$

Suppose that f is locally integrable on $\mathbb{R}^n$, that is, f is integrable on any bounded set of $\mathbb{R}^n$, and we denote it by $f \in L_{loc}(\mathbb{R}^n)$. For each $x \in \mathbb{R}^n$, we define

$$f_x(t) = \frac{1}{\omega_n} \int_{\mathbb{S}^{n-1}} f(x+t\xi)d\sigma(\xi) = \frac{1}{\omega_n t^{n-1}} \int_{|\xi|=t} f(x+\xi)d\sigma(\xi) \quad (t>0).$$

Then equation (2.1.1) can be represented by

$$B_R^\alpha(f;x) = \frac{2^{\alpha+1}\Gamma(\alpha+1)}{2^{n/2}\Gamma(n/2)} R^n \int_0^\infty f_x(t) \frac{J_{\frac{n}{2}+\alpha}(Rt)}{(Rt)^{\frac{n}{2}+\alpha}} t^{n-1} dt. \qquad (2.1.3)$$

Bochner [Bo1] established the localization principle of $B_R^{\alpha_0}(f;x)$ converging at a point, where $f \in L(\mathbb{R}^n)$. That is, if $f \in L(\mathbb{R}^n)$, and there exists a real number $\delta > 0$ such that

$$f(x) = 0, \quad \text{for } |x-x_0| < \delta,$$

then we have

$$\lim_{R\to\infty} B_R^{\alpha_0}(f;x_0) = 0.$$

The above conclusion is included in the following convergence theorem.

Theorem 2.1.1 (Bochner) *Let $f \in L(\mathbb{R}^n)$ and $\alpha = \alpha_0 - \beta$, with $0 \le \beta < 1$. If there exists a number s such that*

$$\lim_{t\to 0_+} t^{-n} \int_0^t \tau^{n-1} |f_x(\tau) - s| d\tau = 0, \qquad (2.1.4)$$

then

$$\lim_{R\to\infty} B_R^\alpha(f;x) = s \qquad (2.1.5)$$

holds if and only if

$$\lim_{R\to\infty} R^\beta \int_{R^{-1}}^\infty (f_x(t) - s) \frac{\cos\left(tR - \frac{(n-\beta)\pi}{2}\right)}{t^{1-\beta}} dt = 0. \qquad (2.1.6)$$

Theorem 2.1.2 *Suppose that (2.1.4) holds. Then*

$$\lim_{R\to\infty} B_R^{\alpha_0}(f;x) = s \qquad (2.1.7)$$

holds if and only if for some positive number η, we have

$$\lim_{R\to\infty} \int_{R^{-1}}^\eta (f_x(t) - s) \frac{\cos\left(tR - \frac{n\pi}{2}\right)}{t} dt = 0. \qquad (2.1.8)$$

2.1 Localization principle and classic results on fixed-point convergence

Proof. Set
$$C_{n,\alpha} = 2^{\alpha+1}\Gamma(\alpha+1)\left(2^{n/2}\Gamma(\tfrac{n}{2})\right)^{-1}.$$

It follows from the equality (2.1.3) that

$$B_R^\alpha(f;x) - s = C_{n,\alpha} \int_0^\infty \left(f_x\left(\tfrac{t}{R}\right) - s\right) \frac{J_{n-\frac{1}{2}-\beta}(t)}{t^{\frac{1}{2}-\beta}} dt$$

$$= C_{n,\alpha} \int_0^1 \left(f_x\left(\tfrac{t}{R}\right) - s\right) t^{\beta-\frac{1}{2}} J_{n-\frac{1}{2}-\beta}(t) dt$$

$$+ C_{n,\alpha} \int_1^\infty \left(f_x\left(\tfrac{t}{R}\right) - s\right) t^{\beta-\frac{1}{2}} J_{n-\frac{1}{2}-\beta}(t) dt. \quad (2.1.9)$$

By $J_{n-\frac{1}{2}-\beta}(t) = O\left(t^{n-\frac{1}{2}-\beta}\right)$ as $t \to 0$, we immediately have

$$\int_0^1 \left(f_x\left(\tfrac{t}{R}\right) - s\right) t^{\beta-\frac{1}{2}} J_{n-\frac{1}{2}-\beta}(t) dt = O\left(\int_0^1 \left|f_x\left(\tfrac{t}{R}\right) - s\right| t^{n-1} dt\right).$$

By (2.1.4), we easily have

$$\int_0^1 \left(f_x\left(\tfrac{t}{R}\right) - s\right) t^{\beta-\frac{1}{2}} J_{n-\frac{1}{2}-\beta}(t) dt = o(1),$$

as $R \to \infty$.

From the equation

$$J_\nu(t) = \sqrt{\frac{2}{\pi t}} \cos\left(t - \frac{\nu\pi}{2} - \frac{\pi}{4}\right) + O\left(t^{-3/2}\right),$$

as $t \to +\infty$, we have

$$\int_1^\infty \left(f_x\left(\tfrac{t}{R}\right) - s\right) t^{\beta-\frac{1}{2}} J_{n-\frac{1}{2}-\beta}(t) dt$$

$$= \int_1^\infty \left(f_x\left(\tfrac{t}{R}\right) - s\right) \sqrt{\tfrac{2}{\pi}} \frac{\cos\left(t - \frac{n-\beta}{2}\pi\right)}{t^{1-\beta}} dt$$

$$+ O\left(\int_1^\infty \left|f_x\left(\tfrac{t}{R}\right) - s\right| \frac{dt}{t^{2-\beta}}\right),$$

where the second term can be estimated as

$$\int_1^\infty \left|f_x\left(\tfrac{t}{R}\right) - s\right| \frac{dt}{t^{2-\beta}} = t^{-(2-\beta+n-1)} \int_0^t \left|f_x\left(\tfrac{\tau}{R}\right) - s\right| \tau^{n-1} d\tau \Big|_1^\infty$$

$$+ O\left(\int_1^\infty t^{-(n+2-\beta)} \int_0^t \left|f_x\left(\tfrac{\tau}{R}\right) - s\right| \tau^{n-1} d\tau dt\right).$$

It follows from $f \in L(\mathbb{R}^n)$ that

$$\int_0^\infty |f_x(t)| t^{n-1} dt = O(1).$$

Together with (2.1.4) and $\beta < 1$, this leads to

$$\int_1^\infty \left| f_x\left(\frac{t}{R}\right) - s \right| \frac{dt}{t^{2-\beta}} = o(1) + O\left(\int_1^\infty \frac{t^n}{t^{n+2-\beta}} dt \right) = o(1),$$

as $R \to \infty$.

Consequently, we get that

$$B_R^\alpha(f;x) - s = C_{n,\alpha} \sqrt{\frac{2}{\pi}} \int_1^\infty \left(f_x\left(\frac{t}{R}\right) - s \right) \frac{\cos\left(t - \frac{n-\beta}{2}\pi\right)}{t^{1-\beta}} dt + o(1), \tag{2.1.10}$$

as $R \to \infty$.

This implies Theorem 2.1.1 holds.

Substituting $\beta = 0$ into (2.1.10), and letting $\eta > 0$, $R > \frac{1}{\eta}$, we conclude that

$$B_R^\alpha(f;x) - s = C_{n,\alpha} \sqrt{\frac{2}{\pi}} \int_1^{\eta R} \left(f_x(\frac{t}{R}) - s \right) \frac{\cos\left(t - \frac{n}{2}\pi\right)}{t} dt$$

$$+ C_{n,\alpha} \sqrt{\frac{2}{\pi}} \int_{\eta R}^\infty \left(f_x(\frac{t}{R}) - s \right) \frac{\cos\left(t - \frac{n}{2}\pi\right)}{t} dt + o(1),$$

as $R \to \infty$.

Since $f_x(t) t^{-1} \in L(\eta, \infty)$, by the Riemann-Lebesgue theorem, we obtain

$$\int_{\eta R}^\infty f_x\left(\frac{t}{R}\right) \frac{\cos\left(t - \frac{n}{2}\pi\right)}{t} dt = o(1)$$

and

$$\int_{\eta R}^\infty \frac{\cos\left(t - \frac{n}{2}\pi\right)}{t} dt = o(1),$$

as $R \to \infty$.

Therefore when $R \to \infty$, we immediately have

$$B_R^\alpha(f;x) - s = C_{n,\alpha} \sqrt{\frac{2}{\pi}} \int_{R^{-1}}^\eta (f_x(t) - s) \frac{\cos\left(Rt - \frac{n}{2}\pi\right)}{t} dt + o(1).$$

This completes the proof of Theorem 2.1.2. ∎

2.2 L^p-convergence

Remark 2.1.1 *In the proof, the notation* $\int_1^\infty$ *denotes* $\lim_{A\to+\infty}\int_1^A$.

Remark 2.1.2 *The condition (2.1.4) is essentially the one for the Lebesgue points.*

Remark 2.1.3 *The localization principle can be deduced from Theorem 2.1.2. If $\alpha < \alpha_0$, the localization property of $B_R^\alpha(f;x)$ at some points, where $f \in L(\mathbb{R}^n)$ is not valid. For the cases when the order is lower than the critical order, Pan [P1] has proven that $B_R^\alpha(f;x)$ of $f \in L_m^1(\mathbb{R}^n)$ still has the localization property of convergence point. More accurately, if $f \in L_m^1(\mathbb{R}^n)$ with $n \geq 2m - 1$, and*

$$\alpha > \frac{n-(2m+1)}{2},$$

then $B_R^\alpha(f;x)$ has the localization property of any point. Here, we denote the Sobolev space $L_m^1(\mathbb{R}^n)$ by

$$L_m^1(\mathbb{R}^n) = \left\{ f \in L(\mathbb{R}^n) : \sum_{|\alpha|\leq m} \left\| \frac{\partial^\alpha f}{\partial x^\alpha} \right\|_1 < \infty \right\}.$$

2.2 L^p-convergence

Let $f \in L^p(\mathbb{R}^n)$. In this section, we consider whether or not

$$\lim_{R\to\infty} \|B_R^\alpha(f) - f\|_p = 0$$

holds for any $f \in L^p(\mathbb{R}^n)$.

Next we first point out that the L^p convergence of the Bochner-Riesz means of the Fourier integral of f is equivalent to the L^p-boundedness of the operator B_R^α.

Theorem 2.2.1 *Let $f \in L^p(\mathbb{R}^n)$. The equality*

$$\lim_{R\to\infty} \|B_R^\alpha(f) - f\|_p = 0$$

holds if and only if there exists a constant C_p such that the inequality

$$\|B_R^\alpha(f)\|_p \leq C_p \|f\|_p$$

holds, where C_p is independent of R.

Proof. Assume
$$\lim_{R\to\infty} \|B_R^\alpha(f) - f\|_p = 0$$
for every $f \in L^p(\mathbb{R}^n)$. Thus, we obtain that $\|B_R^\alpha(f)\|_p$ has bound with respect to R.

It follows from the uniform boundedness theorem in functional analysis that
$$\|B_R^\alpha(f)\|_p \le C_p \|f\|_p.$$

Conversely, we suppose that there exists a constant C_p such that the inequality
$$\|B_R^\alpha(f)\|_p \le C_p \|f\|_p$$
holds for any $f \in L^p(\mathbb{R}^n)$.

Notice that the space $\mathscr{D}(\mathbb{R}^n)$, the class of infinitely differential functions which have compact support, is dense in $L^p(\mathbb{R}^n)$. It suffices to prove that
$$\lim_{R\to\infty} \|B_R^\alpha(f) - f\|_p = 0$$
holds for every $f \in \mathscr{D}(\mathbb{R}^n)$.

Since $f \in \mathscr{D}(\mathbb{R}^n) \subset \mathscr{S}(\mathbb{R}^n)$, we have $\hat{f} \in \mathscr{S}(\mathbb{R}^n)$, where $\mathscr{S}(\mathbb{R}^n)$ denotes the Schwartz class.

Combining Lebesgue's control convergence theorem with the inversion formula of the Fourier transform of f, we have that
$$\lim_{R\to\infty} B_R^\alpha(f;x) = \lim_{R\to\infty} \int_{\mathbb{R}^n} \hat{f}(y) e^{iy\cdot x} \Phi^\alpha\left(\frac{y}{R}\right) dy = f(x),$$
for a.e. $x \in \mathbb{R}^n$.

Therefore, in order to prove
$$\lim_{R\to\infty} \int_{\mathbb{R}^n} |B_R^\alpha(f;x) - f(x)|^p dx = 0,$$
we need to prove that $|B_R^\alpha(f;x) - f(x)|^p$ can be controlled by an integrable function. This can be easily done, (see Pan [P1]). ∎

After summing up the L^p convergence of Bochner-Riesz means of Fourier integral of f to the L^p boundedness of Bochner-Riesz means, we can deduce the question to the L^p-boundedness of another simpler operator. The operator can be defined just by the Fourier transform. Let $f \in \mathscr{S}(\mathbb{R}^n)$, we define the Bochner-Riesz operator T_α
$$\widehat{T_\alpha f}(x) = \Phi^\alpha(x) \hat{f}(x),$$

2.2 L^p-convergence

where $\Phi^\alpha(x)$ is defined in Section 1.2.

The following theorem shows why the L^p boundedness of Bochner-Riesz spherical means can be transformed into the L^p boundedness of the Bochner-Riesz spherical operator.

Theorem 2.2.2 *The operator T_α has a bounded extension on $L^p(\mathbb{R}^n)$ if and only if there exists a constant C_p such that the inequality*

$$\|B_R^\alpha(f)\|_p \leq C_p \|f\|_p$$

holds for every $f \in L^p(\mathbb{R}^n)$.

Proof. Suppose that T_α has a bounded extension on $L^p(\mathbb{R}^n)$. For $f \in \mathscr{S}(\mathbb{R}^n)$, it is easy to see that $T_\alpha f = B^\alpha * f$, where $B^\alpha(x) = \widehat{\Phi^\alpha}(x)$. Therefore, we have

$$\|B_1^\alpha(f)\|_p \leq C_p \|f\|_p, \quad f \in L^p(\mathbb{R}^n).$$

By the change of variables

$$\overline{x} = Rx, \quad u = \frac{y}{R},$$

and letting $g(u) = f(\frac{u}{R})$, we get that

$$B_R^\alpha(f; x) = \int_{|u|<1} \hat{f}(uR)(1-|u|^2)^\alpha e^{i\overline{x} \cdot u} R^n du$$

$$= \int_{|u|<1} \widehat{f\left(\frac{u}{R}\right)}(1-|u|^2)^\alpha e^{i\overline{x} \cdot u} du$$

$$= B_1^\alpha(g; \overline{x}).$$

Hence, we have

$$\|B_R^\alpha(f; x)\|_p^p = \frac{1}{R^n} \int_{\mathbb{R}^n} |B_1^\alpha(g; \overline{x})|^p d\overline{x}$$

$$\leq \frac{C_p}{R^n} \int_{\mathbb{R}^n} |g(\overline{x})|^p d\overline{x}$$

$$= C_p \|f\|_p^p.$$

The proof for the other part of the theorem is trivial.
This completes the proof of Theorem 2.2.2. ∎

2.3 Some basic facts on multipliers

Bochner-Riesz operator is a kind of multipliers. In order to investigate the Bochner-Riesz operator, we begin with establishing some fundamental properties.

Definition 2.3.1 *Let $m \in L^\infty(\mathbb{R}^n)$. The linear operator T_m defined by*

$$\widehat{T_m f}(x) = m(x)\hat{f}(x), \quad f \in L^2(\mathbb{R}^n) \bigcap L^p(\mathbb{R}^n)$$

is called a multiplier. Moreover, if T_m satisfies

$$\|T_m f\|_p \leq C_p \|f\|_p, \quad f \in L^2(\mathbb{R}^n) \bigcap L^p(\mathbb{R}^n),$$

then we call T_m the L^p-bounded multiplier with the multiplier function $m(x)$, and the minimal constant C_p satisfying the above inequality is called the norm of T_m, denoted by $\|T_m\|_p$.

Theorem 2.3.1 *Let $\frac{1}{p} + \frac{1}{p'} = 1$. Then T_m is a L^p-bounded multiplier if and only if T_m is a $L^{p'}$-bounded multiplier and*

$$\|T_m\|_p = \|T_m\|_{p'}.$$

Proof. Assume that T_m is a L^p-bounded multiplier. By the property of the Fourier transform, we have that

$$<T_m f, g> \ = \ <m\hat{f}, \hat{g}> \ = \ <f, T_m g>,$$

which implies

$$\|T_m f\|_{p'} = \sup_{\|g\|_p \leq 1} | <T_m f, g> | \leq \|T_m\|_p \|f\|_{p'}.$$

Therefore, T_m is also a $L^{p'}$-bounded multiplier and

$$\|T_m\|_{p'} \leq \|T_m\|_p.$$

Replacing p by p', similar discussions lead to

$$\|T_m\|_p \leq \|T_m\|_{p'}.$$

This completes the proof. ∎

2.3 Some basic facts on multipliers

Theorem 2.3.2 *If T_m is a L^2-bounded multiplier, then $\|T_m\|_2 = \|m\|_\infty$.*

Proof. Firstly, by the fact $m \in L^\infty$ and the Plancherel theorem, we have that
$$\|T_m f\|_2 = \|m\hat{f}\|_2 \leq \|m\|_\infty \|f\|_2,$$
which yields
$$\|T_m\|_2 \leq \|m\|_\infty.$$
Next, to show the inverse inequality $\|T_m\|_2 \geq \|m\|_\infty$, we denote by
$$E = \{x : |m(x)| > \|T_m\|_2\}.$$
Suppose that $|E| \neq 0$ and let $\hat{f} = \chi_E$. By the inequality
$$\|T_m f\|_2 |E|^{\frac{1}{2}} < \|m\hat{f}\|_2 = \|T_m f\|_2 \leq \|T_m\|_2 |E|^{\frac{1}{2}},$$
we can deduce a contradiction. Hence, we have $|E| = 0$. ∎

Theorem 2.3.3 *Let $1 < p < \infty$. If T_m is a L^p-bounded multiplier, then we have*
$$\|T_m\|_2 \leq \|T_m\|_p.$$

Proof. Without loss of generality, we assume that $1 < p < 2$. By the assumption and Theorem 2.3.1, we obtain
$$\|T_m f\|_p \leq \|T_m\|_p \|f\|_p := k_0 \|f\|_p$$
and
$$\|T_m f\|_{p'} \leq \|T_m\|_{p'} \|f\|_{p'} = \|T_m\|_p \|f\|_{p'} := k_1 \|f\|_{p'}$$
respectively.
According to the Riesz-Thorin convexity theorem, we can get that there exists $0 < t < 1$ such that
$$\|T_m f\|_2 \leq k_0^{1-t} k_1^t \|f\|_2 = \|T_m\|_p \|f\|_2,$$
which leads to
$$\|T_m\|_2 \leq \|T_m\|_p.$$
∎

Theorem 2.3.4 *Let $1 < p < \infty$. If T_m is a $L^p(\mathbb{R}^{n+1})$-bounded multiplier, then T_m is also a $L^p(\mathbb{R}^n)$-bounded multiplier.*

Remark 2.3.1 *To express the meaning of Theorem 2.3.4 more accurately, we rewrite $T_m f(x)$ as $T_{m(x)} f(x)$, where $x = (\xi, \eta) \in \mathbb{R}^{n+1}$, $\xi \in \mathbb{R}$ and $\eta \in \mathbb{R}^n$. Then Theorem 2.3.4 shows that if the multiplier T_m satisfies*

$$\|T_m f\|_{L^p(\mathbb{R}^{n+1})} \leq \|T_m\|_p \|f\|_{L^p(\mathbb{R}^{n+1})},$$

then the following inequality

$$\|T_{m(\xi,\cdot)} g\|_{L^p(\mathbb{R}^n)} \leq C_p \|g\|_{L^p(\mathbb{R}^n)}$$

holds for a.e. $\xi \in \mathbb{R}$ and $C_p \leq \|T_m\|_p$.

Proof. Let $\frac{1}{p} + \frac{1}{q} = 1$. By the property of Fourier transform, we get

$$\int_{\mathbb{R}^{n+1}} T_m f(x) \alpha(x) dx = \int_{\mathbb{R}^{n+1}} m(x) \hat{f}(x) \hat{\alpha}(x) dx.$$

Then we have

$$\left| \int_{\mathbb{R}^{n+1}} m(x) \hat{f}(x) \hat{\alpha}(x) dx \right| \leq \|T_m f\|_{L^p(\mathbb{R}^{n+1})} \|\alpha\|_{L^q(\mathbb{R}^{n+1})}.$$

Let

$$f(x) = h(\xi) g(\eta)$$

and

$$\alpha(x) = \gamma(\xi) \beta(\eta),$$

where $h, \gamma \in \mathscr{S}(\mathbb{R})$ and $g, \beta \in \mathscr{S}(\mathbb{R}^n)$. The above inequality can be rewritten as

$$\left| \int_{\mathbb{R}} \left(\int_{\mathbb{R}^n} m(\xi, \eta) \hat{g}(\eta) \widehat{\beta}(\eta) d\eta \right) \hat{h}(\xi) \hat{\gamma}(\xi) d\xi \right|$$
$$\leq \|T_m\|_p \|h\|_{L^p(\mathbb{R})} \|g\|_{L^p(\mathbb{R}^n)} \|\gamma\|_{L^q(\mathbb{R})} \|\beta\|_{L^q(\mathbb{R}^n)}.$$

Set

$$M(\xi) = \int_{\mathbb{R}^n} m(\xi, \eta) \hat{g}(\eta) \widehat{\beta}(\eta) d\eta$$

and define the operator T_M by

$$\widehat{T_M h}(\xi) = M(\xi) \hat{h}(\xi).$$

Then we rewrite the above inequality as

$$\left|\int_{\mathbb{R}} T_M h(\xi)\gamma(\xi)d\xi\right| \leq \|T_m\|_p \|h\|_{L^p(\mathbb{R})} \|g\|_{L^p(\mathbb{R}^n)} \|\beta\|_{L^q(\mathbb{R}^n)} \|\gamma\|_{L^q(\mathbb{R})},$$

which implies that

$$\|T_M h\|_{L^p(\mathbb{R})} \leq \|T_m\|_p \|g\|_{L^p(\mathbb{R}^n)} \|\beta\|_{L^q(\mathbb{R}^n)} \|h\|_{L^p(\mathbb{R})}.$$

Hence, we have
$$\|T_M\|_p \leq \|T_m\|_p \|g\|_{L^p(\mathbb{R}^n)} \|\beta\|_{L^q(\mathbb{R}^n)}.$$

By Theorem 2.3.2, Theorem 2.3.3 and the above inequality, we obtain

$$\left|\int_{\mathbb{R}^n} m(\xi,\eta)\hat{g}(\eta)\hat{\beta}(\eta)d\eta\right| \leq \|T_m\|_p \|g\|_{L^p(\mathbb{R}^n)} \|\beta\|_{L^q(\mathbb{R}^n)},$$

which can be rewritten as

$$\left|\int_{\mathbb{R}^n} T_{m(\xi,\eta)} g(\eta)\beta(\eta)d\eta\right| \leq \|T_m\|_p \|g\|_{L^p(\mathbb{R}^n)} \|\beta\|_{L^q(\mathbb{R}^n)}. \quad (2.3.1)$$

It obviously follows from the inequality 2.3.1 that the conclusion of this theorem is true. ∎

2.4 The disc conjecture and Fefferman theorem

In Section 2.2, we have defined the Bochner-Riesz operator T_α by

$$\widehat{T_\alpha f}(x) = \Phi^\alpha(x)\hat{f}(x),$$

for $f \in \mathscr{S}(\mathbb{R}^n)$.

We notice that when $\alpha = 0$, $\Phi^0(x)$ is just the characteristic function of the unit ball. The conjecture that the corresponding operator T_0 is bounded on $L^p(\mathbb{R}^n)$ with $\frac{2n}{n+1} < p < \frac{2n}{n-1}$ is just the disc conjecture which is referred for a long time. The conjecture is based on the fact that the conclusion is correct when $n = 1$. However, for the case of high dimension ($n \geq 2$), Fefferman [Fe3] published an amazing result: T_0 is only bounded on $L^2(\mathbb{R}^n)$. Therefore, except for $p = 2$, the disc conjecture is false. In the following, we first introduce the result of Fefferman.

Theorem 2.4.1 *Let $n \geq 2$. The operator T_0 is bounded on $L^p(\mathbb{R}^n)$ if and only if $p = 2$.*

For $p = 2$, by the definition of the multiplier T_0, it is obvious that T_0 is bounded on $L^2(\mathbb{R}^n)$. So the most essential part of the proof of the theorem lies in giving the conclusion that T_0 is unbounded in $L^p(\mathbb{R}^n)$, $p \neq 2$. Before proving, we need some lemmas.

Lemma 2.4.1 *Suppose that $p > 2$, v_j, $1 \leq j \leq k$, is the unit vector in $\mathbb{R}^2$ and*
$$L_j = \{x \in \mathbb{R}^2 : x \cdot v_j \geq 0\}$$
is a half-plane. Define the operator T_j by
$$\widehat{T_j f}(x) = \chi_{L_j}(x) \hat{f}(x), \quad f \in \mathscr{S},$$
where $\chi_E(x)$ is the characteristic function of the set E. If T_0 is bounded on $L^p(\mathbb{R}^2)$, then for any $f_j \in L^p(\mathbb{R}^2)$ ($1 \leq j \leq k$), we have
$$\left\| \left(\sum_{j=1}^k |T_j f_j|^2 \right)^{\frac{1}{2}} \right\|_p \leq C_p \left\| \left(\sum_{j=1}^k |f_j|^2 \right)^{\frac{1}{2}} \right\|_p.$$

Proof. Let D_j^r be a circle, centered at rv_j, with a radius of r. Obviously,
$$D_j^r \to L_j$$
as $r \to \infty$. We define the multiplier T_j^r by
$$\widehat{T_j^r f}(x) = \chi_{D_j^r}(x) \hat{f}(x), \quad f \in \mathscr{S}.$$
For $f \in \mathscr{S}$, by the equation
$$T_j f(x) - T_j^r f(x) = \int_{\mathbb{R}^2} \left(\widehat{T_j f}(y) - \widehat{T_j^r f}(y) \right) e^{ix \cdot y} dy$$
$$= \int_{\mathbb{R}^2} \left(\chi_{L_j}(y) - \chi_{D_j^r}(y) \right) \hat{f}(y) e^{ix \cdot y} dy$$
and Lebesgue's dominated convergence theorem, we obtain that
$$T_j f(x) = \lim_{r \to \infty} T_j^r f(x).$$

2.4 The disc conjecture and Fefferman theorem

Combining the above equation with Fatou lemma, we obtain

$$\left\|\left(\sum |T_j f_j|^2\right)^{\frac{1}{2}}\right\|_p \leq \liminf_{r\to\infty} \left\|\left(\sum |T_j^r f_j|^2\right)^{\frac{1}{2}}\right\|_p.$$

Therefore, in order to prove Lemma 2.4.1, we merely need to prove that

$$\left\|\left(\sum |T_j^r f_j|^2\right)^{\frac{1}{2}}\right\|_p \leq C_p \left\|\left(\sum |f_j|^2\right)^{\frac{1}{2}}\right\|_p,$$

where C_p is independent of r. Similar to the proof of Theorem 2.2.2, it suffices to show that

$$\left\|\left(\sum |T_j^1 f_j|^2\right)^{\frac{1}{2}}\right\|_p \leq C_p \left\|\left(\sum |f_j|^2\right)^{\frac{1}{2}}\right\|_p.$$

Let $D = \{x : |x| < 1\}$. Then $T_j^1 f_j(x)$ can be represented as

$$T_j^1 f_j(x) = \int_{\mathbb{R}^2} \chi_{D_j^1}(y) \widehat{f_j}(y) e^{ix\cdot y} dy$$

$$= \int_{\mathbb{R}^2} \chi_D(y - v_j) \widehat{f_j}(y) e^{ix\cdot y} dy$$

$$= e^{ix\cdot v_j} \int_{\mathbb{R}^2} \chi_D(z) \widehat{f_j}(v_j + z) e^{ix\cdot z} dz$$

$$= e^{ix\cdot v_j} T_0(e^{-iv_j\cdot z} f_j(z))(x).$$

Hence, we have

$$\left\|\left(\sum |T_j^1 f_j|^2\right)^{\frac{1}{2}}\right\|_p = \left\|\left(\sum |T_0(e^{iv_j\cdot y} f_j(y))|^2\right)^{\frac{1}{2}}\right\|_p.$$

If we can prove

$$\left\|\left(\sum |T_0 g_j|^2\right)^{\frac{1}{2}}\right\|_p \leq C_p \left\|\left(\sum |g_j|^2\right)^{\frac{1}{2}}\right\|_p,$$

then letting $g_j(y) = e^{iv_j\cdot y} f_j(y)$, we get that

$$\left\|\left(\sum |T_0(e^{iv_j\cdot y} f_j(y))|^2\right)^{\frac{1}{2}}\right\|_p \leq C_p \left\|\left(\sum |e^{iv_j\cdot y} f_j(y)|^2\right)^{\frac{1}{2}}\right\|_p$$

$$= C_p \left\|\left(\sum |f_j(y)|^2\right)^{\frac{1}{2}}\right\|_p.$$

We complete the proof of this lemma.

Now we prove the inequality

$$\left\|\left(\sum |T_0 g_j|^2\right)^{\frac{1}{2}}\right\|_p \leq C_p \left\|\left(\sum |g_j|^2\right)^{\frac{1}{2}}\right\|_p.$$

Denote by

$$\mathbb{S}^{k-1} = \left\{x \in \mathbb{R}^k : |x| = 1\right\}$$

and let

$$\omega = (\omega_1, \omega_2, \ldots, \omega_k) \in \mathbb{S}^{k-1}.$$

Since T_0 is bounded on $L^p(\mathbb{R}^2)$, we have

$$\int_{\mathbb{R}^2} \left|\sum_{j=1}^k \omega_j T_0 g_j(x)\right|^p dx = \int_{\mathbb{R}^2} \left|T_0\left(\sum_{j=1}^k \omega_j g_j(x)\right)\right|^p dx$$

$$\leq C_p^p \int_{\mathbb{R}^2} \left|\sum_{j=1}^k \omega_j g_j(x)\right|^p dx.$$

Set

$$A(x) = (T_0 g_1(x), T_0 g_2(x), \ldots, T_0 g_k(x))$$

and

$$B(x) = (g_1(x), g_2(x), \ldots, g_k(x)).$$

The above inequality can be rewritten as

$$\int_{\mathbb{R}^2} |<\omega, A(x)>|^p dx \leq C_p^p \int_{\mathbb{R}^2} |<\omega, B(x)>|^p dx,$$

which can also be rewritten as

$$\int_{\mathbb{R}^2} \left|\left(\sum |T_0 g_j(x)|^2\right)^{\frac{1}{2}}\right|^p \left|\left\langle \omega, \frac{A(x)}{(\sum |T_0 g_j(x)|^2)^{\frac{1}{2}}}\right\rangle\right|^p dx$$

$$\leq C_p^p \int_{\mathbb{R}^2} \left|\left(\sum |g_j(x)|^2\right)^{\frac{1}{2}}\right|^p \left|\left\langle \omega, \frac{B(x)}{(\sum |g_j(x)|^2)^{\frac{1}{2}}}\right\rangle\right|^p dx.$$

We notice that

$$\frac{A(x)}{(\sum |T_0 g_j(x)|^2)^{\frac{1}{2}}} \in \mathbb{S}^{k-1}$$

2.4 The disc conjecture and Fefferman theorem

and
$$\frac{B(x)}{(\sum |g_j(x)|^2)^{\frac{1}{2}}} \in \mathbb{S}^{k-1}.$$

Taking the integral about ω on $\mathbb{S}^{k-1}$ to both sides of the above inequality, we can get that

$$\int_{\mathbb{R}^2} \left|\left(\sum |T_0 g_j(x)|^2\right)^{\frac{1}{2}}\right|^p dx \leq C_p^p \int_{\mathbb{R}^2} \left|\left(\sum |g_j(x)|^2\right)^{\frac{1}{2}}\right|^p dx.$$

This finishes the proof Lemma 2.4.1. ∎

Lemma 2.4.2 *Suppose that the longer edge of the rectangle R_j is parallel to the vector v_j. We denote by $\widetilde{R}_j$ two rectangles which have communal edge with the shorter edge of R_j and the same size as R_j. Let $f_j = \mathcal{H}_{R_j}$, then we have*

$$|T_j f_j(x)| \geq C > 0, \quad \text{when } x \in \widetilde{R}_j,$$

where C is some positive constant.

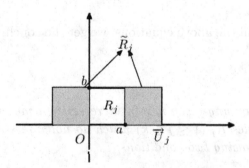

Figure 2.1.

Proof. Without loss of generality, suppose that v_j is horizontal and then construct the coordinate system as Figure 2.1.

Let $R_j = [0, a] \times [0, b]$, and

$$T_j f_j(x) = \int_{L_j} \widehat{f_j}(y) e^{ix \cdot y} dy$$

$$= \int_{L_j} \left(\int_{R^j} e^{-iy \cdot z} dz \right) e^{ix \cdot y} dy$$

$$= \int_0^\infty \frac{1}{iy_1} \left(e^{ix_1 y_1} - e^{i(x_1 - a)y_1} \right) dy_1$$

$$\times \int_{-\infty}^{+\infty} \frac{1}{iy_2} \left(e^{ix_2 y_2} - e^{i(x_2 - b)y_2} \right) dy_2.$$

It is easy to compute that

$$\left| \int_0^\infty \frac{1}{iy_1} \left(e^{ix_1 y_1} - e^{i(x_1 - a)y_1} \right) dy_1 \right| \geq \left| \int_0^\infty \frac{\cos(x_1 y_1) - \cos(x_1 - a)y_1}{y_1} dy_1 \right|$$

$$= \left| \log \frac{x_1}{x_1 - a} \right|$$

$$\geq \log 2,$$

for $x = (x_1, x_2) \in \widetilde{R_j}$, and

$$\int_{-\infty}^{+\infty} \frac{1}{iy_2} \left(e^{ix_2 y_2} - e^{i(x_2 - b)y_2} \right) dy_2 = \int_{-\infty}^{+\infty} \frac{\sin(x_2 y_2) - \sin[(x_2 - b)y_2]}{y_2} dy_2$$

$$= 2\pi.$$

Combining with all the above equations, we get the conclusion of the lemma. ∎

Lemma 2.4.3 *For any given $\eta > 0$, there exists a measurable set $E \subset \mathbb{R}^2$ and finite rectangles R_j ($1 \leq j \leq k$) which do not intersect with each other, satisfying the following two conditions:*

(I) $\left| E \cap \widetilde{R_j} \right| \geq C \left| \widetilde{R_j} \right|$ *(C being a positive constant);*

(II) $|E| \leq \eta \sum_{j=1}^k |R_j|,$

where $\widetilde{R_j}$ are defined as in Lemma 2.4.2.

2.4 The disc conjecture and Fefferman theorem

Proof. In order to construct E, we first introduce a notation. Give a triangle $\triangle ABC$ with the bottom edge AB being horizontal and the height being h. Then we extend AC to A' and BC to B', such that the distances from the point A, B to the edge AB are both equal to h' ($h' > h$). Joining the midpoint D of AB with A', and D with B', we get two triangles $\triangle ADA'$ and $\triangle BDB'$. We call the two triangles as the ones which grow out of the original triangle $\triangle ABC$ from the height of h to h'. Now we proceed to construct the measurable set E. Give an equilateral triangle $\triangle^0$, with the bottom edge $[0,1]$ and the height $h_0 = \sqrt{3}/2$. Let

$$h_0 < h_1 < \cdots < h_m.$$

The triangle obtain $\triangle^0$ grows from the height h_0 to h_1 and we get two triangles $\triangle'$ and $\triangle''$, while $\triangle'$ and $\triangle''$ grow from the height h_1 to h_2 and we get four triangles in all. Thus we make an induction by this way until the m-th time, 2^m triangles in all, denoted by $\triangle_j$, $1 \le j \le 2^m$. Set

$$E = \bigcup_{j=1}^{2^m} \triangle_j,$$

and let

$$h_1 = \frac{\sqrt{3}}{2}\left(1 + \frac{1}{2}\right), \ldots, h_m = \frac{\sqrt{3}}{2}\left(1 + \frac{1}{2} + \cdots + \frac{1}{m+1}\right).$$

It is easy to check that $|E| \le 17$. We next construct finite disjoint rectangles R_j ($1 \le j \le k$). We divide $[0,1]$ equally into 2^m intervals and let I be anyone of them. It is obvious that I must be the bottom edge of some $\triangle_j$. We denote $\triangle_j$ by $\triangle(I)$ and the vertex of $\triangle_j$ as $P(I)$. The region I formed by the extension lines of two edges of $\triangle(I)$ is called the below extension region of $\triangle(I)$. Make a rectangle $R(I)$ in this extension region as in Figure 2.2, such that the longer edge of $R(I)$ is $\log m$.

Obviously, the shorter edge of $R(I)$ is $\sim |I|$. Then, we have

$$|R(I)| \sim 2^{-m}\log m.$$

As in Lemma 2.4.2, $\widetilde{R(I)}$ is the union of two rectangles which share the same edge with the shorter edge of $R(I)$ and has the same size as $R(I)$. Though $\widetilde{R(I)}$ cannot include $\triangle(I)$, by the construction we can see that

$$\left|\widetilde{R(I)} \cap \triangle(I)\right| \ge \frac{1}{10}\left|\widetilde{R(I)}\right|.$$

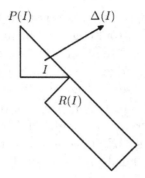

Figure 2.2.

Denote the 2^m equally divided intervals of $[0,1]$ by $I_1, I_2, \ldots, I_{2^m}$. For every I_j with $1 \leq j \leq 2^m$, we get $R(I_j)$ correspondingly. From the last equation above, we have that the finite rectangles

$$\{R(I_j) : 1 \leq j \leq 2^m\}$$

satisfy conclusion I in the lemma. Next, if we choose m big enough such that $\log m > 17/\eta$, then there holds

$$|E| \leq 17 < \eta \log m \sim \eta \sum_{j=1}^{2^m} |R(I_j)|.$$

Thus conclusion II of the lemma is also met. Finally, for different I_i and I_j, we have

$$P(I_i) \neq P(I_j).$$

Then the below extension regions of $\triangle(I_i)$ and $\triangle(I_j)$ are disjoint and so are $R(I_i)$ and $R(I_j)$. This completes the proof. ∎

Next we will give the proof of Theorem 2.4.1.

Proof. By Theorem 2.3.4, it suffices to show that T_0 is unbounded on $L^p(\mathbb{R}^2)$, $p \neq 2$. According to Theorem 2.3.1, we merely need to prove that T_0 is unbounded on $L^p(\mathbb{R}^2)$ with $p > 2$.

If T_0 is bounded on $L^p(\mathbb{R}^2)$. By Lemma 2.4.1, we get that

$$\left\|\left(\sum |T_j f_j|^2\right)^{\frac{1}{2}}\right\|_p \leq C_p \left\|\left(\sum |f_j|^2\right)^{\frac{1}{2}}\right\|_p.$$

2.5 The L^p-boundedness of Bochner-Riesz operator T_α with $\alpha > 0$

Notice that $p > 2$. By Hölder's inequality, we have

$$\int_E \left(\sum_j |T_j f_j(x)|^2 \right) dx \leq C_p |E|^{\frac{p-2}{p}} \left\| \left(\sum_j |f_j|^2 \right)^{\frac{1}{2}} \right\|_p^2,$$

where E is the measurable set in Lemma 2.4.3. Let $f_j = \chi_{R_j}$, where R_j is the rectangle in Lemma 2.4.3. Notice that different R_j disjoint. Then we get from the above inequality and Lemma 2.4.3 that

$$\int_E \left(\sum_j |T_j \chi_{R_j}(x)|^2 \right) dx \leq C_p \eta^{\frac{p-2}{p}} \sum_j |R_j|.$$

On the other hand, by Lemmas 2.4.2 and 2.4.3, we obtain

$$\int_E \left(\sum_j |T_j \chi_{R_j}(x)|^2 \right) dx \geq \sum_j \int_{E \cap \widetilde{R_j}} |T_j \chi_{R_j}(x)|^2 dx \geq C \sum_j |R_j|.$$

Obviously, the last two inequality contradicts with each other and this completes the proof. ∎

2.5 The L^p-boundedness of Bochner-Riesz operator T_α with $\alpha > 0$

In the last section, we obtained the conclusions about L^p-boundedness of the Bochner-Riesz operator T_0. We now turn to investigating the L^p-boundedness of the operators T_α ($\alpha > 0$). In this field, Herz [He1] gave a necessary condition for T_α ($\alpha > 0$) being bounded on $L^p(\mathbb{R}^n)$.

Theorem 2.5.1 Let $0 < \alpha < \frac{n-1}{2}$ and $p > 1$. If T_α is bounded on $L^p(\mathbb{R}^n)$, then we have

$$\frac{2n}{n+1+2\alpha} < p < \frac{2n}{n-1-2\alpha}.$$

Proof. Give any $f_0 \in \mathscr{D}$, satisfying

$$f_0(x) = \begin{cases} 1, & |x| \leq 1; \\ 0, & |x| \geq 2. \end{cases}$$

Then $\widehat{f_0} \in \mathscr{S}$ and by the conditions of the theorem, we get that
$$T_\alpha \widehat{f_0} \in L^p.$$
Let
$$\Phi(x) = T_\alpha \widehat{f_0}(x).$$
Taking the Fourier transform on both sides of the equation, we have
$$\widehat{\Phi}(x) = \Phi^\alpha(x),$$
which leads to
$$\Phi(x) = B^\alpha(x).$$
By the expression of $B^\alpha(x)$, we get that
$$T_\alpha \widehat{f_0} \in L^p$$
if and only if
$$\left(\frac{n}{2} + \alpha + \frac{1}{2}\right) p > n.$$
Therefore, if T_α is bounded on $L^p(\mathbb{R}^n)$, then we must have $p > \frac{2n}{n+1+2\alpha}$. We notice that $\frac{2n}{n+1+2\alpha}$ and $\frac{2n}{n-1-2\alpha}$ are a pair of conjugate indexes. Hence by Theorem 2.3.1, we have that $p < \frac{2n}{n-1-2\alpha}$.

This finishes the proof Theorem 2.5.1. ∎

By the conclusion of Theorem 2.5.1, one naturally raises the following question: if $\alpha > 0$ and
$$\frac{2n}{n+1+2\alpha} < p < \frac{2n}{n-1-2\alpha},$$
is T_α bounded on $L^p(\mathbb{R}^n)$? This question is now called the Bochner-Riesz conjecture. As far as the case of high dimension is concerned, the question is firstly solved by Carleson and Sjölin only if $n = 2$. They make use of the oscillatory integrals. Afterwards, Fefferman and Cordoba acquired the same conclusions using constriction theorem of the Fourier transform and the the method of geometric proof, respectively. In the following, we formulate the theorem due to Carleson-Sjölin [CS1].

Theorem 2.5.2 Let $0 < \alpha < \frac{1}{2}$. T_α is bounded on $L^p(\mathbb{R}^2)$ if and only if $\frac{4}{3+2\alpha} < p < \frac{4}{1-2\alpha}$.

In this section, we will not give the proof of Theorem 2.5.2, and postpone it until we have introduced the oscillatory integral in the next section. It is necessary to be pointed out that the method of proof in Section 2.6.2 basically follows the original ideas of Carleson-Sjölin in [CS1].

2.6 Oscillatory integral and proof of Carleson-Sjölin theorem

2.6.1 Oscillatory integrals.

By the definition of the Bochner-Riesz operator, we rewrite $T_\alpha f(x)$ as

$$T_\alpha f(x) = \int_{\mathbb{R}^n} B^\alpha(x-y) f(y) dy,$$

where

$$B^\alpha(t) = \frac{2^\alpha \Gamma(\alpha+1)}{(2\pi)^{n/2}} \cdot \frac{J_{\frac{n}{2}+\alpha}(|t|)}{|t|^{\frac{n}{2}+\alpha}}.$$

According to the asymptotic property of Bessel function, we have

$$J_m(t) = \left(\frac{2}{\pi t}\right)^{\frac{1}{2}} \cos\left(t - \frac{\pi m}{2} - \frac{\pi}{4}\right) + r(t), \quad t \to \infty,$$

where $r(t) = O(t^{-3/2})$ (when $t \to \infty$). Then $T_\alpha f(x)$ can be represented as

$$T_\alpha f(x) = C \int_{|x-y| \geq 1} \cos\left(|x-y| - \frac{\pi(n+1+2\alpha)}{4}\right) \frac{1}{|x-y|^{\frac{n}{2}+\alpha+\frac{1}{2}}} f(y) dy$$

$$+ C \int_{|x-y| \geq 1} r(|x-y|) \frac{1}{|x-y|^{\frac{n}{2}+\alpha}} f(y) dy$$

$$+ \frac{2^\alpha \Gamma(\alpha+1)}{(2\pi)^{n/2}} \int_{|x-y|<1} \frac{J_{\frac{n}{2}+\alpha}(|x-y|)}{|x-y|^{\frac{n}{2}+\alpha}} f(y) dy$$

$$:= Pf(x) + Qf(x) + Rf(x),$$

where the constant C only depends on n and α.

By the asymptotic estimate $J_m(t) \sim C t^m$, as $t \to 0$, we have

$$|Rf(x)| \leq CMf(x),$$

where Mf is the Hardy-Littlewood maximal function of f. Next, noticing $r(t) = O(t^{-3/2})$, as $t \to \infty$, it follows that

$$|Qf(x)| \leq C \int_{|x-y|\geq 1} \frac{|f(y)|}{|x-y|^{\frac{3}{2}+\frac{n}{2}+\alpha}} dy.$$

Let $p > \frac{2n}{n+1+2\alpha}$. Then by Minkowski's inequality of integral, we can get that

$$\|Qf\|_p \leq C_p \|f\|_p.$$

Therefore, under the assumption that $p > \frac{2n}{n+1+2\alpha}$, the boundedness of T_α on $L^p(\mathbb{R}^n)$ is equivalent to the boundedness of the operator P on $L^p(\mathbb{R}^n)$. Due to the expression of $Pf(x)$, we might as well take it for granted that

$$Pf(x) = \int_{|x-y|\geq 1} e^{i|x-y|}|x-y|^{-(\frac{n}{2}+\alpha+\frac{1}{2})}f(y)dy.$$

But if $p > \frac{2n}{n+1+2\alpha}$, it is easy to see that there holds

$$\left|\int_{|x-y|<1} e^{i|x-y|}|x-y|^{-(\frac{n}{2}+\alpha+\frac{1}{2})}f(y)dy\right| \leq CMf(x).$$

Hence we get that

$$Pf(x) = \int_{\mathbb{R}^n} e^{i|x-y|}|x-y|^{-(\frac{n}{2}+\alpha+\frac{1}{2})}f(y)dy.$$

Obviously, P is a class of convolution operator with oscillatory kernel s. For this reason, we will discuss the operators, called the oscillatory integrals. Suppose that $x \in \mathbb{R}^n$, $\xi \in \mathbb{R}^n$,

$$\nabla_i g(x,\xi) = (\frac{\partial g}{\partial \xi_1}, \frac{\partial g}{\partial \xi_2}, \ldots, \frac{\partial g}{\partial \xi_n}),$$

and

$$\det\left(\frac{\partial^2 g(x,\xi)}{\partial x_i \partial \xi_j}\right)$$

is the determinant with the elements of

$$\left\{\frac{\partial^2 g(x,\xi)}{\partial x_i \partial \xi_j} : 1 \leq i \leq n,\ 1 \leq j \leq n\right\}.$$

In this section, we first study the oscillatory integral defined by

$$P_\lambda f(\xi) = \int_{\mathbb{R}^n} e^{i\lambda\varphi(x,\xi)}\psi(x,\xi)f(x)dx,\quad \xi \in \mathbb{R}^n,$$

where $\psi \in \mathscr{D}(\mathbb{R}^n \times \mathbb{R}^n)$, φ is a real valued smooth function, and

$$\det\left(\frac{\partial^2 \varphi(x,\xi)}{\partial x_i \partial \xi_j}\right) \neq 0.$$

Theorem 2.6.1 *For any $\lambda > 0$, P_λ is of type (2.2) and satisfies*

$$\|P_\lambda f\|_2 \leq C\lambda^{-\frac{n}{2}}\|f\|_2,$$

where the constant C is independent of λ and f.

2.6 Oscillatory integral and proof of Carleson-Sjölin theorem

Proof. Firstly, we rewrite $\|P_\lambda f\|_2^2$ as

$$\int_{\mathbb{R}^n} |P_\lambda f(\xi)|^2 d\xi = \int_{\mathbb{R}^n}\int_{\mathbb{R}^n}\int_{\mathbb{R}^n} e^{i\lambda(\varphi(x,\xi)-\varphi(y,\xi))}\psi(x,\xi)\overline{\psi(y,\xi)}f(x)\overline{f(y)}dxdyd\xi$$

$$= \int_{\mathbb{R}^n}\int_{\mathbb{R}^n} K_\lambda(x,y)f(x)\overline{f(y)}dxdy,$$

where

$$K_\lambda(x,y) = \int_{\mathbb{R}^n} e^{i\lambda(\varphi(x,\xi)-\varphi(y,\xi))}\psi(x,\xi)\overline{\psi(y,\xi)}d\xi.$$

Obviously, we have $|K_\lambda(x,y)| \leq C$, where C is independent of x, y and λ. Therefore, if we can prove

$$|K_\lambda(x,y)| \leq \frac{C}{(\lambda|x-y|)^{(n+1)}}, \qquad (2.6.1)$$

then we can easily obtain that

$$\sup_y \int_{\mathbb{R}^n} |K_\lambda(x,y)|dx \leq \frac{C}{\lambda^n}$$

and

$$\sup_x \int_{\mathbb{R}^n} |K_\lambda(x,y)|dy \leq \frac{C}{\lambda^n}.$$

Consequently, by the Schwarz inequality, we get that

$$\int_{\mathbb{R}^n} |P_\lambda f(\xi)|^2 dx \leq \left(\int_{\mathbb{R}^n}\int_{\mathbb{R}^n} |K_\lambda(x,y)||f(x)|^2 dxdy\right)^{\frac{1}{2}}$$

$$\times \left(\int_{\mathbb{R}^n}\int_{\mathbb{R}^n} |K_\lambda(x,y)||f(y)|^2 dxdy\right)^{\frac{1}{2}}$$

$$\leq C\lambda^{-n}\|f\|_2^2.$$

Thus we complete the proof of the theorem.

Next it suffices to prove the inequality (2.6.1). Choose a vector

$$a = (a_1, a_2, \ldots, a_n)$$

satisfying $|a| \leq 1$. It easily implies that

$$\left(a, \nabla_\xi\right) e^{i\lambda(\varphi(x,\xi)-\varphi(y,\xi))} = i\lambda\left(a, \nabla_\xi(\varphi(x,\xi)-\varphi(y,\xi))\right)e^{i\lambda(\varphi(x,\xi)-\varphi(y,\xi))}.$$

Let

$$M := \left(a, \nabla_\xi\bigl(\varphi(x,\xi)-\varphi(y,\xi)\bigr)\right),$$

then we have
$$K_\lambda(x,y) = \int_{\mathbb{R}^n} \frac{1}{i\lambda M}(a, \nabla_\xi) e^{i\lambda(\varphi(x,\xi)-\varphi(y,\xi))} \psi(x,\xi)\overline{\psi(y,\xi)} d\xi.$$

According to the finite increment formula for functions of multiple variables, we obtain that
$$M = \left(a, \nabla_\xi \{\varphi(x_1,\ldots,x_n;\xi_1,\ldots,\xi_n) - \varphi(y_1,\ldots,y_n;\xi_1,\ldots,\xi_n)\}\right)$$
$$= \left(a, \nabla_\xi \left\{(x_1-y_1)\frac{\partial\varphi}{\partial x_1} + \cdots + (x_n-y_n)\frac{\partial\varphi}{\partial x_n}\right\}_{x=\bar{x}}\right)$$
$$= \left(a, \nabla_\xi \nabla_x \varphi\Big|_{x=\bar{x}} \cdot (x-y)\right),$$

where the component of
$$\bar{x} = (\overline{x_1},\ldots,\overline{x_n}),$$
say, $\overline{x_i}$ is between x_i and y_i. Together with the condition, this implies that the matrix
$$\nabla_\xi \nabla_x \varphi = \begin{pmatrix} \frac{\partial^2 \varphi}{\partial x_1 \partial \xi_1} & \cdots & \frac{\partial^2 \varphi}{\partial x_1 \partial \xi_n} \\ \cdots & \cdots & \cdots \\ \frac{\partial^2 \varphi}{\partial x_n \partial \xi_1} & \cdots & \frac{\partial^2 \varphi}{\partial x_n \partial \xi_n} \end{pmatrix}$$

is a non-degenerate matrix. Due to the smoothness of φ, there exists $C > 0$, such that
$$|\det(\nabla_\xi \nabla_x \varphi)| \geq C.$$

Let
$$b = \nabla_\xi \nabla_x \varphi \cdot (x-y),$$
then we have
$$x - y = (\nabla_\xi \nabla_x \varphi)^{-1} b,$$
which implies
$$|x-y| \leq \|(\nabla_\xi \nabla_x \varphi)^{-1}\| |b|,$$
where the norm of matrix A is defined as
$$\|A\| = \sup_{|x|\neq 0} \frac{|Ax|}{|x|}.$$

For $A = (a_{ij})$, it is obvious that
$$\|A\| \leq C \max_{1\leq j\leq n} \sum_{i=1}^n |a_{ij}|.$$

2.6 Oscillatory integral and proof of Carleson-Sjölin theorem

Now, it follows from the components of the matrix $(\nabla_\xi \nabla_x \varphi)^{-1}$ and

$$|\det(\nabla_\xi \nabla_x \varphi)| \geq C > 0,$$

that

$$\|(\nabla_\xi \nabla_x \varphi)^{-1}\| \leq C.$$

Hence, we have

$$|b| \geq C|x - y|.$$

Choose a with $|a| = 1$ such that

$$M = \left|\nabla_\xi \nabla_x \varphi\Big|_{x=\bar{x}}(x-y)\right|,$$

and thus we have

$$M \geq C|x - y|.$$

By making use of partial integration to the last expression in $K_\lambda(x, y)$ and the fact

$$\frac{1}{M} \leq \frac{C}{|x-y|}, \quad \frac{|\nabla_\xi M|}{M^2} \leq \frac{C}{|x-y|},$$

and $\psi \in \mathscr{D}(\mathbb{R}^n \times \mathbb{R}^n)$, we get that

$$|K_\lambda(x,y)| \leq \frac{C}{\lambda|x-y|}.$$

By repeating the above arguments for $(n+1)$ times, we get the inequality (2.6.1). ∎

Since the operator P_λ is of type $(1, \infty)$, combining Theorem 2.6.1 with the Riesz-Thorin convex theorem, we can get the following theorem.

Theorem 2.6.2 Let $1 \leq p \leq 2$ and $\frac{1}{p} + \frac{1}{q} = 1$. Then P_λ is of type (p, q) and satisfies

$$\|P_\lambda f\|_q \leq C\lambda^{-n/q}\|f\|_p.$$

In order to prove Carleson-Sjölin theorem, we should turn to investigating the $L^p(\mathbb{R}^2)$-boundedness of the oscillatory integral

$$P_\lambda f(\xi) = \int_{\mathbb{R}^2} e^{i\lambda\varphi(x,\xi)} \psi(x,\xi) f(x) dx$$

for $\xi \in \mathbb{R}^2$.

To this end, we begin with the oscillatory integral which is closely related to the above. Let $x = (x_1, x_2) \in \mathbb{R}^2$ and $t \in \mathbb{R}$, we consider the oscillatory integral

$$P_\lambda^{(1)} f(x) = \int_\mathbb{R} e^{i\lambda \Phi(x,t)} \Psi(x,t) f(t) dt,$$

where $\Psi \in \mathscr{D}(\mathbb{R}^2 \times \mathbb{R})$, and Φ is a real and smooth function satisfying

$$\det \begin{pmatrix} \frac{\partial^2 \Phi}{\partial x_1 \partial t} & \frac{\partial^3 \Phi}{\partial x_1 \partial t^2} \\ \frac{\partial^2 \Phi}{\partial x_2 \partial t} & \frac{\partial^3 \Phi}{\partial x_2 \partial t^2} \end{pmatrix} \neq 0.$$

Theorem 2.6.3 Let $\frac{3}{q} + \frac{1}{p} = 1$ and $q > 4$. Then $P_\lambda^{(1)}$ satisfies the inequality

$$\left\| P_\lambda^{(1)} f \right\|_{L^q(\mathbb{R}^2)} \leq C_q \lambda^{-2/q} \|f\|_{L^p(\mathbb{R})}.$$

Proof. We first write that

$$\left(P_\lambda^{(1)} f(x) \right)^2 = \int_\mathbb{R} \int_\mathbb{R} e^{i\lambda \{\Phi(x,s) + \Phi(x,t)\}} \Psi(x,s) \Psi(x,t) f(s) f(t) ds dt,$$

Let $y_1 = s + t$, $y_2 = st$ and set $y = (y_1, y_2)$. We put

$$\varphi(x, y) = \Phi(x, s) + \Phi(x, t),$$

$$\psi(x, y) = \Psi(x, s) \Psi(x, t),$$

and let

$$F(y) = J f(s) f(t),$$

where J is the Jacobi determinant of the transformation, that is,

$$J = \frac{D(t, s)}{D(y_1, y_2)}.$$

Then we have

$$\left(P_\lambda^{(1)} f(x) \right)^2 = \int_{\mathbb{R}^2} e^{i\lambda \varphi(x,y)} F(y) \psi(x,y) dy := P_\lambda F(x).$$

It is easy to see that $\psi \in \mathscr{D}(\mathbb{R}^2 \times \mathbb{R}^2)$. If we can prove that

$$\det(\nabla_x \nabla_y \varphi) \neq 0,$$

2.6 Oscillatory integral and proof of Carleson-Sjölin theorem

then from Theorem 2.6.2, we can get

$$\left(\int_{\mathbb{R}^2}\left|P_\lambda^{(1)}f(x)\right|^{2r'}dx\right)^{1/r'} = \|P_\lambda F\|_{r'} \leq C\lambda^{-2/r'}\|F\|_r,$$

where $\frac{1}{r} + \frac{1}{r'} = 1$ and $1 < r < 2$. From the equation,

$$\|F\|_r = \left(\int_{\mathbb{R}^2}|F(y)|^r dy\right)^{1/r} = \left(\int_\mathbb{R}\int_\mathbb{R}|f(s)f(t)J|^r\frac{D(y_1,y_2)}{D(t,s)}dsdt\right)^{1/r}$$

and

$$\left|\frac{D(y_1,y_2)}{D(t,s)}\right| = |s-t|.$$

We have that

$$\|F\|_r = \left(\int_\mathbb{R}\int_\mathbb{R}|f(s)|^r|f(t)|^r|s-t|^{1-r}dsdt\right)^{1/r}$$

$$= \left\{\int_\mathbb{R}|f(t)|^r\left(\int_\mathbb{R}|f(s)|^r|s-t|^{-1+(2-r)}ds\right)dt\right\}^{1/r}$$

$$\leq C\left(\|f^r\|_u\|I_{2-r}(f^r)\|_{u'}\right)^{1/r},$$

where $\frac{1}{u} + \frac{1}{u'} = 1$ and $I_\alpha(g)$ is the Riesz potential operator. Then by the boundedness of the Riesz potential operator (cf. Stein [St4]), we obtain that

$$\|I_\alpha(g)\|_{u'} \leq C\|g\|_u$$

with $0 < \alpha < 1$. This leads to

$$\|F\|_r \leq C\|f^r\|_u^{2/r}.$$

Consequently, we have

$$\left(\int_{\mathbb{R}^2}|P_\lambda^{(1)}f(x)|^{2r'}dx\right)^{1/2r'} \leq C\lambda^{-1/r'}\left(\int_\mathbb{R}|f(t)|^{ru}dt\right)^{1/ru}.$$

By letting

$$p = ur, \ q = 2r', \text{and } u = \frac{2}{3-r},$$

we get the conclusion of the proof of Theorem.

Hence, it suffices to check

$$\det(\nabla_x\nabla_y\varphi) \neq 0.$$

Let $z = (s,t)$. Then we obtain that
$$|\det(\nabla_y \nabla_x \varphi)| = |\det(\nabla_z \nabla_x \varphi)||t-s|^{-1}$$
and
$$\det(\nabla_z \nabla_x \varphi) = \det \begin{pmatrix} \frac{\partial^2 \Phi(x,s)}{\partial x_1 \partial s} & \frac{\partial^2 \Phi(x,t)}{\partial x_1 \partial t} \\ \frac{\partial^2 \Phi(x,s)}{\partial x_2 \partial s} & \frac{\partial^2 \Phi(x,t)}{\partial x_2 \partial t} \end{pmatrix}$$
$$= \det \begin{pmatrix} \frac{\partial^2 \Phi(x,s)}{\partial x_1 \partial s} & \frac{\partial^2 \Phi(x,t)}{\partial x_1 \partial t} - \frac{\partial^2 \Phi(x,s)}{\partial x_1 \partial s} \\ \frac{\partial^2 \Phi(x,s)}{\partial x_2 \partial s} & \frac{\partial^2 \Phi(x,t)}{\partial x_2 \partial t} - \frac{\partial^2 \Phi(x,s)}{\partial x_2 \partial s} \end{pmatrix}$$
$$\approx (t-s) \det \begin{pmatrix} \frac{\partial^2 \Phi}{\partial x_1 \partial s} & \frac{\partial^3 \Phi}{\partial x_1 \partial s^2} \\ \frac{\partial^2 \Phi}{\partial x_2 \partial s} & \frac{\partial^3 \Phi}{\partial x_2 \partial s^2} \end{pmatrix}_{s=\bar{s}},$$
where $\bar{s}$ is between s and t. Combining the condition of the theorem with the above two equations, we get that
$$\det(\nabla_x \nabla_y \varphi) \neq 0.$$

This completes the proof Theorem 2.6.3. ∎

From the proof of Theorem 2.6.1, we can see that if
$$\operatorname{supp} f \subset S_1, \ \operatorname{supp} \psi \subset S_1 \text{ and } \xi \in S_2,$$
where S_1 and S_2 are both bounded closed sets, then it follows from the condition that
$$\det \left(\frac{\partial^2 \varphi(x,\xi)}{\partial x_i \partial \xi_j} \right) \neq 0$$
on $S_1 \times S_2$ that
$$\|P_\lambda f\|_{L^2(S_2)} \leq C \lambda^{-\frac{n}{2}} \|f\|_{L^2(S_1)}.$$

We denote by S_1 a bounded closed set in $\mathbb{R}$, and by S_2 a bounded closed set in $\mathbb{R}^2$. Let
$$\overline{P_\lambda^{(1)}} f(x) = \int_{S_1} e^{i\lambda \Phi(x,t)} \Psi(x,t) f(t) dt, \quad x \in S_2,$$

2.6 Oscillatory integral and proof of Carleson-Sjölin theorem

where $\Psi \in \mathscr{D}(\mathbb{R}^2 \times \mathbb{R})$, $\mathrm{supp}\Psi \subset S_2 \times S_1$, Φ is real and smooth, which satisfies

$$\det \begin{pmatrix} \frac{\partial^2 \Phi}{\partial x_1 \partial t} & \frac{\partial^3 \Phi}{\partial x_1 \partial t^2} \\ \frac{\partial^2 \Phi}{\partial x_2 \partial t} & \frac{\partial^3 \Phi}{\partial x_2 \partial t^2} \end{pmatrix} \neq 0.$$

on $S_2 \times S_1$. Then we have the following corollary.

Corollary 2.6.1 *Let $\frac{3}{q} + \frac{1}{p} = 1$ and $q > 4$. Then $\overline{P}_\lambda^{(1)}$ satisfies*

$$\left\| \overline{P}_\lambda^{(1)} f \right\|_{L^q(S_2)} \leq C_q \lambda^{-2/q} \|f\|_{L^p(S_1)}.$$

Set $S_1 = I = [0, 1]$ and

$$S_2 = \left[-\frac{3}{2}, \frac{5}{2}\right] \times \left[\frac{3}{2}, \frac{5}{2}\right] \cup \left[-\frac{3}{2}, \frac{5}{2}\right] \times \left[-\frac{3}{2}, -\frac{1}{2}\right].$$

Define

$$P_{\lambda,s}^{(2)} f(x) = \int_I e^{i\lambda \{(x_1-t)^2 + (x_2-s)^2\}^{\frac{1}{2}}} \Psi(x,t) f(t) dt, \ x \in S_2,$$

where $\Psi \in \mathscr{D}(\mathbb{R}^2 \times \mathbb{R})$, and $\mathrm{supp}\Psi \subset S_2 \times I$.

We have the following theorem.

Theorem 2.6.4 *For any $s \in [0, 1]$, the following inequality*

$$\left\| P_{\lambda,s}^{(2)} f \right\|_{L^4(S_2)} \leq C_\varepsilon \lambda^{-\frac{1}{2}+\varepsilon} \|f\|_{L^4(I)}$$

holds for any $\varepsilon > 0$. Here C_ε depends only on ε.

Proof. Let

$$\Phi(x, t) = \{(x_1 - t)^2 + (x_2 - s)^2\}^{\frac{1}{2}}.$$

Then we have

$$P_{\lambda,s}^{(2)} f(x) = \overline{P}_\lambda^{(1)} f(x).$$

Since

$$\det \begin{pmatrix} \frac{\partial^2 \Phi}{\partial x_1 \partial t} & \frac{\partial^3 \Phi}{\partial x_1 \partial t^2} \\ \frac{\partial^2 \Phi}{\partial x_2 \partial t} & \frac{\partial^3 \Phi}{\partial x_2 \partial t^2} \end{pmatrix} = \frac{(x_2 - s)^3}{\Phi^6},$$

$0 \leq s \leq 1$ and $(x_1, x_2) \in S_2$, the above determinant is not equal to 0 on $S_2 \times I$. By Corollary 2.6.1, we directly obtain that

$$\|P^{(2)}_{\lambda,s} f\|_{L^q(S_2)} \leq C_q \lambda^{-2/q} \|f\|_{L^p(I)},$$

which follows the conclusion of the theorem. This completes the proof. ∎

If we denote by

$$S_2^{(1)} = \left[-\frac{3}{2}, \frac{5}{2}\right] \times \left[\frac{3}{2}, \frac{5}{2}\right] \bigcup \left[-\frac{3}{2}, \frac{5}{2}\right] \times \left[-\frac{1}{2}, \frac{3}{2}\right],$$

and

$$S_2^{(2)} = \left[-\frac{3}{2}, -\frac{1}{2}\right] \times \left[-\frac{3}{2}, \frac{5}{2}\right] \bigcup \left[\frac{3}{2}, \frac{5}{2}\right] \times \left[-\frac{3}{2}, \frac{5}{2}\right],$$

and let

$$S_1 = I^2, \quad S_2 = S_2^{(1)} \bigcup S_2^{(2)},$$

then it is easy to check that S_2 consists of the union of 12 squares each in size of I^2, and for $x \in S_1$ and $y \in S_2$, we have

$$|x - y| \geq \frac{1}{2}.$$

For convenience, we denote by $S_2 = F(I^2)$ and set

$$P^{(3)}_\lambda f(x) = \int_{I^2} e^{i\lambda|x-y|} \Psi(x, y) f(y) dy,$$

for $x \in F(I^2)$, where $\Psi \in \mathscr{D}(\mathbb{R}^2 \times \mathbb{R}^2)$ and $\operatorname{supp}\Psi \subset F(I^2) \times I^2$.

We have the following theorem.

Theorem 2.6.5 *The following inequality*

$$\left\|P^{(3)}_\lambda f\right\|_{L^4(F(I^2))} \leq C_\varepsilon \lambda^{-\frac{1}{2}+\varepsilon} \|f\|_{L^4(I^2)}$$

holds for any $\varepsilon > 0$.

2.6 Oscillatory integral and proof of Carleson-Sjölin theorem

Proof. Let $x = (x_1, x_2) \in F(I^2)$ and $y = (t, \tau) \in I^2$. Then by the Minkowski inequality of integral, we get that

$$\left\| P_\lambda^{(3)} f \right\|_{L^4(F(I^2))}$$
$$\leq \left\| P_\lambda^{(3)} f \right\|_{L^4(S_2^{(1)})} + \left\| P_\lambda^{(3)} f \right\|_{L^4(S_2^{(2)})}$$
$$\leq \int_I \left\{ \int_{S_2^{(1)}} \left| \int_I e^{i\lambda[(x_1-t)^2+(x_2-\tau)^2]^{\frac{1}{2}}} \Psi(x,t,\tau) f(t,\tau) dt \right|^4 dx \right\}^{\frac{1}{4}} d\tau$$
$$+ \int_I \left\{ \int_{S_2^{(2)}} \left| \int_I e^{i\lambda[(x_1-t)^2+(x_2-\tau)^2]^{\frac{1}{2}}} \Psi(x,t,\tau) f(t,\tau) d\tau \right|^4 dx \right\}^{\frac{1}{4}} dt.$$

Applying Theorem 2.6.4 to the two terms on the right side of the above equation, we get that

$$\left\| P_\lambda^{(3)} f \right\|_{L^4(F(I^2))}$$
$$\leq C_\varepsilon \lambda^{-\frac{1}{2}+\varepsilon} \left\{ \int_I \left[\int_I |f(t,\tau)|^4 dt \right]^{\frac{1}{4}} d\tau + \int_I \left[\int_I |f(t,\tau)|^4 dt \right]^{\frac{1}{4}} dt \right\}.$$

Hence from Hölder's inequality we get the conclusion of the theorem. ∎

For the operator

$$P_\lambda^{(4)} f(x) = \int_{I^2} e^{i\lambda|x-y|} \frac{1}{|x-y|^{\frac{3}{2}+\alpha}} f(y) dy,$$

for $x \in F(I^2)$, we can use the operator

$$\overline{P}_\lambda^{(3)} f(x) = \int_{I^2} e^{i\lambda|x-y|} \frac{1}{|x-y|^{\frac{3}{2}+\alpha}} \psi_0(x,y) f(y) dy$$

to approximate it arbitrarily (with different ψ_0 chosen), where

$$\operatorname{supp}\psi_0 \subset F(I^2) \times I^2.$$

For $x \in D_1 \subset F(I^2)$ and $y \in D_2 \subset I^2$, we let

$$\psi_0(x,y) = 1,$$

and D_1 and D_2 can arbitrarily approximate to $F(I^2)$ and I^2, respectively. Noticing the smoothness of
$$\frac{1}{|x-y|^{\frac{3}{2}+\alpha}}$$
on $F(I^2) \times I^2$, we see that if $P_\lambda^{(3)}$ is replaced by $P_\lambda^{(4)}$, Theorem 2.6.5 still holds.

Now let
$$S_\lambda f(x) = \lambda^{\frac{1}{2}-\alpha} P_\lambda^{(4)} f(x),$$
for $x \in F(I^2)$.

By Theorem 2.6.5, we have that
$$\|S_\lambda f\|_{L^4(F(I^2))} \leq C_\varepsilon \lambda^{-\alpha+\varepsilon} \|f\|_{L^4(I^2)}.$$

Hence, by taking ε small enough, we obtain the following corollary.

Corollary 2.6.2 *There exists a positive constant $\delta > 0$ such that the inequality*
$$\|S_\lambda f\|_{L^4(F(I^2))} \leq C\lambda^{-\delta} \|f\|_{L^4(I^2)}$$
holds.

Remark 2.6.1 *It is easy to see that*
$$F(I^2) = [-1.5, 2.5]^2 \setminus [-0.5, 1.5]^2.$$
If $F(I^2)$ is replaced by the set
$$\overline{F}(I^2) = [-2.5, 3.5]^2 \setminus [-1.5, 2.5]^2,$$
we can still obtain the same result as Corollary 2.6.2.

2.6.2 Proof of Carleson-Sjölin theorem.

In Section 2.6.1, we mentioned that, under the condition $p > \frac{4}{3+2\alpha}$, the boundedness of T_α is equivalent to the L^p-boundedness of the operator
$$Pf(x) = \int_{|x-y|\geq 1} e^{i|x-y|} \frac{1}{|x-y|^{\frac{3}{2}+\alpha}} f(y) dy,$$
for $x \in \mathbb{R}^2$.

2.6 Oscillatory integral and proof of Carleson-Sjölin theorem

It is not difficult to prove that the integral

$$\int_{|x-y|<1} e^{i|x-y|} \frac{1}{|x-y|^{\frac{3}{2}+\alpha}} f(y) dy$$

can be controlled by some constant times of $Mf(x)$. Thus, we might as well take it for granted that

$$T_\alpha f(x) = \int_{\mathbb{R}^2} e^{i|x-y|} \frac{1}{|x-y|^{3/2+\alpha}} f(y) dy, \quad x \in \mathbb{R}^2.$$

In order to prove the boundedness of T_α on $L^p(\mathbb{R}^2)$, we find the key is the case that $p = 4$. By the boundedness of T_α on $L^4(\mathbb{R}^2)$ and the theory of interpolation and duality, we can easily obtain that T_α is bounded on $L^p(\mathbb{R}^2)$ where

$$\frac{4}{3+2\alpha} < p < \frac{4}{1-2\alpha}.$$

Next, to prove the boundedness of T_α on $L^4(\mathbb{R}^2)$, it suffices to show

$$\int_{(\lambda I)^2} \left| \int_{(\lambda I)^2} e^{i|x-y|} \frac{1}{|x-y|^{\frac{3}{2}+\alpha}} f(y) dy \right|^4 dx \leq C \int_{(\lambda I)^2} |f(x)|^4 dx,$$

where $\lambda I = [0, \lambda]$. By some change of variables, the inequality above turns out to be

$$\int_{I^2} \left| \lambda^{\frac{1}{2}-\alpha} \int_{I^2} e^{i\lambda |x-y|} \frac{1}{|x-y|^{\frac{3}{2}+\alpha}} F(y) dy \right|^4 dx \leq C \int_{I^2} |F(x)|^4 dx.$$

Then, the proof of Carleson-Sjölin theorem can be reduced to the proof of the following basic facts.

Proposition 2.6.1 *There exists a constant $C > 0$ such that the inequality*

$$\|S_\lambda f\|_{L^4(I^2)} \leq C \|f\|_{L^4(I^2)}$$

holds.

Proof. Before the proof of the proposition, we recall the meaning of the set $F(I^2)$ in section 6.1. Let F_1 and F_2 be two squares whose centers are the same as I^2 and edges both parallel to the axes, with lengths 2 and 4. Then we have $F(I^2) = F_2 \setminus F_1$. In the following, for any square ω with edges paralleling to the axes, we similarly define $F(\omega)$. We denote by $F_1(\omega)$

and $F_2(\omega)$ two squares, whose centers are the same as ω and edges are both parallel to the axes, with lengths 2 and 4 times of the one of ω. Then, we define
$$F(\omega) = F_2(\omega) \setminus F_1(\omega).$$
Obviously, for $x \in \omega$ and $y \in F(\omega)$, we have $|x - y| \geq \frac{1}{2} \times$ (the length of ω). In addition, any square ω whose edges are parallel to the axes, we denote by $\frac{1}{2}\omega$ the square which has the same center as ω and edges paralleling to the axes, with half length of ω. Similarly, we define
$$\overline{F}(\omega) = \overline{F_2}(\omega) \setminus \overline{F_1}(\omega),$$
where $\overline{F_1}(\omega)$ and $\overline{F_2}(\omega)$ are two squares, whose centers are the same as ω and edges are both parallel to the axes, with lengths 4 and 6 times of the one of ω.

Let Ω_k be all of the binary squares in the square of $[-2, 2] \times [-2, 2]$ with length 2^{-k} and Ω_k^* be a collection of all the squares which consist of all four squares belonging to Ω_k. It is obvious that the edge length of the square in Ω_k^* is 2^{-k+1}. For any $x \in I^2$, but not in the boundary of any binary squares, noticing that the measure of the set consisted by all those boundaries is zero in two dimensional sense, obviously there exists an unique square in Ω_k^*, such that the half small square which is concentrically contracted from the original square contains x. We denote the unique square by $\omega_k^*(x)$. Then, $x \in \frac{1}{2}\omega_k^*(x)$. Since
$$I^2 = \frac{1}{2}\Big([-2,2] \times [-2,2]\Big),$$
we denote by
$$\omega_{-1}^*(x) = [-2, 2] \times [-2, 2].$$
Set
$$A(x, D) = \lambda^{\frac{1}{2}-\alpha} \int_D e^{i\lambda|x-y|} \frac{f(y)}{|x-y|^{3/2+\alpha}} dy, \quad x \in I^2,$$
and
$$A_k(x) = A\Big(x, [\omega_{k-1}^*(x) \setminus \omega_k^*(x)] \cap I^2\Big), \quad k \geq 0.$$
Then we have
$$S_\lambda f(x) = \sum_{k=0}^{k_N} A_k(x) + A(x, \omega_{k_N}^*(x)),$$
where k_N satisfies
$$2^{-(k_N+1)} < \lambda^{-1} \leq 2^{-k_N}.$$

2.6 Oscillatory integral and proof of Carleson-Sjölin theorem

We notice that if k is large enough, then we have

$$\omega_{k-1}^*(x) \setminus \omega_k^*(x) \subset I^2.$$

Thus we might as well take it for granted that

$$A_k(x) = A(x, \omega_{k-1}^*(x) \setminus \omega_k^*(x)).$$

By the construction of $\omega_k^*(x)$, we have that $\omega_{k-1}^*(x) \setminus \omega_k^*(x)$ is the union of 12 squares in the same size, with length 2^{-k}. The arrangement of the 12 squares at least falls into one of the following three cases.
Case 1. (see Figure 2.3)

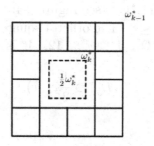

Figure 2.3: Case 1.

The 12 squares share either the same edges or points with $\omega_k^*(x)$, which are denoted by D_i, $1 \leq i \leq 12$. Obviously, for any i, $1 \leq i \leq 12$, we have

$$F(D_i) \supset \frac{1}{2}\omega_k^*(x).$$

Thus, we have $x \in F(D_i)$. Then we have

$$A_k(x) = \sum_{i=1}^{12} A(x, D_i) \chi_{F(D_i)}(x).$$

It follows that

$$|A_k(x)|^4 \leq C \sum_{i=1}^{12} |A(x, D_i)|^4 \chi_{F(D_i)}(x).$$

and

$$\int_{I^2} |A_k(x)|^4 dx \leq C \sum_{i=1}^{12} \int_{F(D_i)} |A(x, D_i)|^4 dx.$$

Applying Corollary 2.6.2 to each integral on the right side of the above inequality and making change of variables, we conclude that

$$\int_{I^2} |A_k(x)|^4 dx \leq C \sum_{i=1}^{12} (\lambda 2^{-k})^{-4\delta} \int_{D_i} |f(x)|^4 dx$$

$$\leq C\lambda^{-4\delta} 2^{4k\delta} \int_{I^2} |f(x)|^4 dx.$$

Case 2. Among all the 12 squares, there are 8 squares sharing either the same edges or points with $\omega_k^*(x)$ (see Figure 2.4: Case 2 and Case 3), which are denoted by D_i, $1 \leq i \leq 8$. The other 4 squares have no common points with $\omega_k^*(x)$, denoted by D_j, $9 \leq j \leq 12$. It is evident that for any D_j, $9 \leq j \leq 12$, we have

$$\overline{F}(D_j) \supset \frac{1}{2}\omega_k^*(x).$$

Then we have

$$A_k(x) = \sum_{i=1}^{8} A(x, D_i)\chi_{F(D_i)}(x) + \sum_{j=9}^{12} A(x, D_j)\chi_{\overline{F}(D_j)}(x).$$

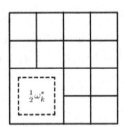

Figures 2.4: Case 2 and Case 3.

Noticing the remark at the end of Section 2.6.1 and by the similar discussion in Case 1, we have that

$$\int_{I^2} |A_k(x)|^4 dx \leq C\lambda^{-4\delta} 2^{4k\delta} \int_{I^2} |f(x)|^4 dx.$$

2.6 Oscillatory integral and proof of Carleson-Sjölin theorem

Case 3. Among all the 12 squares, there are only 5 ones sharing either the same edges or points with $\omega_k^*(x)$, (see Figure 2.4: Case 2 and Case 3).

In this case, the proof is the same as that in Case 2. Consequently, for any case, we always have that

$$\|A_k\|_{L^4(I^2)} \leq C\lambda^{-\delta} 2^{k\delta} \|f\|_{L^4(I^2)}.$$

Then we have

$$\|S_\lambda f\|_{L^4(I^2)} \leq C\left\{\lambda^{-\delta}\sum_{k=0}^{k_N} 2^{k\delta}\|f\|_{L^4(I^2)} + \left(\int_{I^2} |A(x,\omega_{k_N}^*(x))|^4 dx\right)^{\frac{1}{4}}\right\}$$

$$\leq C\left\{\|f\|_{L^4(I^2)} + \left(\int_{I^2} |A(x,\omega_{k_N}^*(x))|^4 dx\right)^{\frac{1}{4}}\right\}.$$

Therefore, it suffices to prove that

$$\left(\int_{I^2} |A(x,\omega_{k_N}^*(x))|^4 dx\right)^{\frac{1}{4}} \leq C\|f\|_{L^4(I^2)}.$$

In fact, we have

$$|A(x,\omega_{k_N}^*(x))| \leq \lambda^{\frac{1}{2}-\alpha} \int_{\omega_{k_N}^*} \frac{|f(y)|}{|x-y|^{3/2+\alpha}} dy$$

$$\leq \lambda^{\frac{1}{2}-\alpha} \sum_{i=0}^{\infty} \int_{\omega_{k_N+i}^* \setminus \omega_{k_N+i+1}^*} \frac{|f(y)|}{|x-y|^{3/2+\alpha}} dy.$$

We notice that for

$$y \in \omega_{k_N+i}^*(x) \setminus \omega_{k_N+i+1}^*(x)$$

and $x \in \frac{1}{2}\omega_{k_N+i+1}^*(x)$, there holds

$$|x-y| \geq 2^{-(k_N+i+2)},$$

which yields

$$|A(x,\omega_{k_N}^*(x))| \leq \lambda^{\frac{1}{2}-\alpha} \sum_{i=0}^{\infty} 2^{(3/2+\alpha)(k_N+i+2)} \int_{\omega_{k_N+i}^*(x)} |f(y)| dy$$

$$\leq C \sum_{i=0}^{\infty} 2^{-(\frac{1}{2}-\alpha)i} Mf(x)$$

$$\leq CMf(x).$$

This finishes the proof of Theorem 2.5.2. ∎

Remark 2.6.2 *Let us recall the proof of Carleson-Sjölin theorem in this section. Firstly, the conclusions of Theorems 2.6.1 to 2.6.3 cannot be applied directly to the proof of Carleson-Sjölin theorem. The reason for that lies in the fact that when*

$$\varphi(x,y) = |x - y|$$

is restricted, the condition

$$\det\left(\frac{\partial^2 \varphi(x,y)}{\partial x_i \partial y_i}\right) \neq 0$$

does not hold. But when $n = 2$, a special method could be utilized to conquer the difficulty. Therefore, the method of proving in this section is not suitable for the case $n > 2$. To this end, ones are trying to find out new methods of Carleson-Sjölin theorem, such that the new one is feasible in solving the following speculation: T_α, $\alpha > 0$, is bounded on $L^p(\mathbb{R}^n)$, where $n > 2$ and and

$$\frac{2n}{n+1+2\alpha} < p < \frac{2n}{n-1-2\alpha}.$$

Here, what needs to be mentioned is the work done by Cordoba, Tomas and Stein. Cordoba used the geometric method which combined the proof of Carleson-Sjölin theorem with the Kakeya maximal function. Making use of the restriction theorem of the Fourier transform, Tomas and Stein put forward a general result on the boundedness of T_α on $L^p(\mathbb{R}^n)$ (cf. [Fe4]).

However, the two kinds of new methods still fail to settle the above conjecture under the case $n > 2$, yet they offer a powerful tool in solving other problems. For this reason, we will introduce the methods of Cordoba and Tomas-Stein.

2.7 Kakeya maximal function

To investigate the L^p-boundedness of the operator T_α, we can use the method for proving the L^p-boundedness of partial sum operator in the case of one dimension.

It is well known that in the case of one dimension, the Hilbert transform plays an important role in solving questions and the estimate of Hilbert transform can be reduced to Hardy-Littlewood maximal function. However it is different when it comes to the cases of high dimension. For instance, when Cordoba considered the case of two-dimension, what he met with was not Hardy-Littlewood maximal function but the maximal function about the

2.7 Kakeya maximal function

rectangle with some eccentricity, which is called Kakeya maximal function. The Kakeya maximal function is defined as follows. If given $\delta > 0$ and $N > 0$, we denote by $\mathscr{R} = \{\text{rectangle } R : |R| = N\delta \times \delta\}$, that is, the length of the longer edge of R is $N\delta$ and the shorter is δ. Let the direction of the long edge be the direction of the rectangle. Then we define

$$M_N f(x) = \sup_{R \ni x} \frac{1}{|R|} \int_R |f(y)|dy,$$

where sup is taken over all the $R \in \mathscr{R}$ including x, and N is the eccentricity. Cordoba [Co1] has proven the following basic estimate on M_N.

Theorem 2.7.1 *There exists a constant C independent of δ and N, such that*

$$\|M_N f\|_2 \leq C(\log 3N)^{\frac{1}{2}} \|f\|_2$$

for all $f \in L^2(\mathbb{R}^2)$.

Proof. Since the proof of Theorem 2.7.1 is rather long, it will be convenient to divide it into seven steps.

Step I. Restrict the direction of the rectangle.

Divide the range of the direction of rectangle $[0, 2\pi)$ into eight parts as

$$[0, 2\pi) = \left[0, \frac{\pi}{4}\right) \cup \left[\frac{\pi}{4}, \frac{\pi}{2}\right) \cup \cdots \cup \left[\frac{7\pi}{4}, 2\pi\right) = \bigcup_{i=1}^{8} I_i.$$

Similar to the definition of $M_N f(x)$, we define $M_N^i f(x)$, with the requirement that the direction of rectangle R must be inside I_i. Then we have

$$M_N f(x) \leq \sum_{i=1}^{8} M_N^i f(x).$$

Since the method of dealing with all these $M_N^i f(x)$ is the same, we merely need to prove that

$$\|M_N^1 f\|_2 \leq C(\log 3N)^{\frac{1}{2}} \|f\|_2$$

for any $f \in L^2(\mathbb{R}^2)$. For convenience, we still denote $M_N^1 f$ by $M_N f$.

Step II. Restrict the support of f.

Dividing the plane into squares with the length of the edge of $N\delta$ along the axas as
$$\mathbb{R}^2 = \bigcup Q_\alpha,$$
where the length of the edge of Q_α is $N\delta$ and $Q_\alpha \cap Q_\beta = \emptyset$ for $\alpha \neq \beta$.

Set
$$f_\alpha(x) = f(x)\chi_{Q_\alpha}(x).$$
It is obvious that
$$f(x) = \sum f_\alpha(x),$$
for a.e. $x \in \mathbb{R}^2$.

Denoted by $Q^* = 3Q$, the length of the edge of Q^* is 3 times of Q and the two have the same center.

When $x \in Q_\alpha^*$, and $Q_\alpha^* \cap Q_\beta^* = \emptyset$, we have
$$M_N f_\beta(x) = 0.$$

Consequently, there exist at most 5^2 terms in the sum $\sum_\alpha M_N f_\alpha(x)$. Therefore, it follows that

$$|M_N f(x)|^2 \leq \left|\sum_\alpha M_N f_\alpha(x)\right|^2 \leq 5^4 \sum_\alpha |M_N f_\alpha(x)|^2. \qquad (2.7.1)$$

For each f_α, if the inequality
$$\|M_N f_\alpha\|_2 \leq C(\log 3N)^{\frac{1}{2}} \|f_\alpha\|_2$$
holds, then the conclusion of the theorem immediately follows from (2.7.1).

Consequently, the proof of the theorem can be reduced to prove that if $\operatorname{supp} f \subset Q$ and the edge length of Q is $N\delta$, then
$$\|M_N f\|_2 \leq C(\log 3N)^{\frac{1}{2}} \|f\|_2 \qquad (2.7.2)$$
holds.

It is obvious that, for such f, we must have
$$M_N f(x) = 0 \qquad (2.7.3)$$
for $x \in (Q^*)^c$.

Step III. The linearization of the maximal operator.

2.7 Kakeya maximal function

Suppose that $\mathrm{supp} f \subset$ the square Q, the edge length of Q is $N\delta$, and $Q^* = 3Q$. Again by the straight line along the axas, we divide Q^* into $9N^2$ squares $Q_{i,p}$ $(1 \leq i \leq 3N, 1 \leq p \leq 3N)$ of the edge length δ. For any $x \in Q_{i,p}$, by the definition of $M_N^1 f$, there exists a rectangle $R_{i,p}(x)$ such that

$$M_N f(x) \leq \frac{2}{|R_{i,p}(x)|} \int_{R_{i,p}(x)} |f(y)|dy$$

for $x \in Q_{i,p}$, where the direction of the rectangle $R_{i,p}(x)$ belongs to $I_1 = [0, \frac{\pi}{4})$, and $|R_{i,p}(x)| = N\delta \times \delta$. By the definition of the supremum, there exists a

$$R_{i,p} \in \{R_{i,p}(x) : x \in Q_{i,p}\},$$

such that

$$\sup_{x \in Q_{i,p}} \int_{R_{i,p}(x)} |f(y)|dy \leq 2 \int_{R_{i,p}} |f(y)|dy,$$

where the direction of $R_{i,p}$ belongs to I_1, and $|R_{i,p}| = N\delta \times \delta$. It follows from the above discussion that

$$M_N f(x) \leq \frac{4}{|R_{i,p}|} \int_{R_{i,p}} |f(y)|dy,$$

for $x \in Q_{i,p}$. Together with (2.7.3), the above inequality leads to

$$M_N f(x) \leq 4 \sum_{i,p} \left(\frac{1}{|R_{i,p}|} \int_{R_{i,p}} |f(y)|dy \right) \chi_{Q_{i,p}}(x).$$

It is obvious that there holds, $R_{i,p} \cap Q_{i,p} \neq \emptyset$. For fixed f, if we define the linear operator $g \to T_f(g)$ as

$$T_f(g)(x) = \sum_{i,p} \left(\frac{1}{|R_{i,p}|} \int_{R_{i,p}} g(y)dy \right) \chi_{Q_{i,p}}(x),$$

then, in order to prove (2.7.2), it suffices to show that

$$\|T_f(g)\|_2 \leq C(\log 3N)^{\frac{1}{2}} \|g\|_2, \quad \text{for any } g \in L^2(Q^*). \tag{2.7.4}$$

Step IV. Transfer into the estimate for the conjugate operator T_f^*.

Let $h, g \in L^2(Q^*)$. If we denote the conjugate operator of T_f by T_f^*, then by the equation

$$\int_{Q^*} g(y) T_f^*(h)(y) dy = \int_{Q^*} T_f(g)(x) h(x) dx.$$

We easily have

$$T_f^*(h)(y) = \sum_{i,p} \left(\frac{1}{|R_{i,p}|} \int_{Q_{i,p}} h(y)dy \right) \chi_{R_{i,p}}(x).$$

By the fact that $\|T_f^*\| = \|T_f\|$, to get (2.7.4), it suffices to show

$$\|T_f^*(h)\|_2 \leq C(\log 3N)^{\frac{1}{2}}\|h\|_2, \quad for\ any\ h \in L^2(Q^*). \tag{2.7.5}$$

Step V. The proof of the equation (2.7.5).

Let $h \in L^2(Q^*)$. Now with straight lines parallel to the y-axis, we divide Q^* into $3N$ rectangles, that is,

$$Q^* = \bigcup_{i=1}^{3N} E_i, \quad |E_i| = \delta \times 3N\delta.$$

Let

$$h_i(x) = h(x)\chi_{E_i}(x),$$

then we have

$$h(x) = \sum_{i=1}^{3N} h_i(x).$$

It is easy to prove that if the inequality

$$\|T_f^*(h_i)\|_2 \leq CN^{-\frac{1}{2}}(\log 3N)^{\frac{1}{2}}\|h_i\|_2, \quad (1 \leq i \leq 3N) \tag{2.7.6}$$

holds, then, (2.7.5) is also holds. In fact, this can be obtained from the following inequality

$$\|T_f^*(h)\|_2 \leq \sum_{i=1}^{3N} \|T_f^*(h_i)\|_2$$

$$\leq CN^{-\frac{1}{2}}(\log 3N)^{\frac{1}{2}} \sum_{i=1}^{3N} \|h_i\|_2$$

$$\leq CN^{-\frac{1}{2}}(\log 3N)^{\frac{1}{2}} \left(\sum_{i=1}^{3N} \int_{E_i} |h(x)|^2 dx \right)^{\frac{1}{2}} \left(\sum_{i=1}^{3N} 1 \right)^{\frac{1}{2}}$$

$$\leq C(\log 3N)^{\frac{1}{2}}\|h\|_2.$$

2.7 Kakeya maximal function

Step VI. The proof of the equation (2.7.6).

We divide E_i into $3N$ squares with the edge length δ:

$$E_i = \bigcup_{p=1}^{3N} Q_{i,p}.$$

Let

$$h_{i,p}(x) = h_i(x)\chi_{Q_{i,p}}.$$

We easily have

$$h_i(x) = \sum_{p=1}^{3N} h_{i,p}(x).$$

Since there holds

$$T_f^*(h_i)(x) = \sum_p \left(\frac{1}{|R_{i,p}|} \int_{Q_{i,p}} h_{i,p}(y) dy \right) \chi_{R_{i,p}}(x),$$

we have

$$|T_f^*(h_i)(x)| \leq \frac{1}{N\delta} \sum_p \|h_{i,p}\|_2 \chi_{R_{i,p}}(x),$$

which yields

$$\|T_f^*(h_i)\|_2^2 \leq \frac{1}{(\delta N)^2} \int \left(\sum_p \|h_{i,p}\|_2 \chi_{R_{i,p}}(x) \right)^2 dx$$

$$= \frac{1}{(\delta N)^2} \sum_{p,q} \|h_{i,p}\|_2 \|h_{i,q}\|_2 \left| R_{i,p} \bigcap R_{i,q} \right|.$$

It is not difficult to prove that if the inequality

$$\left| R_{i,p} \bigcap R_{i,q} \right| \leq CN\delta^2/(|p-q|+1) \qquad (2.7.7)$$

holds, then we can get (2.7.6). This can be obtained from the following equation

$$\|T_f^*(h_i)\|_2^2 \leq \frac{C}{N} \sum_{1 \leq p,\, q \leq 3N} \frac{\|h_{i,p}\|_2 \|h_{i,q}\|_2}{|p-q|+1}$$

$$= \frac{C}{N} \left(\sum_{p=1}^{3N} \|h_{i,p}\|_2^2 + \sum_{\nu=1}^{3N-1} \frac{1}{1+\nu} \sum_{|p-q|=\nu} \|h_{i,p}\|_2 \|h_{i,q}\|_2 \right)$$

$$\leq \frac{C}{N} \left\{ \|h_i\|_2^2 + \sum_{\nu=1}^{3N-1} \frac{1}{1+\nu} \left(\sum_{p=1}^{3\nu} \|h_{i,p}\|_2^2 \right) \right\}$$

$$\leq \frac{C}{N} (\log 3N) \|h_i\|_2^2.$$

Step VII. The proof of (2.7.7).

Here, $R_{i,p} \cap Q_{i,p} \neq \emptyset$, the direction of $R_{i,p}$ is inside I_1, and

$$|R_{i,p}| = \delta \times N\delta.$$

We might as well prove (2.7.7) for $q = 1$. Suppose that the direction of $R_{i,1}$ and $R_{i,p}$ forms an angle of θ, and $R_{i,1} \cap R_{i,p}$ is a parallelogram $ABCD$ (see Figure 2.5).

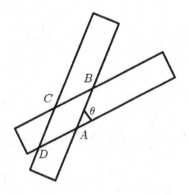

Figure 2.5.

Then we have

$$\left| R_{i,1} \cap R_{i,p} \right| = \delta \times |AB| = \frac{\delta^2}{\sin \theta}.$$

2.7 Kakeya maximal function

We consider the case: $\theta > \alpha$, where α is just the angle leaning toward the edge EF in the triangle EFG in Figure 2.6. Here, FG and HF represent $R_{i,p}$ and $R_{i,1}$, respectively.

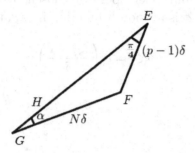

Figure 2.6.

Hence, we have

$$|R_{i,1} \cap R_{i,p}| \leq \frac{\delta^2}{\sin\alpha} = \frac{N\delta^2}{(p-1)\sin\frac{\pi}{4}}.$$

Thus (2.7.7) is valid (when $q = 1$), then we could duplicate the proof for q in general case.

This finishes the proof of Theorem 2.7.1. ∎

Cordoba (see [Co1] and [Co2]) gave another proof of Carleson-Sjölin theorem by the method of geometric proof and Kakeya maximal function. We only point out why Cordoba's proof is related with the Kakeya maximal function briefly. It is well known that the key point of proving Carleson-Sjölin theorem lies in the estimate for the case $p = 4$, while for other p, we can prove by the methods of interpolation and duality. The estimate for the case when $p = 4$ is carried out in the following way. Firstly, by segmentation along the radial diameters, we divide the unit disc $\{|x| < 1\}$ into the union of a series of loops $\bigcup_k D_k$ and give some smooth decomposition of $(1-|x|^2)_+^\alpha$ as

$$(1-|x|^2)_+^\alpha = \sum_k \overline{m}_k(x),$$

where $\overline{m}_k$ is smooth and supp $\overline{m}_k \subset D_k$.

Define

$$\widehat{T_\alpha^k f} = \overline{m}_k \hat{f}.$$

Then the estimate for T_α can be transfered into the one for T_α^k. These T_α^k are all multiplier operators in corresponding with the smooth multiplier function whose support is in a loop. By dilations, we merely need to estimate one of the T_α^ks. The multiplier operator chosen by Cordoba is defined as follows. Suppose that $\widetilde{\phi}$ is a smooth function in $\mathbb{R}$, and supp $\widetilde{\phi} \subset [-1, 1]$,

$$\phi(\xi) = \widetilde{\phi}\left(\frac{|\xi| - 1}{\delta}\right)$$

for $\xi \in \mathbb{R}^2$.

It is easy to see that

$$\operatorname{supp} \phi \subset \{\xi \in \mathbb{R}^2 : 1 - \delta \leq |\xi| \leq 1 + \delta\}.$$

Define

$$\widehat{S_\delta f}(\xi) = \phi(\xi)\hat{f}(\xi)$$

for $f \in \mathscr{D}$.

Then the estimate for $\|T_\alpha f\|_4$ is transferred into the one for $\|S_\delta f\|_4$. Secondly, let $\xi = |\xi|e^{i\theta}$ and through the partition of the argument θ, we give a smooth decomposition of $\phi(\xi)$ as

$$\phi(\xi) = \sum_j m_j(\xi).$$

We also define the multiplier T_j by

$$\widehat{T_j f}(\xi) = m_j(\xi)\hat{f}(\xi), \quad (0 \leq j \leq [\delta^{-\frac{1}{2}}] - 1),$$

and then we transfer the estimate for $\|S_\delta f\|_4$ into the one for $\|\sum_j T_j f\|_4$. We notice that suppm_j is a small part of the loop

$$\{\xi : 1 - \delta \leq |\xi| \leq 1 + \delta\}$$

and the central angle which is leaning toward the small part is $2\pi\delta^{\frac{1}{2}}$. Now, we write T_j as a convolution operator, that is,

$$T_j f = \widehat{m_j} * f.$$

Since

$$\widehat{m_j}(x_1, x_2) \approx \int_{1-\delta}^{1+\delta} \int_{2\pi j \delta^{\frac{1}{2}}}^{2\pi(j+1)\delta^{\frac{1}{2}}} m_j(\xi_1, \xi_2) e^{-i(\xi_1 x_1 + \xi_2 x_2)} d\xi_1 d\xi_2,$$

2.7 Kakeya maximal function

by partial integration, we obtain a control function of $\widehat{m_j}$

$$|\widehat{m_j}(x_1, x_2)| \leq C_{p,q} \delta^{3/2} |\delta x_1|^{-p} |\delta^{\frac{1}{2}} x_2|^{-q} \quad (p \geq 0, \ q \geq 0).$$

From the above inequality, we verify that

$$|\widehat{m_j}(x_1, x_2)| \leq C \sum_{\nu=0}^{\infty} 2^{-\nu} \frac{1}{|R_{\nu,j}|} \chi_{R_{\nu,j}}(x_1, x_2),$$

where

$$R_{\nu,j} = \left\{ (x_1, x_2) : |x_1| \leq 2^\nu \delta^{-1}, |x_2| \leq 2^\nu \delta^{-\frac{1}{2}} \right\}.$$

Obviously, $R_{\nu,j}$ is a rectangle with the eccentricity being $\delta^{-\frac{1}{2}}$. Then, by the method for the case of L^2 and the inequality

$$\left\| \sum_j T_j f \right\|_4^4 \leq C \left\| \left(\sum_j |T_j f|^2 \right)^{\frac{1}{2}} \right\|_4^4,$$

we have that

$$\left\| \sum_j T_j f \right\|_4^4 \leq C \int_{\mathbb{R}^2} \left(\sum_j |(\widehat{m_j} * f)(x)|^2 \right)^2 dx$$

$$\leq C \int_{\mathbb{R}^2} \left\{ \sum_j \left(\sum_\nu 2^{-\nu} \frac{\chi_{R_{\nu,j}}}{|R_{\nu,j}|} * |f|(x) \right)^2 \right\}^2 dx$$

$$\leq C \sum_\nu 2^{-\nu} \left\| \left\{ \sum_j \left(\frac{\chi_{R_{\nu,j}}}{|R_{\nu,j}|} * |f| \right)^2 \right\}^{\frac{1}{2}} \right\|_4^4.$$

Consequently, we face the estimate of the norm on the right side of the above inequality. It is easy to see that

$$\left\|\left\{\sum_j \left(\frac{\chi_{R_{\nu,j}}}{|R_{\nu,j}|}*|f|\right)^2\right\}^{\frac{1}{2}}\right\|_4^2 = \left\|\sum_j \left(\frac{\chi_{R_{\nu,j}}}{|R_{\nu,j}|}*|f|\right)^2\right\|_2$$

$$= \sup_{\|\omega\|_2 \le 1} \int_{\mathbb{R}^2} \sum_j \left(\frac{\chi_{R_{\nu,j}}}{|R_{\nu,j}|}*|f|(x)\right)^2 \omega(x)dx$$

$$\le \sup_{\|\omega\|_2 \le 1} \sum_j \int_{\mathbb{R}^2} |f(y)|^2 \left(\frac{\chi_{R_{\nu,j}}}{|R_{\nu,j}|}*\omega\right)(y)dy$$

$$\le \sup_{\|\omega\|_2 \le 1} \sum_j \int_{\mathbb{R}^2} |f(y)|^2 M_{\delta^{-1/2}}(\omega)(y)dy,$$

where $M_{\delta^{-\frac{1}{2}}}(\omega)(y)$ is the Kakeya maximal function with eccentricity being $\delta^{-\frac{1}{2}}$. In this way, in Cordoba's proof, the solution to the problem is closely related with the Kakeya maximal function.

It is worthy of being pointed out that Cordoba and Lopez-Melero both used the same methods in [CoL1] to extend the Carleson-Sjölin theorem to the case of vector-valued functions, and get the following theorem.

Theorem 2.7.2 Let $0 < \alpha < \frac{1}{2}$. If

$$\frac{4}{3+2\alpha} < p < \frac{4}{1-2\alpha},$$

then we have

$$\left\|\left(\sum_j |B_{R_j}^\alpha f_j|^2\right)^{\frac{1}{2}}\right\|_{L^p(\mathbb{R}^2)} \le C_{p,\alpha} \left\|\left(\sum_j |f_j|^2\right)^{\frac{1}{2}}\right\|_{L^p(\mathbb{R}^2)},$$

for all $f \in \mathscr{S}$.

Remark 2.7.1 Adding into Theorem 2.7.2 the condition that $\{R_j\}_1^\infty$ is a lacunary sequence, Igari [Ig1] independently acquired the results as well.

When $\alpha = 0, n \ge 2$, the following result similarily to Theorem 2.7.2 is obtained by Carbey, Rubio de Francia, Vega [CRV1] and independently by Ma, Liu, Lu [MLL1].

Theorem 2.7.3 *Let $2 \leq p < \frac{2n}{n-1}$. If $\{R_j\}_1^\infty$ is a lacunary sequence, then*

$$\left\| \left(\sum_j |B_{R_j}^0 f_j|^2 \right)^{\frac{1}{2}} \right\|_{L^p(\mathbb{R}^n)} \leq C_{p,\alpha} \left\| \left(\sum_j |f_j|^2 \right)^{\frac{1}{2}} \right\|_{L^p(\mathbb{R}^n)}$$

for $f \in \mathscr{S}(\mathbb{R}^n)$.

2.8 The restriction theorem of the Fourier transform

For any n with $n \geq 3$ and

$$\frac{2n}{n+1+2\alpha} < p < \frac{2n}{n-1-2\alpha},$$

it still has not been verified that whether T_α is bounded on $L^p(\mathbb{R}^n)$. However, Stein offered a general result by making use of the restriction theorem of the the Fourier transform (see Fefferman [Fe4]). To the end, we first give the definition of the restriction theorem of the Fourier transform.

Definition 2.8.1 *Suppose that there exists $p \geq 1$ such that*

$$\left(\int_{\mathbb{S}^{n-1}} |\hat{f}(\xi)|^2 d\sigma(\xi) \right)^{\frac{1}{2}} \leq C_p \|f\|_{L^p(\mathbb{R}^n)},$$

for $f \in \mathscr{S}$. We say that the (L^p, L^2) restriction theorem of the Fourier transform holds.

The result due to Stein illustrates that: under the condition that the (L^p, L^2) restriction theorem of the Fourier transform holds, the Fefferman's conjecture above is valid. By the method of duality, we merely need to consider the case when $\frac{2n}{n+1+2\alpha} < p < 2$.

Theorem 2.8.1 *Suppose that*

$$\frac{2n}{n+1+2\alpha} < p < 2,$$

if the (L^p, L^2) restriction theorem of the Fourier transform holds, then

$$\|B_R^\alpha(f)\|_p \leq C_{p,\alpha} \|f\|_p.$$

Proof. It is easy to prove that there exists a radial function $\psi_0, \psi \in \mathscr{D}(\mathbb{R}^n)$, where supp$\psi$ is around $|x| = 1$, satisfying

$$\psi_0(x) + \sum_{j=1}^{\infty} \psi(2^{-j}x) = 1.$$

Let
$$\psi_j(x) = \psi(2^{-j}x) \ (j = 1, 2, \ldots)$$
and
$$B_{j,R}^{\alpha}(x) = B_R^{\alpha}(x) \cdot \psi_j(x) \ (j = 0, 1, \ldots).$$

Then we have
$$B_R^{\alpha}(f)(x) = \sum_j (B_{j,R}^{\alpha} * f)(x) := \sum_j (T_j f)(x).$$

Obviously, it suffices to prove for some $\varepsilon > 0$, there holds

$$\int_{\mathbb{R}^n} |T_j f(x)|^p dx \leq C^p 2^{-jp\varepsilon} \int_{\mathbb{R}^n} |f(x)|^p dx, \qquad (2.8.1)$$

where C is independent of f, ε.

Now we separate $\mathbb{R}^n$ into the union of countable n-dimensional cubes, with the edges parallel to the axas and the edge length being 2^j. Denote any one of the cubes by Q and the constant times extension of Q by $\widetilde{Q}$, which share the same center with Q. According to the fact that $B_{j,R}^{\alpha}$ is around $|x| = 2^j$, then in order to prove (2.8.1), we merely need to prove that

$$\int_{\widetilde{Q}} |T_j f(x)|^p dx \leq C^p 2^{-jp\varepsilon} \int_Q |f(x)|^2 dx. \qquad (2.8.2)$$

By the translation invariance, we might as well take it for granted that Q is centered at O. It is not difficult to see that we might as well deem Q as a ball, centered at O, with a diameter being 2^j. To this end, we suppose that
$$\mathrm{supp} f \subset Q = \{x : |x| < 2^j\}.$$

Let $\frac{2}{p}$ and r be a pair of conjugate indexes, that is,

$$\frac{1}{\frac{2}{p}} + \frac{1}{r} = 1.$$

By Hölder's inequality, we have

2.8 The restriction theorem of the Fourier transform

$$\int_{\widetilde{Q}} |T_j f(x)|^p dx \leq \left(\int_{\widetilde{Q}} |T_j f(x)|^2 dx \right)^{p/2} \cdot |\widetilde{Q}|^{1/r}$$

$$\leq C 2^{\frac{nj}{r}} \left(\int_{\mathbb{R}^n} |T_j f(x)|^2 dx \right)^{p/2}. \qquad (2.8.3)$$

It follows from the Plancherel theorem that

$$\int_{\mathbb{R}^n} |T_j f(x)|^2 dx = \int_{\mathbb{R}^n} |\widehat{B}^\alpha_{j,R}(x)|^2 \cdot |\hat{f}(x)|^2 dx$$

$$= \int_{|x| \leq \alpha} |\widehat{B}^\alpha_{j,R}(x)|^2 \cdot |\hat{f}(x)|^2 dx$$

$$+ \int_{|x| > \alpha} |\widehat{B}^\alpha_{j,R}(x)|^2 |\hat{f}(x)|^2 dx,$$

where α is to be determined.

Next, we make estimates of the two terms on the right side of the above inequality. Firstly, we have

$$\int_{|x| \leq \alpha} |\widehat{B}^\alpha_{j,R}(x)|^2 |\hat{f}(x)|^2 dx \leq \sup_{|x| \leq \alpha} |\widehat{B}^\alpha_{j,R}(x)|^2 \int_{|x| \leq \alpha} |\hat{f}(x)|^2 dx.$$

According to the fact

$$\widehat{B}^\alpha_{j,\alpha}(x) = (\widehat{B}_R * \widehat{\psi}_j)(x)$$

and

$$\psi(0) = \int_{\mathbb{R}^n} \psi(u) du = 0,$$

we obtain

$$\widehat{B}^\alpha_{j,\alpha}(x) = \int_{\mathbb{R}^n} \left\{ \widehat{B}_R(y) - 1 \right\} \widehat{\psi} \left(2^j (x-y) \right) 2^{nj} dy.$$

Since

$$\lim_{y \to 0} \widehat{B}^\alpha_R(y) = 1$$

and

$$\psi \in \mathscr{D}(\mathbb{R}^n) \subset \mathscr{S}(\mathbb{R}^n),$$

there exists $\delta > 0$ small enough such that

$$|\widehat{B}^\alpha_{j,R}(x)| \leq C \left| \int_{|y| \geq \delta} \widehat{B}^\alpha_R(y) \widehat{\psi} \left(2^j (x-y) \right) 2^{nj} dy \right|$$

$$\leq C \int_{|y| \geq \delta} \left| \widehat{\psi}(2^j (x-y)) \right| 2^{nj} dy.$$

By letting $\alpha \leq \frac{\delta}{2}$, we obtain

$$\sup_{|x|\leq \alpha} |\widehat{B}^\alpha_{j,R}(x)| \leq C \int_{|u|\geq 2^{j-1}\delta} |\widehat{\psi}(u)|du \leq C2^{-j(\alpha+\frac{1}{2})}.$$

The reason being on hold for the last inequality on the right side of the above inequalities is $\widehat{\psi} \in \mathscr{S}$. Choosing $p' > 2$ such that $1/p + 1/p' = 1$, by Hausdorff-Young's inequality, we have

$$\left(\int_{|x|\leq \alpha} |\hat{f}(x)|^2 dx\right)^{\frac{1}{2}} \leq \left(\int_{|x|\leq \alpha} |\hat{f}(x)|^{p'} dx\right)^{1/p} \leq C\|f\|_p,$$

which implies

$$\int_{|x|\leq \alpha} \left|\widehat{B}^\alpha_{j,R}(x)|^2|\hat{f}(x)\right|^2 dx \leq C2^{-2j(\alpha+\frac{1}{2})}\|f\|_p^2.$$

Secondly, for $\widehat{B}^\alpha_{j,R}(x)$ is a radial function, then we get

$$\int_{|x|\geq \alpha} |\widehat{B}^\alpha_{j,R}(x)|^2 |\hat{f}(x)|^2 dx = \int_\alpha^\infty \rho^{n-1}|\widehat{B}^\alpha_{j,R}(\rho)|^2 \left(\int_{|\xi|=1} |\hat{f}(\rho\xi)|^2 d\xi\right) d\rho.$$

Thus, by the (L^p, L^2) restriction theorem of the Fourier transform, we have

$$\int_{|x|\geq \alpha} |\widehat{B}^\alpha_{j,R}(x)|^2 |\hat{f}(x)|^2 dx$$

$$\leq C \int_\alpha^\infty \rho^{-n-1}|\widehat{B}^\alpha_{j,R}(\rho)|^2 \left(\int_{\mathbb{R}^n} \left|f\left(\frac{y}{\rho}\right)\right|^p dy\right)^{2/p} d\rho$$

$$= C \left(\int_\alpha^\infty \rho^{-n-1+\frac{2n}{p}}|\widehat{B}^\alpha_{j,R}(\rho)|^2 d\rho\right) \|f\|_p^2$$

$$\leq C \left(\int_\alpha^\infty |\widehat{B}^\alpha_{j,R}(\rho)|^2 \rho^{n-1} d\rho\right) \|f\|_p^2$$

$$\leq C \left(\int_{\mathbb{R}^n} |B^\alpha_{j,R}(x)|^2 dx\right) \|f\|_p^2$$

$$\leq C \left(\int_{|x|\geq 2^j} |B^\alpha_R(x)|^2 dx\right) \|f\|_p^2$$

$$\leq C \left(\int_{|x|\geq 2^j} \frac{dx}{|x|^{n+2\alpha+1}}\right) \|f\|_p^2$$

$$\leq C2^{-2j(\alpha+\frac{1}{2})}\|f\|_p^2.$$

2.8 The restriction theorem of the Fourier transform

It follows
$$\int_{\mathbb{R}^n} |T_j f(x)|^2 dx \leq C 2^{-2j(\alpha+\frac{1}{2})} \|f\|_p^2.$$

Together with (2.8.3), the above inequality leads to
$$\int_Q |T_j f(x)|^p dx \leq C^p 2^{nj/r} \cdot 2^{-pj(\alpha+\frac{1}{2})} \|f\|_p^p$$
$$= C^p 2^{-n(\frac{n+1+2\alpha}{2n}-1)pj} \|f\|_p^p.$$

Since
$$\frac{2n}{n+1+2\alpha} < p,$$

we take
$$\varepsilon = n\left(\frac{n+1+2\alpha}{2n} - \frac{1}{p}\right),$$

and get (2.8.2), which completes the proof of the theorem. ∎

Remark 2.8.1 *The converse of Theorem 2.8.1 is also valid. This is proven by Tao [Ta1] in 1999. That means the Bochner-Riesz conjecture is equivalent to the restriction theorem of the Fourier transform.*

By Theorem 2.8.1, we are interested in the following problem. For what kind of p, (L^p, L^2)-restriction of the Fourier transforms holds. The following result is due to Tomas [To1].

Theorem 2.8.2 *Let $1 \leq p < \frac{2n+2}{n+3}$. Then we have*
$$\left(\int_{|\xi|=1} |\hat{f}(\xi)|^2 d\sigma(\xi)\right)^{\frac{1}{2}} \leq A_p \|f\|_{L^p(\mathbb{R}^n)}, \quad f \in \mathscr{S}.$$

Proof. Actually, we have
$$\int_{|\xi|=1} |\hat{f}(\xi)|^2 d\sigma(\xi) = \int_{\mathbb{R}^n} (f * \overline{f})(x) \int_{|\xi|=1} e^{-ix\cdot\xi} d\sigma(\xi) dx.$$

Set
$$\widehat{d\xi}(x) = \int_{|\xi|=1} e^{-ix\cdot\xi} d\sigma(\xi).$$

Then we have

$$\int_{|\xi|=1} |\hat{f}(\xi)|^2 d\sigma(\xi) = \int_{\mathbb{R}^n} \overline{f}(x) \cdot (\widehat{d\xi} * f)(x) dx$$

$$\leq \|f\|_p \|\widehat{d\xi} * f\|_{p'} \quad \left(\frac{1}{p} + \frac{1}{p'} = 1\right).$$

It suffices to prove that

$$\left\|\widehat{d\xi} * f\right\|_{p'} \leq C_p \|f\|_p. \tag{2.8.4}$$

Now suppose that the radial function $K(x) \in \mathscr{S}$, satisfying

$$K(x) = 1, \quad \text{when } |x| \leq 100,$$

and set

$$T_k(x) = \left\{K(\frac{x}{2^k}) - K(\frac{x}{2^{k-1}})\right\} \widehat{d\xi}(x).$$

In order to prove (2.8.4), it suffices to show that there exists a positive number $\varepsilon = \varepsilon(\rho)$, such that

$$\|T_k * f\|_{p'} \leq C_p 2^{-\varepsilon k} \|f\|_p. \tag{2.8.5}$$

We notice that there holds

$$\widehat{d\xi}(x) = \int_{|\xi|=1} e^{-ix\cdot\xi} d\sigma(\xi) = C_n |x|^{-\frac{(n-2)}{2}} J_{\frac{n-2}{2}}(|x|),$$

and get that

$$\left|\int_{\mathbb{R}^n} T_k(y) f(x-y) dy\right|$$

$$\leq C_n \int_{|y| \geq 100 \times 2^{k-1}} \left|K\left(\frac{y}{2^k}\right) - K\left(\frac{y}{2^{k-1}}\right)\right| |y|^{-\frac{n-1}{2}} |f(x-y)| dy$$

$$\leq C_n 2^{-\frac{k(n-1)}{2}} \|f\|_1.$$

Hence, we obtain

$$\|T_k * f\|_\infty \leq C_n 2^{-\frac{k(n-1)}{2}} \|f\|_1. \tag{2.8.6}$$

In addition, since

$$\|T_k * f\|_2 \leq \|\widehat{T_k}\|_\infty \|f\|_2,$$

then we have

$$\|T_k * f\|_2 \leq C 2^k \|f\|_2. \tag{2.8.7}$$

2.8 The restriction theorem of the Fourier transform

By (2.8.6), (2.8.7) and Riesz-Thorin's convexity theorem, we get the conclusion that

$$\|T_k * f\|_p \leq C \left(2^{-\frac{k(n-1)}{2}}\right)^{1-t} \left(2^k\right)^t \|f\|_p$$

$$= C 2^{-\left[\frac{(n+1)(1-t)}{2} - 1\right]k} \|f\|_p,$$

where

$$\frac{1}{p} = 1 - t + \frac{t}{2}.$$

We notice that if

$$1 \leq p < \frac{2n+2}{n+3},$$

then

$$\frac{(n+1)(1-t)}{2} - 1 = \frac{2n+2}{p} - n - 2 > 0.$$

Let $\varepsilon = \frac{2n+2}{p} - n - 2$. Then (2.8.5) holds and this finishes the proof. ∎

Remark 2.8.2 The condition in Theorem 2.8.2 that $1 \leq p < \frac{2n+2}{n+3}$ has been improved by Stein to $1 \leq p \leq \frac{2n+2}{n+3}$.

Combining Theorem 2.8.2 with Theorem 2.8.1, we can get the next corollary.

Corollary 2.8.1 Let

$$\frac{2n}{n+1+2\alpha} < p < \frac{2n}{n-1-2\alpha}.$$

If

$$\alpha > \frac{n-1}{2n+2},$$

then we have

$$\|B_R^\alpha(f)\|_p \leq C_{p,\alpha} \|f\|_p.$$

Proof. By the method of duality, we merely need to consider the case when

$$\frac{2n}{n+1+2\alpha} < p < 2.$$

When

$$\alpha > \frac{n-1}{2n+2},$$

we have
$$\frac{2n}{n+1+2\alpha} < \frac{2n+2}{n+3}.$$
Then, by Theorems 2.8.1 and 2.8.2, we know that when
$$\frac{2n}{n+1+2\alpha} < p < \frac{2n+2}{n+3},$$
the operator $B_R^\alpha : f \to B_R^\alpha(f)$, is bounded on $L^p(\mathbb{R}^n)$. In addition, by the interpolation theorem of operators, we get that when
$$\frac{2n+2}{n+3} \leq p < 2,$$
the conclusion of the boundedness of the operator
$$f \to B_R^\alpha(f)$$
on $L^p(\mathbb{R}^n)$ is obtained, and this finishes the proof. ∎

Remark 2.8.3 *We point out that after some proper revisions about the proof of Theorem 2.8.1 in this section, Fefferman [Fe4] gave a new proof of Carleson-Sjölin theorem. Limited by the space, here we omit the proof in details.*

Many researchers take the restriction theorem of the Fourier transform as the starting point of the study on the Bochner-Riesz conjecture (see [TVV1], [Le1],[BG]). Other researchers regard the Kakeya maximal function as the starting point to study the Bochner-Riesz conjecture (see [Bou1], [Wo1]). Here the remarkable result that we state is due to Bourgain and Guth [BG]. If $n \geq 3$, $0 < \alpha < \frac{n-1}{2}$ and
$$max(p, p') > \begin{cases} 2\frac{4n+3}{4n-3}, & n = 0 \ (mod\, 3), \\ \frac{2n+1}{n-1}, & n = 1 \ (mod\, 3), \\ \frac{4n+4}{2n-1}, & n = 2 \ (mod\, 3), \end{cases}$$
then the Bochner-Riesz conjecture is true.

2.9 The case of radial functions

In this section, we will use the notation $L^p(\mathbb{R}^n, r)$ to denote the collection of all the functions f which satisfy $f \in L^p(\mathbb{R}^n)$ and
$$f(x) = \varphi(|x|),$$
and we call it the radial functions class of $L^p(\mathbb{R}^n)$. In Section 2.4, we have proven that Bochner-Riesz operator T_0 is unbounded on $L^p(\mathbb{R}^n)$, where
$$\frac{2n}{n+1} < p < \frac{2n}{n-1}$$
and $p \neq 2$.

However, it is different for $L^p(\mathbb{R}^n, r)$. Herz [He1] proved that T_0 is bounded in $L^p(\mathbb{R}^n, r)$, with
$$\frac{2n}{n+1} < p < \frac{2n}{n-1}.$$
Similarly, since Carleson-Sjölin theorem holds for the operator T_α with $\alpha > 0$, then it is natural to guess that when $n > 2$, T_α should be bounded on $L^p(\mathbb{R}^n)$ for
$$\frac{2n}{n+1+2\alpha} < p < \frac{2n}{n-1-2\alpha}.$$
The problem still has not been completely solved till now. For the radial function class $L^p(\mathbb{R}^n, r)$, the above conclusion is correct, which was proven by Welland [We1]. The two results above about $L^p(\mathbb{R}^n, r)$ can be formulated as follows.

Theorem 2.9.1 *If $f \in L^p(\mathbb{R}^n, r)$ and*
$$\frac{2n}{n+1} < p < \frac{2n}{n-1},$$
then the inequality
$$\|B_R^0 f\|_p \leq C_{p,n} \|f\|_p$$
holds.

Theorem 2.9.2 *Suppose that $f \in L^p(\mathbb{R}^n, r)$, $0 < \alpha < \frac{n-1}{2}$, and*
$$\frac{2n}{n+1+2\alpha} < p < \frac{2n}{n-1-2\alpha}.$$
Then there holds
$$\|B_R^\alpha f\|_p \leq C_{p,\alpha,n} \|f\|_p.$$

We merely give the proof of Theorem 2.9.2. From the proof of Theorem 2.2.2 in Section 2.2, it suffices to prove that

$$\|B_1^\alpha(f)\|_p \le C_p \|f\|_p,$$

for $f \in L^p(\mathbb{R}^n, r)$, where p, α satisfy the assumptions of Theorem 2.9.2. Similarly, since the class of radial simple functions having compact support is dense in $L^p(\mathbb{R}^n, r)$, we merely need to prove the above inequality for any simple function $f \in L^p(\mathbb{R}^n, r)$. Assume that $f \in L^p(\mathbb{R}^n, r)$ is a simple function. By the equation (2.1.1) in Section 2.1, we have

$$B_1^\alpha(f; x) = C_\alpha \int_{\mathbb{R}^n} f(t) \frac{J_{\frac{n}{2}+\alpha}(|x-t|)}{|x-t|^{\frac{n}{2}+\alpha}} dt.$$

We consider a set of operators $\{B_1^{\alpha(z)}\}$, where

$$\alpha(z) = \alpha_0(1-z) + \varepsilon, \quad \varepsilon > 0, \quad \alpha_0 = \frac{n-1}{2}$$

and

$$0 < \text{Re}(z) < 1.$$

According to the following three properties of the Bessel functions,

(i) $\quad J_\zeta(t) = \dfrac{\left(\frac{t}{2}\right)^\zeta}{\Gamma\left(\frac{1}{2}\right)\Gamma\left(\zeta+\frac{1}{2}\right)} \int_0^1 (1-u^2)^{\zeta-\frac{1}{2}} \cos ut \, du, \quad \text{Re}(\zeta) > -\dfrac{1}{2},$

(ii) $\quad |J_{\xi+i\eta}(t)| \le A_\xi e^{\pi|\eta|} t^{-1/2}, \quad t \ge 0, \, \xi \ge 0,$

(iii) $\quad |J_{\xi+i\eta}(t)| \le A_\xi e^{\pi|\eta|} t^\xi, \quad t > 0, \, \xi \ge 0.$

It is easy to check that the operator family $\{B_1^{\alpha(z)}\}$ is an admissible analytic family (see Stein and Weiss [SW1]). In order to use the interpolation theorem of the analytic family of operators, we take

$$p_0 = q_0 = 1, \quad p_1 = q_1 > \frac{2n}{n+1},$$

and point out that the inequality

$$|B_1^{\alpha(iy)} f|_1 \le A_\xi e^{\pi \eta} \|f\|_1 \tag{2.9.1}$$

holds, where

$$\xi + i\eta = \alpha(iy) + \frac{\eta}{2}.$$

2.9 The case of radial functions

In fact, we have
$$\left(B_1^{\alpha(iy)}f\right)(x) = \left(f * B_1^{\alpha(iy)}\right)(x),$$
where
$$B_1^{\alpha(iy)}(x) = C|x|^{-\alpha(iy)-\frac{n}{2}} J_{\alpha(iy)+\frac{n}{2}}(|x|).$$
Obviously, there holds
$$\|B_1^{\alpha(iy)}\|_1 \leq A_\xi e^{\pi|\eta|}.$$
Then we have
$$\left\|B_1^{\alpha(iy)}f\right\|_1 \leq \left\|B_1^{\alpha(iy)}\right\|_1 \|f\|_1 \leq A_\xi e^{\pi|\eta|}\|f\|_1,$$
which leads to (2.9.1).

Next, we need to prove
$$\left\|B_1^{\alpha(1+iy)}f\right\|_{p_1} \leq A_p \left(|\eta|\varepsilon^{-1}+1\right)\|f\|_{p_1}, \qquad (2.9.2)$$
for $\frac{2n}{n+1} < p_1 \leq 2$, where $\eta = -\alpha_0 y$.

Since $\alpha(1+iy) = \varepsilon + i\eta$, then (2.9.2) is transfered into
$$\|B_1^{\varepsilon+i\eta}f\|_{p_1} \leq A_{p_1}\left(|\eta|\varepsilon^{-1}+1\right)\|f\|_{p_1}. \qquad (2.9.3)$$

For any $f \in L^{p_1}(\mathbb{R}^n, r)$, we first give an expression of $(B_1^{\varepsilon+i\eta}f)(x)$ as

$$\left(B_1^{\varepsilon+i\eta}f\right)(x) \qquad\qquad\qquad\qquad\qquad\qquad\qquad\qquad (2.9.4)$$
$$= C|x|^{-\frac{n-2}{2}}\int_0^\infty f(t)t^{\frac{n}{2}}\int_0^1 (1-r^2)^{\varepsilon+i\eta} J_{\frac{n-2}{2}}(r|x|) J_{\frac{n-2}{2}}(rt) r\, dr\, dt.$$

In fact, we have that if $f(x)$ is a radial function, then so is $\hat{f}(x)$. Denote by
$$\hat{f}(x) = F(|x|) = F(r),$$
and through the computation in polar coordinates, we can get that
$$F(r) = (n-1)\omega_{n-1}\int_0^\infty f(t)\left\{\int_0^\pi e^{-irt\cos\theta}\sin^{n-2}\theta\, d\theta\right\} t^{n-1}\, dt,$$
where ω_{n-1} is the surface area of the unit sphere in $\mathbb{R}^n$. Since there holds
$$\left(B_1^{\varepsilon+i\eta}f\right)(x) = \frac{1}{(2\pi)^n}\int_{|y|<1}(1-|y|^2)^{\varepsilon+i\eta}\hat{f}(y)e^{ix\cdot y}\, dy,$$

if we also put the computation for the integral on the right side of the above equation under the polar coordinates, we have that

$$\left(B_1^{\varepsilon+i\eta}f\right)(x) = \frac{(n-1)\omega_{n-1}}{(2\pi)^n}\int_0^1 (1-r^2)^{\varepsilon+i\eta} r^{n-1} F(r)$$

$$\times \left\{\int_0^\pi e^{i|x|r\cos\varphi}\sin^{n-2}\varphi d\varphi\right\} dr$$

$$= \frac{(n-1)^2\omega_{n-1}^2}{(2\pi)^n}\int_0^1 (1-r^2)^{\varepsilon+i\eta} r^{n-1}$$

$$\times \left\{\int_0^\infty f(t)\left(\int_0^\pi e^{-irt\cos\theta}\sin^{n-2}\theta d\theta\right) t^{n-1}dt\right\}$$

$$\times \left\{\int_0^\pi e^{i|x|r\cos\varphi}\sin^{n-2}\varphi d\varphi\right\} dr.$$

Thus, by the deformation of the property (i) of the Bessel function, we have

$$J_\zeta(z) = \frac{\left(\frac{z}{2}\right)^\zeta}{\Gamma\left(\zeta+\frac{1}{2}\right)\Gamma\left(\frac{1}{2}\right)}\int_0^\pi e^{\pm iz\cos\varphi}\sin^{2\zeta}\varphi d\varphi,$$

and

$$\operatorname{Re}(\zeta+\frac{1}{2}) > 0,$$

we immediately have (2.9.4).

By the relationship of two Bessel functions (see Welland [We1])

$$rJ_\nu(|x|r)J_\nu(tr) = \frac{1}{|x|^2-t^2}\frac{d}{dr}\left\{rtJ_\nu(|x|r)\left(\frac{d}{dy}J_\nu(y)\right)\bigg|_{y=tr}\right.$$

$$\left.-r|x|J_\nu(tr)\left(\frac{d}{dy}J_\nu(y)\right)\bigg|_{y=|x|r}\right\}$$

and

$$\frac{d}{dy}J_\nu(y) = \frac{\nu J_\nu(y)}{y} - J_{\nu+1}(y),$$

we conclude that

$$\int_0^1 (1-r)^{\varepsilon+i\eta} J_{\frac{n-2}{2}}(|x|r) J_{\frac{n-2}{2}}(tr) r dr = \frac{1}{|x|^2-t^2}(I_2-I_1),$$

where

$$I_1 = \int_0^1 (1-r^2)^{\varepsilon+i\eta}\frac{d}{dr}\{rtJ_\nu(|x|r)J_{\nu+1}(tr)\}dr$$

2.9 The case of radial functions

and

$$I_2 = \int_0^1 (1-r^2)^{\varepsilon+i\eta} \frac{d}{dr}\{r|x|J_\nu(tr)J_{\nu+1}(|x|r)\}dr.$$

By the integration by part, we have

$$I_2 = 2(\varepsilon+i\eta)\int_0^1 r^2|x|(1-r^2)^{\varepsilon-1+i\eta} J_\nu(tr)J_{\nu+1}(|x|r)dr.$$

Due to the property (ii) of Bessel function, we easily have that

$$|I_2| \le C\varepsilon^{-1}(|\eta|+\varepsilon)\sqrt{\frac{|x|}{t}},$$

where C is a constant.

Similarly, we have

$$|I_1| \le C\varepsilon^{-1}(|\eta|+\varepsilon)\sqrt{\frac{t}{|x|}}.$$

Thus, it follows from (2.9.4) and the estimate above that

$$\left|(B_1^{\varepsilon+i\eta}f)(x)\right| \le C\varepsilon^{-1}(|\eta|+\varepsilon)|x|^{-\frac{n-2}{2}}\Bigg\{\int_0^{|x|} \frac{|f(t)|t^{n/2}}{|x|^2-t^2}\left(\sqrt{\frac{|x|}{t}}+\sqrt{\frac{t}{|x|}}\right)dt$$

$$+ \int_{|x|}^\infty \frac{|f(t)|t^{n/2}}{t^2-|x|^2}\left(\sqrt{\frac{|x|}{t}}+\sqrt{\frac{t}{|x|}}\right)dt\Bigg\}$$

$$\le C\varepsilon^{-1}(|\eta|+\varepsilon)\Bigg\{|x|^{-\frac{n-3}{2}}\int_0^{|x|} \frac{|f(t)|t^{\frac{n-1}{2}}}{|x|^2-t^2}dt$$

$$+ |x|^{-\frac{n-3}{2}}\int_{|x|}^\infty \frac{|f(t)|t^{\frac{n-1}{2}}}{t^2-|x|^2}dt + |x|^{-\frac{n-1}{2}}\int_0^{|x|} \frac{|f(t)|t^{\frac{n-1}{2}}}{|x|^2-t^2}dt$$

$$+ |x|^{-\frac{n-1}{2}}\int_{|x|}^\infty \frac{|f(t)|t^{\frac{n+1}{2}}}{t^2-|x|^2}dt\Bigg\}.$$

For any integral on the right side of the inequality above, if we take it for granted that f equals to zero in the complement of the region of the integration, then in order to prove (2.9.3), we merely need to prove that for

$$\frac{2n}{n+1} < p_1 \le 2,$$

there holds
$$\|T_i f\|_{p_1} \leq A_{p_1} \|f\|_{p_1}, \tag{2.9.5}$$
for $i = 1, 2$ and $f \in L^{p_1}(\mathbb{R}^n, r)$, where
$$T_1 f(x) = |x|^{-\frac{n-3}{2}} \int_0^\infty \frac{|f(t)| t^{\frac{n-1}{2}}}{|x|^2 - t^2} dt,$$
and
$$T_2 f(x) = |x|^{-\frac{n-1}{2}} \int_0^\infty \frac{|f(t)| t^{\frac{n-1}{2}}}{|x|^2 - t^2} dt.$$

Since there holds
$$\|T_1 f\|_{p_1}^{p_1} = \int_{\mathbb{R}^n} |x|^{-\frac{n-3}{2} p_1} \left| \int_0^\infty \frac{|f(t)| t^{\frac{n-1}{2}}}{|x|^2 - t^2} dt \right|^{p_1} dx$$
$$= C \int_0^\infty r^{n-1-\frac{(n-3)p_1}{2}} \left| \int_0^\infty \frac{|f(t)| t^{\frac{n-1}{2}}}{r^2 - t^2} dt \right|^{p_1} dr,$$
by the change of variables $r^2 = \sigma$, $t^2 = \tau$, we get that
$$\|T_1 f\|_{p_1}^{p_1} = C \int_0^\infty \left| \int_0^\infty \frac{\tau^{\frac{n-2}{2p}} |f(\tau^{\frac{1}{2}})|}{\sigma - \tau} \left(\frac{\tau}{\sigma}\right)^{\frac{n-1}{4} - \frac{n-2}{2p}} d\tau \right|^{p_1} d\sigma.$$

By another change of variables $r^2 = \sigma$, we have
$$\|f\|_{p_1}^{p_1} = C \int_0^\infty |f(r)|^{p_1} r^{n-1} dr = C \int_0^\infty \left\{ \tau^{\frac{n-2}{2p}} |f(\tau^{\frac{1}{2}})| \right\}^{p_1} d\tau.$$

Let
$$\phi(\tau) = \tau^{\frac{n-2}{2p}} \left| f(\tau^{\frac{1}{2}}) \right|$$
and
$$\psi(\sigma) = \int_0^\infty \frac{\phi(\tau)}{\sigma - \tau} \left(\frac{\tau}{\sigma}\right)^{\frac{n-3}{4} - \frac{n-2}{2p}} d\tau.$$
Then the inequality $\|T_1 f\|_{p_1} \leq A_{p_1} \|f\|_{p_1}$ can be reduced to prove
$$\|\psi\|_{p_1} \leq A_{p_1} \|\phi\|_{p_1}.$$
Now we separate ψ into $\psi = \psi_1 + \psi_2$, where
$$\psi_1(\sigma) = \int_0^\infty \frac{\phi(\tau)}{\sigma - \tau} d\tau$$

2.9 The case of radial functions

and
$$\psi_2(\tau) = \int_0^\infty \phi(\tau) \frac{\left(\frac{\tau}{\sigma}\right)^{\frac{n-3}{4} - \frac{n-2}{2p}} - 1}{\sigma - \tau} d\tau.$$

By the L^{p_1}-boundedness of the Hilbert transform with $p_1 > 1$, it suffices to estimate ψ_2.

Let
$$K(\sigma, \tau) = (\sigma - \tau)^{-1} \left\{ \left(\frac{\tau}{\sigma}\right)^{\frac{n-3}{4} - \frac{n-2}{2p_1}} - 1 \right\},$$

we have
$$\psi_2(\sigma) = \int_0^\infty \phi(\tau) K(\sigma\tau) d\tau.$$

It is obvious that in order to prove
$$\|\psi_2\|_{p_1} \leq A_{p_1} \|\phi\|_{p_1},$$

we merely need to prove that, for any $h \in L^{p_1}$ and $\|h\|_{p_1'} \leq 1$,
$$\left| \int_0^\infty \psi_2(\sigma) h(\sigma) d\sigma \right| \leq A_{p_1} \|\phi\|_{p_1}$$

holds.

In fact, it follows that
$$\left| \int_0^\infty \psi_2(\sigma) h(\sigma) d\sigma \right| = \left| \int_0^\infty \int_0^\infty K(\sigma, \tau) \phi(\tau) h(\sigma) d\tau d\sigma \right|$$
$$\leq \int_0^\infty \int_0^\infty |\phi(\tau)| \left(\frac{\tau}{\sigma}\right)^{\frac{1}{p_1 p_1'}} \left| K^{\frac{1}{p_1}}(\sigma, \tau) \right|$$
$$\times |h(\sigma)| \left(\frac{\sigma}{\tau}\right)^{\frac{1}{p_1 p_1'}} \left| K^{\frac{1}{p_1'}}(\sigma, \tau) \right| d\tau d\sigma$$
$$\leq \left\{ \int_0^\infty |\phi(\tau)|^{p_1} \int_0^\infty |K(\sigma, \tau)| \left(\frac{\tau}{\sigma}\right)^{\frac{1}{p_1'}} d\tau d\sigma \right\}^{1/p_1}$$
$$\times \left\{ \int_0^\infty |h(\sigma)|^{p_1'} \int_0^\infty |K(\sigma, \tau)| \left(\frac{\sigma}{\tau}\right)^{\frac{1}{p_1}} d\tau d\sigma \right\}^{1/p_1'}.$$

By a change of variables $\alpha = \frac{\tau}{\sigma}$, we obtain that
$$\int_0^\infty |K(\sigma, \tau)| \left(\frac{\tau}{\sigma}\right)^{\frac{1}{p_1'}} d\sigma = \int_0^\infty \left| \frac{\left(\frac{\tau}{\sigma}\right)^{\frac{n-3}{4} - \frac{n-2}{2p_1}} - 1}{\sigma - \tau} \right| \left(\frac{\tau}{\sigma}\right)^{\frac{1}{p_1'}} d\sigma$$
$$= \int_0^\infty \left| \frac{\alpha^{\frac{n-3}{4} - \frac{n-2}{2p_1}} - 1}{\alpha - 1} \right| \alpha^{-\frac{1}{p_1}} d\alpha.$$

It is easy to check that the integral on the right side of the above equation converges when
$$\frac{2n}{n+1} < p_1 \leq 2.$$
Similarly, when
$$\frac{2n}{n+1} < p_1 \leq 2,$$
we have that
$$\int_0^\infty |K(\sigma,\tau)| \left(\frac{\sigma}{\tau}\right)^{\frac{1}{p_1}} d\tau < c < \infty.$$
Hence we get
$$\left|\int_0^\infty \psi_2(\sigma) h(\sigma) d\sigma\right| \leq A_{p_1} \|\phi\|_{p_1},$$
provided that $\|h\|_{p_1'} \leq 1$. This implies
$$\|T_1 f\|_{p_1} \leq A_{p_1} \|f\|_{p_1},$$
for
$$\frac{2n}{n+1} < p_1 \leq 2.$$

Similarly, we can get the similar inequality for T_2. Consequently, (2.9.3) holds and then (2.9.2) is valid.

Now we choose
$$p_0 = q_0 = 1, \quad p_1 = q_1 = \frac{2n}{n+1} + \varepsilon$$
and
$$\alpha(z) = \frac{n-1}{2}(1-z) + \varepsilon.$$
By the interpolation theorem on analytic families of operators, (2.9.1) and (2.9.2), we get that
$$\left\|B_1^{\alpha(t)} f\right\|_p \leq A_p \|f\|_p,$$
where
$$\frac{1}{p} = \frac{1-t}{p_0} + \frac{t}{p_1}, \quad 0 < t < 1.$$
Thus we have
$$\frac{1}{p} = (1-t)\left(1 - \frac{1}{p_1}\right) + \frac{1}{p_1} = \frac{\alpha(t) - \varepsilon}{\frac{n-1}{2}}\left(1 - \frac{1}{p_1}\right) + \frac{1}{p_1}$$

and
$$\varepsilon \leq \alpha(t) \leq \frac{n-1}{2} + \varepsilon.$$
Since ε is arbitrary, it is not hard to get the conclusion of Theorem 2.9.2.

2.10 Almost everywhere convergence

In this section, we will study the almost everywhere convergence of the Bochner-Riesz means of Fourier integral of f. That is to say whether or not
$$\lim_{R \to \infty} B_R^\alpha(f)(x)$$
exists and is finite for almost every $x \in \mathbb{R}^n$ and for all $f \in L^p(\mathbb{R}^n)$, $1 \leq p < \infty$.

Since the operator
$$f \to B_R^\alpha(f)$$
is linear, thus we discuss the property of the almost everywhere convergence of any family of linear operators $f \to T_R f$.

Suppose that $\{T_R : R > 0\}$ is a family of linear operators and each T_R maps $L^p(\mathbb{R}^n)$, $1 \leq p \leq \infty$, to a measurable function space $\mathscr{M}(\mathbb{R}^n)$. Besides, we also assume that, for any $g \in \mathscr{D}(\mathbb{R}^n)$,
$$\lim_{R \to \infty} T_R g(x)$$
exits and is finite for a.e. $x \in \mathbb{R}^n$. To guarantee the almost everywhere convergence of $T_R f(x)$ when $R \to \infty$, it suffices to prove that
$$\left| \left\{ x \in \mathbb{R}^n : \limsup_{R_1, R_2 \to \infty} |T_{R_1} f(x) - T_{R_2} f(x)| > \lambda \right\} \right| = 0$$
is valid for every $\lambda > 0$. However, for any $f \in L^p(\mathbb{R}^n)$, we choose $g \in \mathscr{D}(\mathbb{R}^n)$ and set $h = f - g$. Obviously, for any $\lambda > 0$, we have
$$\left| \left\{ x \in \mathbb{R}^n : \limsup_{R_1, R_2 \to \infty} |T_{R_1} f(x) - T_{R_2} f(x)| > \lambda \right\} \right|$$
$$\leq \left| \left\{ x \in \mathbb{R}^n : \limsup_{R_1, R_2 \to \infty} |T_{R_1} h(x) - T_{R_2} h(x)| > \lambda \right\} \right|.$$

Thus, in order to prove the a.e. convergence of $T_R f(x)$, it suffices to show that
$$\left|\left\{x \in \mathbb{R}^n : \limsup_{R_1, R_2 \to \infty} |T_{R_1} h(x) - T_{R_2} h(x)| > \lambda \right\}\right| \to 0,$$
as $\|h\|_p \to 0$.

We denote by
$$T_* f(x) = \sup_{R > 0} |T_R f(x)|,$$
and notice that
$$\left|\left\{x \in \mathbb{R}^n : \limsup_{R_1, R_2 \to \infty} |T_{R_1} h(x) - T_{R_2} h(x)| > \lambda \right\}\right| \leq \left|\left\{x \in \mathbb{R}^n : T_* h(x) > \frac{\lambda}{2} \right\}\right|.$$

Thus, to guarantee the a.e. convergence of $T_R f(x)$, we merely need to prove that
$$|\{x \in \mathbb{R}^n : T_* f(x) > \lambda\}| \to 0,$$
as $\|f\|_p \to 0$ for any fixed $\lambda > 0$. It is necessary to point out that the validity of the last relationship above is closely related to the type of the operator T_*. In the following, we begin with giving the definition of the type of the operator T.

Definition 2.10.1 Let $1 \leq p, q \leq \infty$. The operator T maps $L^p(\mathbb{R}^n)$ to $L^q(\mathbb{R}^n)$.

(i) If the operator T satisfies the inequality
$$\|T f\|_q \leq C_{p,q} \|f\|_p,$$
then we call T to be of type (p, q) or strong type (p, q).

(ii) If the operator T satisfies the inequality
$$|\{x \in \mathbb{R}^n : T f(x) > \lambda\}| \leq C_{p,q} \left(\frac{\|f\|_p}{\lambda}\right)^q,$$
then we call T an operator of weak type (p, q), where $C_{p,q}$ is independent of f.

By Definition 2.10.1, if T_* is of weak type (p, q), then we have that
$$|\{x \in \mathbb{R}^n : T_* f(x) > \lambda\}| \to 0,$$
as $\|f\|_p \to 0$.

Thus, we obtain the following important facts.

2.10 Almost everywhere convergence

Theorem 2.10.1 *Suppose that $\{T_R : R > 0\}$ is a family of linear operators and each T_R maps $L^p(\mathbb{R}^n)$ ($1 \le p \le \infty$) to the measurable function space $\mathscr{M}(\mathbb{R}^n)$. Moreover, we also assume that for any $g \in \mathscr{D}(\mathbb{R}^n)$, $\lim_{R \to \infty} T_R g(x)$ exits and is finite for a.e x. If the maximal operator*

$$T_* f(x) = \sup_{R > 0} |T_R f(x)|$$

is of weak type (p, q), then for every $f \in L^p(\mathbb{R}^n)$, $\lim_{R \to \infty} T_R f(x)$ exists and is finite almost everywhere. Here $q \ge 1$.

Corollary 2.10.1 *Let $\alpha > \frac{n-1}{2}$. If $f \in L^p(\mathbb{R}^n)$ ($1 \le p < \infty$), then*

$$\lim_{R \to \infty} B_R^\alpha f(x) = f(x)$$

holds for a.e. $x \in \mathbb{R}^n$.

Proof. For $\alpha > \frac{n-1}{2}$, let

$$B_*^\alpha(f; x) = \sup_{R > 0} |B_R^\alpha(f; x)|,$$

we first prove

$$B_*^\alpha(f; x) \le C M f(x),$$

where Mf is the Hardy-Littlewood maximal function of f and C is independent of f. In fact, the equation (2.1.3) int Section 2.1 of this chapter reveals that

$$B_R^\alpha f(x) = C_{n,\alpha} R^n \int_0^\infty f_x(t) \frac{J_{\frac{n}{2}+\alpha}(Rt)}{(Rt)^{\frac{n}{2}+\alpha}} t^{n-1} dt.$$

Separating the integration region on the the right side of the above integral as

$$(0, \infty) = \left(0, \frac{1}{R}\right) \bigcup \left[\frac{1}{R}, \infty\right),$$

and on $(0, \frac{1}{R})$ and $[\frac{1}{R}, \infty)$, using $J_m(f) \sim C r^m$, as $r \to 0$, and $J_m(r) = O(r^{-\frac{1}{2}})$, when $r \to \infty$, we have that

$$|B_R^\alpha f(x)| \le C \left\{ R^n \int_0^{\frac{1}{R}} |f_x(t)| t^{n-1} dt + R^{\frac{n-1}{2} - \alpha} \int_{\frac{1}{R}}^\infty |f_x(t)| t^{\frac{n-1}{2} - \alpha - 1} dt \right\}.$$

It easily follows that

$$R^n \int_0^{\frac{1}{R}} |f_x(t)| t^{n-1} dt \leq \frac{Mf(x)}{n}.$$

By partial integration, we get that

$$R^{\frac{n-1}{2}-\alpha} \int_{\frac{1}{R}}^{\infty} |f_x(t)| t^{\frac{n-1}{2}-\alpha-1} dt \leq \frac{C}{\alpha - \frac{n-1}{2}} Mf(x).$$

Hence, we get the conclusion that

$$B_*^\alpha(f;x) \leq CMf(x).$$

According to the well-known property of the Hardy-Littlewood maximal function, B_*^α is of type (p,p) for $1 < p < \infty$ and B_*^α is of weak type $(1,1)$. Besides, Section 2.2 in this chapter has pointed out that: for any $f \in \mathscr{D}(\mathbb{R}^n)$,

$$\lim_{R \to \infty} B_R^\alpha(f;x) = f(x)$$

holds for a.e x. Hence, from Theorem 2.10.1, we can directly deduce that for any $f \in L^p(\mathbb{R}^n)$ with $1 \leq p < \infty$,

$$\lim_{R \to \infty} B_R^\alpha(f;x)$$

exits and is finite almost everywhere. To prove

$$\lim_{R \to \infty} B_R^\alpha(f;x) = f(x) \text{ a.e.}$$

holds, it suffices to show that there exists a subsequence $\{R_k\}$, such that

$$\lim_{k \to \infty} B_k^\alpha(f;x) = f(x)$$

for a.e. $x \in \mathbb{R}^n$.

In fact, when $\alpha > \frac{n-1}{2}$, the kernel $B_R^\alpha(x) \in L(\mathbb{R}^n)$ and $\|B_R^\alpha\|$ is independent of R.

By the property of convolution, we have

$$\|B_R^\alpha(f)\|_p \leq C_p \|f\|_p.$$

Due to Theorem 2.2.1, we have

$$\lim_{R \to \infty} \|B_R^\alpha(f) - f\|_p = 0.$$

2.10 Almost everywhere convergence

Thus there exists a subsequence $\{R_k\}$ such that

$$\lim_{k\to\infty} B_{R_k}^\alpha(f;x) = f(x), \quad \text{a.e.}$$

and this completes the proof of the corollary. ∎

For the case when $\alpha \leq \frac{n-1}{2}$, there are some basic results due to Stein as follow.

Theorem 2.10.2 *Let $1 < p < \infty$, $n \geq 2$ and*

$$(n-1)\left|\frac{1}{2} - \frac{1}{p}\right| < \alpha \leq \frac{n-1}{2}.$$

Then we have

$$\|B_*^\alpha(f)\|_p \leq C_{p,n,\alpha}\|f\|_p$$

and

$$\lim_{R\to\infty} B_R^\alpha(f;x) = f(x)$$

for a.e. $x \in \mathbb{R}^n$.

Obviously, the second assertion is directly obtained from Theorem 2.10.1 and the first assertion. As far as the first assertion is concerned, its proof is the same as the case of the series in the following and thus we omit the proof here. One can see Theorems 3.4.1 and 3.4.3 in Section 3.4 of Chapter 3.

In Theorem 2.10.2, when $n = 2$, the first assertion shows that if

$$\frac{2}{1+2\alpha} < p < \frac{2}{1-2\alpha}$$

with $0 < \alpha < \frac{1}{2}$, then we have

$$\|B_*^\alpha(f)\|_p \leq C_{p,\alpha}\|f\|_p. \tag{2.10.1}$$

Noticing that

$$\left(\frac{2}{1+2\alpha}, \frac{2}{1-2\alpha}\right) \subset \left(\frac{4}{3+2\alpha}, \frac{4}{1-2\alpha}\right),$$

and applying Carleson-Sjölin theorem, that is Theorem 2.5.2, in Section 2.5, one naturally anticipates the following results. When

$$\frac{4}{3+2\alpha} < p < \frac{4}{1-2\alpha},$$

the inequality (2.10.1) should hold. Though the conjecture is still not proven completely, yet Carbery [Ca1] has promoted the conclusion of Theorem 2.10.2 in an essential step for the case $n = 2$. If we replace the normal maximal operator

$$\sup_{R>0} |(B_R^\alpha f)(x)|$$

by the following maximal operator with the lacunary form

$$\sup_j |(B_{R_j}^\alpha f)(x)|,$$

then the conjecture has been verified by Igari [Ig1] and independently by Cordoba and Lopez-Melero [CoL1].

Theorem 2.10.3 *Suppose that there exists a positive number $q > 1$ such that the sequence $\{R_j\}_{j=1}^\infty$ satisfies*

$$\frac{R_{j+1}}{R_j} \geq q$$

for any $j \in \mathbb{N}$. If

$$\frac{4}{3+2\alpha} < p < \frac{4}{1-2\alpha},$$

we have

$$\left\| \sup_j |B_{R_j}^\alpha(f)| \right\|_p \leq C_{p,\alpha} \|f\|_p.$$

Particularly,

$$\lim_{j \to \infty} B_{R_j}^\alpha(f)(x) = f(x)$$

holds for a.e. $x \in \mathbb{R}^n$.

Proof. The proof of Theorem 2.10.3 mainly depends on Theorem 2.7.2 about the vector valued function and the Littlewood-Paley decomposition theorem (see Stein [St4]).

Firstly, it is easy to prove that there exists $\phi \in \mathscr{D}(\mathbb{R})$ such that

$$\mathrm{supp}\phi \subset (1,3)$$

and

$$\sum_{n=-\infty}^\infty \phi\left(2^{-n}\rho\right) = 1$$

2.10 Almost everywhere convergence

with $\rho > 0$.

Set
$$\phi_j(x) = \phi\left(\frac{2|x|}{R_j}\right)$$

and
$$\psi_j(x) = \begin{cases} 1 - \phi_j(x), & |x| \le R_j, \\ 0, & \text{otherwise.} \end{cases}$$

Clearly we obtain that
$$B_{R_j}^\alpha(x) = \widehat{\left(1 - \frac{|x|^2}{R_j^2}\right)_+^\alpha}$$
$$= \widehat{\left(1 - \frac{|x|^2}{R_j^2}\right)_+^\alpha} * \widehat{\phi_j}(x) + \widehat{\left(1 - \frac{|x|^2}{R_j^2}\right)_+^\alpha} * \widehat{\psi_j}(x).$$

Put
$$\eta\left(\frac{|x|}{R_j}\right) = \widehat{\left(1 - \frac{|x|^2}{R_j^2}\right)_+^\alpha} * \widehat{\psi_j}(x),$$

then we have $\eta \in \mathscr{D}(\mathbb{R})$ and
$$\widehat{\left(1 - \frac{|x|^2}{R_j^2}\right)_+^\alpha} * \widehat{\psi_j}(x) = \eta\left(\frac{|x|}{R_j}\right) = (\widehat{\eta})_{\frac{1}{R_j}}(x).$$

Thus, we get
$$\widehat{\left(1 - \frac{|x|^2}{R_j^2}\right)_+^\alpha} * \widehat{\psi_j}(x) * f(x) = (\widehat{\eta})_{\frac{1}{R_j}} * f(x).$$

By the property of the convolution operator ([SW1]), it follows that
$$\sup_j \left| \widehat{\left(1 - \frac{|x|^2}{R_j^2}\right)_+^\alpha} * \widehat{\psi_j}(x) * f(x) \right| \le CMf(x).$$

Hence we conclude that

$$\sup_j |B_{R_j}^\alpha f(x)| \le CMf(x) + \sup_j \left|\left(1 - \frac{|x|^2}{R_j^2}\right)_+^\alpha \cdot \phi_j * f(x)\right|$$

$$\le CMf(x) + \left(\sum_j \left|B_{R_j}^\alpha(\widehat{\phi}_j * f)(x)\right|^2\right)^{\frac{1}{2}}.$$

From the inequality above and Theorem 2.7.2 in Section 2.7, we can obtain that when

$$\frac{4}{3+2\alpha} < p < \frac{4}{1-2\alpha},$$

we have

$$\left\|\sup_j |B_R^\alpha(f)|\right\|_p \le C\left(\|f\|_p + \left\|\left(\sum_i |\widehat{\phi}_j * f|^2\right)^{\frac{1}{2}}\right\|_p\right).$$

Consequently, by the theorem with the Littlewood-Paley decomposition (see Stein [St4]), we can deduce the conclusion of the theorem. This completes the proof. ∎

From the conditions of Theorem 2.10.3, we also know that when $p = 2$, we have

$$\|B_*^\alpha(f)\|_2 \le C_{n,\alpha}\|f\|_2 \quad (\alpha > 0),$$

where we require the order $\alpha > 0$. Naturally, one hopes the inequality above can still hold for the case when $\alpha = 0$. However, the problem is still an open problem until now. Similarly, if we replace the normal maximal operator

$$\sup_{R>0} |B_R^0 f(x)|$$

with the maximal operator $\sup_j |B_{R_j}^0 f(x)|$ in the lacunary form, then, the above guess is correct, Lu [Lu8] gives a proof as follows.

Theorem 2.10.4 *Let $f \in L^2(\mathbb{R}^n)$. If there exists a positive number $q > 1$, such that the positive number sequence $\{R_j\}_1^\infty$ satisfies $\frac{R_{j+1}}{R_j} \ge q$, $\forall\, j \in N$, then we have*

$$\|\sup_j |B_{R_j}^0(f)|\|_2 \le C\|f\|_2.$$

In particular, we have

$$\lim_{j\to\infty} B_{R_j}^0(f)(x) = f(x), \quad \text{a.e. } x.$$

2.10 Almost everywhere convergence

The proof of the theorem is similar to the one in the case of the series, which can be checked out at Theorem 3.5.6 in Chapter 3.

Remark 2.10.1 *In 1988, Carbery, Rubio de Francia and Vega carried Theorem 2.10.4' to the final form in [CRV1]. Let $f \in L^p(\mathbb{R}^n), 2 \leq p < \frac{2n}{n-1}$. If there exists a positive number $q > 1$, such that the positive number sequence $\{R_j\}_1^\infty$ satisfies $\frac{R_{j+1}}{R_j} \geq q$ for all $j \in \mathbb{N}$, then we have*

$$\lim_{j \to \infty} B_{R_j}^0(f)(x) = f(x)$$

for a.e. $x \in \mathbb{R}^n$. In addition, it should be pointed out that the range of p above is best (see [CSo1]).

As pointed out as previous chapters, the results in Theorem 2.10.3 are not best. When $n > 1$ and $0 < \alpha < \frac{n-1}{2}$, the convergence of the Bochner-Riesz means almost everywhere in $\mathbb{R}^n$ has a conjecture as follows: If $n > 1$, $< \alpha < \frac{n-1}{2}$, $f \in L^p(\mathbb{R}^n)$, and

$$\frac{2n}{n+1+2\alpha} < p < \frac{2n}{n-1-2\alpha},$$

then we have

$$\lim_{R \to \infty} B_R^\alpha f(x) = f(x),$$

for a.e. $x \in \mathbb{R}^n$.

The Bochner-Riesz conjecture has not been solved yet as of now. But when $p \geq 2$, Carbery-Rubio de Francia-Vega [CRV1] gave an affirmative answer to the conjecture above.

Let us now state this result as follows.

Theorem 2.10.5 *Let $0 < \lambda \leq \frac{1}{2}(n-1)$, $n \geq 2$. If $f \in L^p(\mathbb{R}^n)$ and $2 \leq p < p_\lambda$, then*

$$\lim_{R \to \infty} B_R^\lambda f(x) = f(x) \qquad (2.10.2)$$

holds for a.e. $x \in \mathbb{R}^n$, where $p_\lambda = \frac{2n}{n-1-2\lambda}$.

Denote B_*^λ as the corresponding maximal operator of B_R^λ, that is,

$$B_*^\lambda f(x) = \sup_{R>0} \left| B_R^\lambda f(x) \right|.$$

Consequently, the proof of Theorem 2.10.2 can be attributed to doing research on the maximal operator of B_*^λ. But now, we will study it through the decomposition of space, not discussing the boundedness of the maximal operator of type (p,p) as usual. Thus we convert the consideration of the boundedness of the operator into the problem of the boundedness of the square integral space with a power weighted function. For the sake of conciseness, we denote

$$L^2(|x|^{-\alpha}) = L^2(\mathbb{R}^n, |x|^{-\alpha}dx),$$

and $L^p = L^p(\mathbb{R}^n)$ for $1 < p < \infty$.

Thus when $2 \leq p < p_\lambda$ for any $f \in L^p(\mathbb{R}^n)$, there exists

$$0 \leq \alpha < 1 + 2\lambda = n\left(1 - \frac{2}{p_\lambda}\right),$$

such that

$$L^p \subset L^2 + L^2(|x|^{-\alpha}).$$

In fact, we set $\chi_{\{x:|x|<1\}}$ to be the characteristic function of the unit ball, then

$$f = f\chi_{\{x:|x|<1\}} + f(1 - \chi_{\{x:|x|<1\}}). \tag{2.10.3}$$

By Hölder's inequality, it is easy to check that $f\chi_{\{x:|x|<1\}} \in L^2(\mathbb{R}^n)$ and $f(1 - \chi_{|x|<1}) \in L^2(|x|^{-\alpha})$. Thus Theorem 2.10.5 can be derived from the boundedness of the maximal operator with a weighted function.

Theorem 2.10.6 *Let $\lambda > 0$ and $0 \leq \alpha < 1 + 2\lambda \leq n$. For any Schwartz function f, we have*

$$\int_{\mathbb{R}^n} \left(B_*^\lambda f(x)\right)^2 |x|^{-\alpha}dx \leq C_{\alpha,\lambda} \int_{\mathbb{R}^n} |f(x)|^2 |x|^{-\alpha}dx. \tag{2.10.4}$$

It is easy to see that Theorem 2.10.6 includes the consequences of Theorem 2.10.5. Actually, because of the decomposition formula (2.10.3), for any $f \in L^p$, we have

$$f = f_1 + f_2,$$

where $f_1 \in L^2(\mathbb{R}^n)$ and $f_2 \in L^2(|x|^{-\alpha})$.

If Theorem 2.10.6 has been proven, then it follows from Theorem 2.10.6 that

$$\lim_{R \to \infty} B_R^\lambda f_1(x) = f_1(x)$$

holds for almost everywhere $x \in \mathbb{R}^n$.

2.10 Almost everywhere convergence

Since the measure $|x|^{-\alpha}dx$ is absolutely continuous with respect to the Lebesgue measure, we have

$$\lim_{R\to\infty} B_R^\lambda f_2(x) = f_2(x)$$

for almost any $x \in \mathbb{R}^n$.

Consequently, next our main task is to show Theorem 2.10.6.

For a given small positive real number $\delta > 0$, we denote D as the differential operator. Set $m^\delta \in C^\infty$ satisfying

$$\text{supp } m^\delta \subset [1-\delta, 1],$$

and

$$0 \leq m^\delta(t) \leq 1, \quad \left|D^k m^\delta(t)\right| \leq C\delta^{-k} \qquad (2.10.5)$$

for every positive integer k. We define the operator S_t^δ as

$$\widehat{S_t^\delta f}(\xi) = m^\delta(t|\xi|)\hat{f}(\xi).$$

Next we give a fundamental estimate of the operator S_t^δ.

Lemma 2.10.1 *For any $\delta > 0$ and $0 \leq \alpha < n$, we have that*

$$\int_{\mathbb{R}^n} \int_1^2 \left|S_t^\delta f(x)\right|^2 \leq C_\alpha A_\alpha(\delta) \int_{\mathbb{R}^n} |f(x)|^2 \frac{dx}{|x|^\alpha}, \qquad (2.10.6)$$

where C_α is a constant which is independent of δ, and

$$A_\alpha(\delta) = \begin{cases} \delta^{2-\alpha}, & 1 < \alpha < n, \\ \delta|\log\delta|, & \alpha = 1, \\ \delta, & 0 \leq \alpha < 1. \end{cases}$$

Assume the inequality (2.10.6) holds. Then we will prove Theorem 2.10.6 is valid. We denote the corresponding maximal operator of S_t^δ by S_*^δ, that is,

$$S_*^\delta f(x) = \sup_{t>0} \left|S_t^\delta f(x)\right|.$$

Similar to Section 2.7 in this chapter, we have the following decomposition

$$(1-|\xi|^2)_+^\lambda = \sum_{k=0}^\infty 2^{-k\lambda} m^{2^{-k}}(|\xi|),$$

thus we have
$$B_*^\lambda f(x) \le \sum_{k=0}^\infty 2^{-k\lambda} S_*^{2-k} f(x).$$

Clearly we have that the terms about $k=0,1$ can be controlled by the Hardy-Littlewood maximal function, and when $-n < \alpha < n$, the Hardy-Littlewood maximal function operator is bounded on $L^2(|x|^\alpha)$. Thus we mainly pay attention to the terms that $k \ge 1$.

To research the above maximal operator, we introduce the square operator
$$G^\delta f(x) = \left(\int_0^\infty |S_t^\delta f(x)|^2 \frac{dt}{t}\right)^{\frac{1}{2}}$$
and the operator $\widetilde{G}^\delta$, whose corresponding multiplier is
$$\widetilde{m}^\delta(t) = \delta t \frac{d}{dt} m^\delta(t)$$
instead of $m^\delta(t)$. It is easy to check that $\widetilde{m}^\delta(t)$ satisfies (2.10.5) similar to $m^\delta(t)$. By the differential integral fundamental theorem, we have
$$(S_*^\delta f(x))^2 \le \int_0^\infty 2\left|S_t^\delta f(x)\frac{d}{dt}S_t^\delta f(x)\right|dt \le 2\delta^{-1} G^\delta f(x) \widetilde{G}^\delta f(x).$$

Next we will consider the estimate about G^δ. Let a function ψ satisfy
$$\widehat{\psi}(0) = 0,$$
and when $\frac{1}{2} \le |\xi| \le 2$, we have
$$\widehat{\psi}(\xi) = 1.$$
Then, we denote
$$\psi_k(x) = 2^{-kn} \psi\left(2^{-k} x\right).$$
Then when
$$2^{k-1} \le t \le 2^k,$$
the support of $m^\delta(t|\xi|)$ is the subset of the set
$$\left\{\xi : 2^{-k}(1-\delta) \le |\xi| \le 2^{-k+1}\right\},$$
and when ξ is in the set, we have
$$\widehat{\psi_k}(\xi) = \widehat{\psi}\left(2^k \xi\right) = 1.$$

2.10 Almost everywhere convergence

Therefore, we have

$$\widehat{S_t^\delta f}(\xi) = m^\delta(t|\xi|)\hat{f}(\xi)$$
$$= m^\delta(t|\xi|)\widehat{\psi_k}(\xi)\hat{f}(\xi)$$
$$= \widehat{S_t^\delta(\psi_k * f)}(\xi).$$

Because of Lemma 2.10.1, we obtain

$$\int_{\mathbb{R}^n}\int_{2^{k-1}}^{2^k} \left|S_t^\delta f(x)\right|^2 \frac{dt}{t}\frac{dx}{|x|^\alpha} \leq C_\alpha A_\alpha(\delta) \int_{\mathbb{R}^n} |\psi_k * f(x)|^2 \frac{dx}{|x|^\alpha}.$$

About the summation of $k \in \mathbb{Z}$, using the fact that when $-n < \alpha < n$, the operator

$$f \to \left(\sum |\psi_k * f|^2\right)^{\frac{1}{2}}$$

is bounded on $L^2(|x|^{-\alpha})$ (see [GR1]), then we get

$$\int_{\mathbb{R}^n} \left(G^\delta f(x)\right)^2 \frac{dx}{|x|^\alpha} \leq C_\alpha A_\alpha(\delta) \int_{\mathbb{R}^n} |f(x)|^2 \frac{dx}{|x|^\alpha}.$$

Consequently we get

$$\int_{\mathbb{R}^n} \left(S_*^\delta f(x)\right)^2 \frac{dx}{|x|^\alpha} \leq C_\alpha A_\alpha(\delta)\delta^{-1} \int_{\mathbb{R}^n} |f(x)|^2 \frac{dx}{|x|^\alpha}. \qquad (2.10.7)$$

Set $\delta = 2^{-k}$, then we get the boundedness of B_*^λ on $L^2(|x|^{-\alpha})$.

Here we just choose $0 \leq \alpha \leq 1$ when $\lambda > 0$; or $1 < \alpha < n$ when $\lambda > \frac{1}{2}(\alpha - 1)$, thus we have the conclusion of Theorem 2.10.6.

Now we turn to the proof of the fundamental estimate.

Here we consider the duality operator of S_t^δ. In Lemma 2.10.1, we need to show the boundedness of the operator

$$S_t^\delta : L^2(|x|^{-\alpha}) \to L^2\left([1,2] \times \mathbb{R}^n, \frac{dt}{t}\frac{dx}{|x|^{-\alpha}}\right).$$

It is easy to note that the duality spaces of $L^2(|x|^{-\alpha})$ and

$$L^2\left([1,2] \times \mathbb{R}^n, \frac{dt}{t}\frac{dx}{|x|^{-\alpha}}\right)$$

are $L^2(|x|^\alpha)$ and

$$L^2\left([1,2] \times \mathbb{R}^n, \frac{dt}{t}\frac{dx}{|x|^\alpha}\right).$$

For
$$g_t \in L^2\left([1,2] \times \mathbb{R}^n, \frac{dt}{t}\frac{dx}{|x|^\alpha}\right),$$
and $f \in L^2(|x|^{-\alpha})$, denote
$$\widetilde{K}_t^\delta(x) = K_t^\delta(-x).$$

We conclude that
$$(f, (S_t^\delta)^* g_t) = (S_t^\delta f, g_t)$$
$$= \int_{\mathbb{R}^n} \int_1^2 S_t^\delta f(x) g_t(x) \frac{dt}{t} dx$$
$$= \int_{\mathbb{R}^n} \int_1^2 \int_{\mathbb{R}^n} K_t^\delta(x-y) f(y) dy\, g_t(x) \frac{dt}{t} dx$$
$$= \int_{\mathbb{R}^n} \int_{\mathbb{R}^n} \int_1^2 K_t^\delta(x-y) g_t(x) \frac{dt}{t} dx\, f(y) dy$$
$$= \int_{\mathbb{R}^n} \int_{\mathbb{R}^n} \int_1^2 \widetilde{K}_t^\delta(y-x) g_t(x) \frac{dt}{t} dx\, f(y) dy,$$

then we obtain
$$(S_t^\delta)^* g_t(x) = \int_1^2 S_t^\delta g_t(x) \frac{dt}{t}.$$

Because of the property of the duality, the proof of Lemma 2.10.1 is equivalent to proving the following inequality,

$$\int_{\mathbb{R}^n} \left|\int_1^2 S_t^\delta g_t(x) \frac{dt}{t}\right|^2 |x|^\alpha dx \leq C_\alpha A_\alpha(\delta) \int_{\mathbb{R}^n} \int_1^2 |g_t(x)|^2 \frac{dt}{t} |x|^\alpha dx. \quad (2.10.8)$$

In order to prove (2.10.8), we now introduce the Sobolev spaces with the fractional order. When $0 < \beta < 1$, we define the differential operator with the fractional order as follows

$$\mathcal{D}^\beta f(x) = \left(\int_{\mathbb{R}^n} \frac{|f(x+y) - f(x)|^2}{|y|^{2\beta}} \frac{dy}{|y|^n}\right)^{\frac{1}{2}}.$$

When $\beta \geq 1$, we define $\mathcal{D}^\beta$ by the difference with a higher order which is not smaller than $[\beta] + 1$. We also define the Sobolev spaces with the homogeneous order $L_\beta^2(\mathbb{R}^n)$ as follow

$$L_\beta^2(\mathbb{R}^n) = \left\{f : \left\|\mathcal{D}^\beta \hat{f}\right\|_2 < \infty\right\}.$$

2.10 Almost everywhere convergence

Due to the Plancherel theorem, we have that

$$\begin{aligned}\left\|\mathcal{D}^\beta f\right\|_2^2 &= \int_{\mathbb{R}^n}\int_{\mathbb{R}^n}\frac{|f(x+y)-f(x)|^2}{|y|^{2\beta}}\frac{dy}{|y|^n}dx\\ &= \int_{\mathbb{R}^n}\int_{\mathbb{R}^n}|f(x+y)-f(x)|^2 dx\frac{1}{|y|^{2\beta}}\frac{dy}{|y|^n}\\ &= \int_{\mathbb{R}^n}\int_{\mathbb{R}^n}\frac{|e^{-i2\pi y\cdot\xi}-1|^2}{|y|^{2\beta+n}}dy\left|\hat{f}(\xi)\right|^2 d\xi. \end{aligned} \qquad (2.10.9)$$

Now we consider the integral in (2.10.9)

$$I(\xi) = \int_{\mathbb{R}^n}\frac{|e^{-i2\pi y\cdot\xi}-1|^2}{|y|^{2\beta+n}}dy.$$

Set ρ is a rotational transform with the original point as the circle, and ρ^t as its transpose.

Since a rotational transform is orthonormal transform with the determinant 1, then we have

$$\det(\rho) = \det\left(\rho^t\right) = 1.$$

Thus we have

$$\begin{aligned}I(\rho\xi) &= \int_{\mathbb{R}^n}\frac{|e^{-i2\pi y\cdot\rho\xi}-1|^2}{|y|^{2\beta+n}}dy\\ &= \int_{\mathbb{R}^n}\frac{|e^{-i2\pi\rho^t y\cdot\xi}-1|^2}{|y|^{2\beta+n}}dy\\ &= \int_{\mathbb{R}^n}\frac{|e^{-i2\pi y\cdot\xi}-1|^2}{|y|^{2\beta+n}}dy\\ &= I(\xi).\end{aligned}$$

This means that $I(\xi)$ is rotation-invariant. Thus $I(\xi) = I_0(|\xi|)$.

Next set $\xi = |\xi|\eta$, where η is a unit vector. We have

$$\begin{aligned}I(\xi) &= \int_{\mathbb{R}^n}\frac{|e^{-i2\pi y\cdot|\xi|\eta}-1|^2}{|y|^{2\beta+n}}dy\\ &= \int_{\mathbb{R}^n}\frac{|e^{-i2\pi|\xi|y\cdot\eta}-1|^2}{|y|^{2\beta+n}}dy\\ &= |\xi|^{2\beta}\int_{\mathbb{R}^n}\frac{|e^{-i2\pi y\cdot\eta}-1|^2}{|y|^{2\beta+n}}dy\\ &= |\xi|^{2\beta}I(\eta).\end{aligned}$$

Then we will prove that $I(\eta)$ is a constant.

Since $\left|e^{-i2\pi y\cdot\eta}\right|\leq 2$ and $\left|e^{-i2\pi y\cdot\eta}-1\right|\leq C|y|$, with the polar coordinate expression, we have

$$|I(\eta)|\leq \left|\int_{|y|\leq 1}\frac{\left|e^{-i2\pi y\cdot\eta}-1\right|^2}{|y|^{2\beta+n}}dy\right|+\left|\int_{|y|\geq 1}\frac{\left|e^{-i2\pi y\cdot\eta}-1\right|^2}{|y|^{2\beta+n}}dy\right|$$

$$\leq C\int_0^1 \frac{r^2}{r^{2\beta+n}}r^{n-1}dr+C\int_1^\infty \frac{1}{r^{2\beta+n}}r^{n-1}dr$$

$$<+\infty.$$

Now let us turn to (2.10.9). We obtain

$$\left\|\mathcal{D}^\beta f\right\|_2^2 = C\int_{\mathbb{R}^n}|\hat{f}(\xi)|^2|\xi|^{2\beta}d\xi. \tag{2.10.10}$$

Due to the Plancherel theorem and (2.10.10), (2.10.8) is equivalent to

$$\left\|\mathcal{D}^{\frac{1}{2}\alpha}\int_1^2 m^\delta(t|\cdot|)\hat{g}_t(\cdot)\frac{dt}{t}\right\|_2^2 \leq C_\alpha A_\alpha(\delta)\int_{\mathbb{R}^n}\int_1^2 |\mathcal{D}^{\frac{1}{2}\alpha}\hat{g}_t(\xi)|^2\frac{dt}{t}d\xi. \tag{2.10.11}$$

If we calculate the difference with the definition of $\mathcal{D}^\beta$ and use the Cauchy-Schwartz's inequality with the variable t, the integral of the nonzero part of the function with the measure $\frac{dt}{t}$ being at least $C\delta$. Thus the left side of (2.10.11) is controlled by

$$C\delta\int_1^2\int_{\mathbb{R}^n}\left|\mathcal{D}^{\frac{1}{2}\alpha}(m^\delta(t|\xi|)\hat{g}_t(\xi))\right|^2 d\xi\frac{dt}{t}.$$

Consequently, if we can prove that $m^\delta(t\cdot)$ is a multiplier in the pointwise in the homogeneous Sobolev spaces, and the constant is uniformly controlled by $C_\alpha\delta^{-\frac{1}{2}}A_\alpha^{\frac{1}{2}}(\delta)$, provided that $1\leq t\leq 2$. Thus Lemma 2.10.1 will be proven.

Notice the homogeneity property, we merely need to prove the inequality when $t=1$.

Proposition 2.10.1 *If $0\leq\beta<\frac{n}{2}$, we have*

$$\left\|m^\delta f\right\|_{L^2_\beta}^2 \leq C_\beta A_{2\beta}(\delta)\delta^{-1}\|f\|_{L^2_\beta}^2.$$

Equivalently, if $0\leq\alpha<n$, we have

$$\int_{\mathbb{R}^n}\left|S^\delta f(x)\right|^2\frac{dx}{|x|^\alpha}\leq C_\alpha A_\alpha(\delta)\delta^{-1}\int_{\mathbb{R}^n}|f(x)|^2\frac{dx}{|x|^\alpha}.$$

2.10 Almost everywhere convergence

To prove Proposition 2.10.1, we need the following estimate.

Proposition 2.10.2 *If $0 < \varepsilon < \frac{1}{2}$, we have*

$$\int_{||x|-1|\leq \varepsilon} |f(x)|^2 dx \leq C_\beta B_\beta(\varepsilon) \|f\|_{L^2_\beta}^2,$$

where

$$B_\beta(\varepsilon) = \begin{cases} \varepsilon^{2\beta}, & 0 \leq \beta < \frac{1}{2}, \\ \varepsilon |\log \varepsilon|, & \beta = \frac{1}{2}, \\ \varepsilon, & \frac{1}{2} < \beta < \frac{1}{2}n. \end{cases}$$

Proof. Taking the Fourier transform, Proposition 2.10.2 is equivalent to

$$\int_{||x|-1|\leq \varepsilon} \left|\hat{f}(x)\right|^2 dx \leq C_\beta B_\beta(\varepsilon) \int_{\mathbb{R}^n} |f(x)|^2 |x|^{2\beta} dx.$$

By the property of the duality, for the function g whose support set is a subset of $\{x : ||x| - 1| \leq \varepsilon\}$, then the above inequality is also equivalent to

$$\int_{\mathbb{R}^n} |\hat{g}(x)|^2 \frac{dx}{|x|^{2\beta}} \leq C_\beta B_\beta(\varepsilon) \int_{||x|-1|\leq \varepsilon} |g(x)|^2 dx.$$

Using the formula of the inner product, the convolution formula and Hölder's inequality, when $0 < \beta < \frac{1}{2}n$, it follows

$$\widehat{\left(\frac{1}{|\cdot|^{2\beta}}\right)}(y) = C \frac{1}{|y|^{n-2\beta}}.$$

Thus we have

$$\int_{\mathbb{R}^n} |\hat{g}(x)|^2 \frac{dx}{|x|^{2\beta}} = \int_{\mathbb{R}^n} \widehat{(g * \tilde{g})}(x) \frac{dx}{|x|^{2\beta}}$$

$$= \int_{\mathbb{R}^n} g * \tilde{g}(x) \frac{1}{|x|^{n-2\beta}} dx$$

$$= \int_{||x|-1|\leq \varepsilon} \int_{||y|-1|\leq \varepsilon} g(x)\overline{g(y)} \frac{1}{|x-y|^{n-2\beta}} dxdy$$

$$\leq C \int_{||x|-1|\leq \varepsilon} |g(x)|dx \int_{||y|-1|\leq \varepsilon} |\overline{g(y)}| \sup_x \frac{1}{|x-y|^{n-2\beta}} dy$$

$$\leq C \int_{||x|-1|\leq \varepsilon} |g(x)|dx \|g\|_2 \left\|\sup_x |x-\cdot|^{2\beta-n}\right\|_{L^1(||y|-1|\leq \varepsilon)}$$

$$\leq C \|g\|_2^2 \left\|\sup_x |x-\cdot|^{2\beta-n}\right\|_{L^1(||y|-1|\leq \varepsilon)}.$$

By the change of the variable $v = (v_1, v')$, $v_1 \in \mathbb{R}$, $v' \in \mathbb{R}^{n-1}$, we have the following estimate

$$\int_{|v_1| \leq \varepsilon, |v'| \leq 1} \frac{dv}{|v|^{n-2\beta}} \leq C_\beta B_\beta(\varepsilon).$$

It is easy to have that

$$\int_{|v_1| \leq \varepsilon, |v'| \leq \varepsilon} \frac{dv}{|v|^{n-2\beta}} \approx \int_0^\varepsilon \frac{r^{n-1} dr}{r^{n-2\beta}} \approx \varepsilon^{2\beta} \leq B_\beta(\varepsilon),$$

and

$$\int_{|v_1| \leq \varepsilon, \varepsilon \leq |v'| \leq 1} \frac{dv}{|v|^{n-2\beta}} \approx \int_0^\varepsilon dv_1 \int_{|v'| \geq \varepsilon} \frac{dv'}{|v'|^{n-2\beta}} \approx \varepsilon \int_\varepsilon^1 \frac{r^{n-2} dr}{r^{n-2\beta}} \approx B_\beta(\varepsilon).$$

∎

Next we will give some estimates about the kernel of the operator S^δ. Denote $K(x) = K^\delta(x)$ as its kernel, such that

$$\widehat{K}(\xi) = m^\delta(|\xi|).$$

Suppose $\varphi \in C^\infty(\mathbb{R})$, whose support set is $[-1, 2]$, and satisfies $\varphi(t) = 1$ when $0 \leq t \leq 1$. Set

$$h_j(x) = \begin{cases} \varphi(|x|), & j = 0, \\ \varphi(2^j |x|) - \varphi(2^{j-1}|x|), & j \geq 1. \end{cases}$$

Thus we get the binary unit decomposition $\{h_j(x)\}$.

Denote $K_j(x) = K(x) h_j(\delta x)$. We have the decomposition

$$K(x) = \sum_{j=0}^\infty K_j(x).$$

Since K_j and $\widehat{K}_j$ are radial functions, when $|\xi| = r$, we can denote

$$\widehat{K}_j(r) = \widehat{K}_j(\xi).$$

The following lemma will give an elaborate estimate about the kernel.

2.10 Almost everywhere convergence

Proposition 2.10.3 *For any positive integer m, if $0 \leq \alpha < n$, we have*

$$\int_0^\infty |\widehat{K}_j(r)| r^{\alpha-1} dr \leq C 2^{-mj} \delta.$$

Proof. Denote $h(x) = \varphi(2|x|) - \varphi(|x|)$. Since $\widehat{K}_j = m^\delta * \widehat{h_j(\delta \cdot)}$, we have that
$$\begin{aligned}\widehat{K}_j(\xi) &= 2^{nj} \delta^{-n} m^\delta * \widehat{h}\left(2^j \delta^{-1} \cdot\right)(\xi)\\ &= 2^{nj}\delta^{-n} \int_{\mathbb{R}^n} m^\delta(\xi - \eta) \widehat{h}(2^j \delta^{-1} \eta) d\eta\\ &= \int_{\mathbb{R}^n} m^\delta(\xi - 2^{-j}\delta\eta) \widehat{h}(\eta) d\eta.\end{aligned}$$

Due to the fact that h equals 0 in the neighborhood of the origin, for any multi index $a = (a_1, a_2, \ldots, a_n)$, we have

$$\int_{\mathbb{R}^n} \eta^a \widehat{h}(\eta) d\eta = 0.$$

Expanding m^δ at ξ in Taylor's series, we have

$$\widehat{K}_j(\xi) = \int_{\mathbb{R}^n} R_m(\xi, \eta) \widehat{h}(\eta) d\eta,$$

where the remaining term R_m satisfies

$$|R_m(\xi, \eta)| \leq \sum_{|a|=m} \left\|D^a m^\delta\right\|_\infty |2^{-j}\delta\eta|^m \leq C 2^{-mj} |\eta|^m.$$

Here $\widehat{h}$ is a Schwartz function, the integral of the product of $\widehat{h}$ and $|\eta|^m$ is bounded, then for any positive integer m, $\xi \in \mathbb{R}^n$, we have

$$\left|\widehat{K}_j(\xi)\right| \leq C 2^{-mj}. \tag{2.10.12}$$

On the other hand, in the integral of the definition of $\widehat{K}_j(\xi)$, if $|\xi| < \frac{1}{2}$, then we have

$$|\eta| > 2^{j-2} \delta^{-1}.$$

Using the fact that $\widehat{h}$ is a Schwartz function again, for any positive integer m and $|\xi| < \frac{1}{2}$, we have

$$\left|\widehat{K}_j(\xi)\right| \leq C 2^{-mj} \delta^m.$$

Consequently, we obtain

$$\int_0^{\frac{1}{2}} |\widehat{K}_j(r)| r^{\alpha-1} dr \leq C 2^{-mj} \delta^m \int_0^{\frac{1}{2}} r^{\alpha-1} dr \leq C 2^{-mj} \delta.$$

For the other part, set $S = \{\xi \in \mathbb{R}^n : 1 - 2\delta < |\xi| < 1 + 2\delta\}$, we have

$$\int_{\frac{1}{2}}^{\infty} \left|\widehat{K}_j(r)\right| r^{\alpha-1} dr \leq C \int_{\mathbb{R}^n} \left|\widehat{K}_j(\xi)\right| d\xi$$

$$= C \left(\int_S + \int_{\mathbb{R}^n \setminus S}\right) \left|\widehat{K}_j(\xi)\right| d\xi$$

$$\leq C 2^{-mj} \delta + C \int_{|\eta| \geq 2^j} \left|\widehat{h}(\eta)\right| \int_{\mathbb{R}^n \setminus S} m^\delta(\xi - 2^{-j}\delta\eta) d\xi d\eta$$

$$\leq C 2^{-mj} \delta + C \left\|m^\delta\right\|_1 \int_{|\eta| \geq 2^j} \left|\widehat{h}(\eta)\right| d\eta$$

$$\leq C 2^{-mj} \delta.$$

∎

Proposition 2.10.4 *If $\beta > 0$, we have*

$$|\mathcal{D}^\beta m^\delta(|\xi|)| \leq C_\beta \begin{cases} \delta^{-\beta}, & ||\xi| - 1| \leq 2\delta, \\ \delta^{\frac{1}{2}} \left||\xi| - 1\right|^{-(\beta + \frac{1}{2})}, & 0 \leq |\xi| \leq 2, \\ \delta^{\frac{1}{2}} |\xi|^{-(\beta + \frac{1}{2}n)}, & |\xi| \geq 2. \end{cases}$$

Proof. We just prove the case of $0 < \beta < 1$, the case $\beta \geq 1$ one can get the similar estimate about the difference with the higher order.

Firstly, considering the case when $||\xi| - 1| \leq 2\delta$, we divide the integral about $\mathcal{D}^\beta$ in its definition into two parts: $|\eta| \leq \delta$ and $|\eta| > \delta$. We have

$$\left|m^\delta(|\xi + \eta|) - m^\delta(|\xi|)\right| \leq |\eta| \|\nabla m^\delta\|_\infty \leq \frac{|\eta|}{\delta}.$$

Thus we conclude that

$$\int_{|\eta| \leq \delta} \frac{\left|m^\delta(|\xi + \eta|) - m^\delta(|\xi|)\right|^2}{|\eta|^{2\beta}} \frac{d\eta}{|\eta|^n} \leq \delta^{-2} \int_{|\eta| \leq \delta} \frac{|\eta|^2}{|\eta|^{2\beta+n}} d\eta$$

$$= C\delta^{-2} \int_0^\delta r^{-2\beta+1} dr$$

$$= C\delta^{-2\beta}.$$

2.10 Almost everywhere convergence

When $|\eta| \geq \delta$, we have

$$\int_{|\eta|\geq\delta} \frac{|m^\delta(|\xi+\eta|) - m^\delta(|\xi|)|^2}{|\eta|^{2\beta}} \frac{d\eta}{|\eta|^n} \leq C \int_{|\eta|\geq\delta} \frac{d\eta}{|\eta|^{n+2\beta}}$$

$$= C \int_\delta^\infty \frac{dr}{r^{1+2\beta}}$$

$$= C\delta^{-2\beta}.$$

Secondly, we consider the case of $|\xi| \geq 2$. Noting that the support set of m^δ is $[1-\delta, 1]$, we conclude that

$$\left(\mathcal{D}^\beta m^\delta(|\xi|)\right)^2 = \int_{\mathbb{R}^n} \frac{|m^\delta(|\xi+\eta|) - m^\delta(|\xi|)|^2}{|\eta|^{2\beta}} \frac{d\eta}{|\eta|^n}$$

$$= \int_{\mathbb{R}^n} \frac{|m^\delta(|\xi+\eta|)|^2}{|\eta|^{n+2\beta}} d\eta$$

$$= \int_{\mathbb{R}^n} \frac{|m^\delta(|\eta|)|^2}{|\eta-\xi|^{n+2\beta}} d\eta$$

$$= \int_{1-\delta\leq|\eta|\leq 1} \frac{|m^\delta(|\eta|)|^2}{|\eta-\xi|^{n+2\beta}} d\eta$$

$$\leq C \frac{1}{|\xi|^{n+2\beta}} \int_{1-\delta}^1 r^{n-1} dr$$

$$\leq C \frac{1}{|\xi|^{n+2\beta}} \int_{1-\delta}^1 dr$$

$$\leq C \frac{\delta}{|\xi|^{n+2\beta}}.$$

Finally we consider the case when $|\xi| \leq 2$ and $||\xi|-1| \geq 2\delta$. Think of the support set, we should estimate the integral with the variable ξ

$$\int_{\{\eta:||\xi-\eta|-1|\leq\delta\}} \frac{d\eta}{|\eta|^{n+2\beta}}.$$

If $|\xi| \approx 2^{-j}$, we consider the integral on the rings $|\eta| \approx 2^{-j+r} (r \geq 0)$ respectively. The integral area has a measure $O(\delta 2^{-(j-r)(n-1)})$, the integral function has the size about $O(2^{(j-r)(n+2\beta)})$, then the integral on this ring is controlled by $O(\delta 2^{(j-r)(2\beta+1)})$. Then take the summation about r, we get

$$|\mathcal{D}^\beta m^\delta(|\xi|)| \leq C\delta^{\frac{1}{2}} 2^{j(\beta+\frac{1}{2})} \approx C\delta^{\frac{1}{2}} ||\xi|-1|^{-(\beta+\frac{1}{2})}.$$

∎

Let us come to the proof of Proposition 2.10.1.
We consider the two cases $1 < \alpha < n$ and $0 < \alpha < 2$, respectively. Firstly, we think about the first case.

Assume $j \geq 0$. By Proposition 2.10.2, we have

$$\int_{\mathbb{R}^n} |K_j * f(x)|^2 dx = \int_{\mathbb{R}^n} \left|\widehat{K}_j(x)\right|^2 |\hat{f}(\xi)|^2 d\xi$$

$$= \int_0^\infty r^{n-1} \left|\widehat{K}_j(r)\right|^2 \int_{S^{n-1}} |\hat{f}(r\omega)|^2 d\omega dr$$

$$\leq C \int_{\mathbb{R}^n} |f(x)|^2 |x|^\alpha dx \int_0^\infty r^{\alpha-n} r^{n-1} \left|\widehat{K}_j(r)\right|^2 dr.$$

Due to $\|\widehat{K}_j\|_\infty \leq 1$, and Proposition 2.10.3 for any positive integer m, we obtain

$$\int_{\mathbb{R}^n} |K_j * f(x)|^2 dx \leq C 2^{-mj} \delta \int_{\mathbb{R}^n} |f(x)|^2 |x|^\alpha dx.$$

Using the property of the duality, we have

$$\int_{\mathbb{R}^n} |K_j * f(x)|^2 \frac{dx}{|x|^\alpha} \leq C 2^{-mj} \delta \int_{\mathbb{R}^n} |f(x)|^2 dx.$$

Applying the above inequality to f, who has the support $\{x : |x| \leq C 2^j \delta^{-1}\}$, we have

$$\int_{\mathbb{R}^n} |K_j * f(x)|^2 \frac{dx}{|x|^\alpha} \leq C 2^{(\alpha-m)j} \delta^{1-\alpha} \int_{\mathbb{R}^n} |f(x)|^2 \frac{dx}{|x|^\alpha}.$$

Choose $m \geq \alpha + 1$, then we have

$$\int_{\mathbb{R}^n} |K_j * f(x)|^2 \frac{dx}{|x|^\alpha} \leq C 2^{-j} \delta^{1-\alpha} \int_{\mathbb{R}^n} |f(x)|^2 \frac{dx}{|x|^\alpha}. \qquad (2.10.13)$$

We will prove (2.10.13) is true for any f. Notice that $K_j(x) = K(x) h_j(x)$, and the support of h_j is in the ball with the origin as its center and $2^{j+2} \delta^{-1}$ as its radius, then K_j has the similar ball. We decompose f as

$$f = \sum_{i \in \mathbb{Z}} f_i$$

where $f_i = f \chi_{Q_i}$, and the center of Q_0 is the origin, Q_i has the side length $10 \cdot 2^j \delta^{-1}$, which does not intersect with each other and $\{Q_i\}$ covers $\mathbb{R}^n$.

2.10 Almost everywhere convergence

Thus the support of $K_j * f_{Q_i}$ shares the same center with Q_i, but its radius is $C2^j \delta^{-1}$, where the constant C is independent of j, i, δ. Then for any i, $K_j * f_{Q_i}$ has compact support, and they satisfy

$$K_j * f_i(x) \cdot K_j * f_{i'}(x) = 0,$$

provided that $|i - i'| > C$.

Consequently we have

$$\int_{\mathbb{R}^n} |K_j * f(x)|^2 \frac{dx}{|x|^\alpha} = \int_{\mathbb{R}^n} \left|\sum_i K_j * f_i(x)\right|^2 \frac{dx}{|x|^\alpha}$$

$$= \int_{\mathbb{R}^n} \left|\sum_{i,i'} (K_j * f_i(x))(K_j * f_{i'}(x))\right| \frac{dx}{|x|^\alpha}$$

$$= \int_{\mathbb{R}^n} \left|\sum_i \sum_{|i-i'|\le C} (K_j * f_i(x))(K_j * f_{i'}(x))\right| \frac{dx}{|x|^\alpha}$$

$$\le \int_{\mathbb{R}^n} \sum_i \sum_{|i-i'|\le C} \frac{|K_j * f_i(x)|^2 + |K_j * f_{i'}(x)|^2}{2} \frac{dx}{|x|^\alpha}$$

$$\le C \int_{\mathbb{R}^n} \sum_i |K_j * f_i(x)|^2 \frac{dx}{|x|^\alpha}.$$

It suffices to prove that (2.10.13) is valid for any $i \in \mathbb{N}$. The case of $i = 0$ has been proven. When $i \ge 1$, the term $|x|^{-\alpha}$ can be roughly regarded as a constant in the integral. So we merely need to prove

$$\int_{\mathbb{R}^n} |K_j * f(x)|^2 dx \le C 2^{-j} \delta^{1-\alpha} \int_{\mathbb{R}^n} |f(x)|^2 dx.$$

Noting that $\alpha > 1$, $\delta \le \frac{1}{2}$ and (2.10.12), we have

$$\left\|\widehat{K_j}\right\|_\infty \le C 2^{-\frac{1}{2}j} \delta^{\frac{1}{2}(1-\alpha)}.$$

Applying the Plancherel theorem, we get the proof of Proposition 2.10.1 for the case of $1 < \alpha < n$.

For the case $0 < \alpha < 2$, we can use the similar method as before. But we choose another technique to consider it. When $0 < \beta < 1$, by the definition of $\mathcal{D}^\beta$, it implies from Leibniz's law that

$$\left|\mathcal{D}^\beta(gh)(x)\right| \le \|g\|_\infty \left|\mathcal{D}^\beta h(x)\right| + |h(x)| \left|\mathcal{D}^\beta g(x)\right|.$$

Then we have

$$\left\|m^\delta f\right\|_{L^2_\beta} \leq C \left(\left\|m^\delta\right\|_\infty \left\|\mathcal{D}^\beta f\right\|_2 + \left\|f\mathcal{D}^\beta m^\delta\right\|_2 \right).$$

Next we merely need to show

$$\int_{\mathbb{R}^n} |f(x)|^2 \left|\mathcal{D}^\beta m^\delta(x)\right|^2 dx \leq C A_{2\beta}(\delta) \delta^{-1} \|f\|^2_{L^2_\beta}. \qquad (2.10.14)$$

Due to Proposition 2.10.4, the left side of (2.10.14) can be controlled by the following estimate

$$C\delta^{-2\beta} \int_{||x|-1|\leq 2\delta} |f(x)|^2 dx + \delta \sum_{k=1}^{\log \delta} 2^{2k(\beta+\frac{1}{2})} \int_{||x|-1|\leq 2^{-k}} |f(x)|^2 dx$$
$$+ \delta \int_{|x|\geq 2} |f(x)|^2 \frac{dx}{|x|^{n+2\beta}}$$
$$\leq C \left(\delta^{-2\beta} B_\beta(\delta) + \delta \sum_{k=1}^{\log \delta} 2^{2k(\beta+\frac{1}{2})} B_\beta\left(2^{-k}\right) \right) \|f\|^2_{L^2_\beta}$$
$$+ C\delta \sum_{k=1}^\infty \int_{|x|\sim 2^k} |f(x)|^2 dx$$
$$\leq C A_{2\beta}(\delta) \delta^{-1} \|f\|^2_{L^2_\beta} + C\delta \sum_{k=1}^\infty 2^{-k(n+2\beta)} 2^{2k\beta} \|f\|^2_{L^2_\beta}$$
$$\leq C A_{2\beta}(\delta) \delta^{-1} \|f\|^2_{L^2_\beta}.$$

Remark 2.10.2 *Theorem 2.10.5 only confirms the conjecture on a.e. convergence of Bochner-Riesz means of Fourier integrals below the critical index for $p \geq 2$. The conclusion of Theorem 2.10.5 is also obtained by Christ [Chr1] under the restrict condition $\alpha > \frac{n-1}{2(n+1)}$. For the case $p < 2$, the conjecture is still open. However, we have to mention two remarkable results in this direction. In 2002, Tao [Ta2] proved that if $n = 2, 1 < p < 2$ and $\alpha > \max\{\frac{3}{4p} - \frac{3}{8}, \frac{7}{6p} - \frac{2}{3}\}$, then the conclusion of the conjecture holds. In 2004, Lee [Le1] proved that if $n \geq 3$ and $\frac{2n+4}{n} \leq p < \frac{2n}{n-1-2\alpha}$, then the conclusion of the conjecture also holds.*

Remark 2.10.3 *Let $\Omega \subset \mathbb{R}^n$, and $|\Omega| < \infty$. We also assume that $\{T_R : R > 0\}$ is a family of linear operators, with each of T_R mapping $L^p(\Omega)$ with $1 \leq p \leq \infty$ to the measurable function space $\mathscr{M}(\Omega)$. With the method*

completely the same as the one in Theorem 2.10.1, we know that, if the maximal operator
$$T^*f(x) = \sup_{R>0} |T_R f(x)|$$
is of weak type (p,p), then for any $f \in L^p(\Omega)$,
$$\lim_{R\to\infty} T_R f(x)$$
exists almost every $x \in \mathbb{R}^n$. Stein [St3] pointed out that, for the case $1 \leq p \leq 2$, the converse proposition also holds. That is to say, if for any $f \in L^p(\Omega)$ with $1 \leq p \leq 2$,
$$\lim_{R\to\infty} T_R f(x)$$
exists almost every $x \in \mathbb{R}^n$. Thus T_* is of weak type (p,p). The fact can be applied to the case of the series.

2.11 Commutator of Bochner-Riesz operator

After the work on commutators of singular integrals, which is due to Coifman, Rochberg and Weiss [CRW1], was published, it has largely promoted the research of this field. However, without united method for L^p-boundedness of commutators of other operators, its commutator for each operator is studied independently. In 1993, Alvarez, Bagby, Kurtz and Perez [ABKP1] got a judge rule for the boundedness of the commutators generated by general linear operator
$$T: L^p_\omega \to L^p_\omega \Rightarrow [b, T]: L^p \to L^p.$$
for $\omega \in A_q$ and $1 < p, q < \infty$.

Thus, it is easy to know that when the weighted norm inequality of a linear operator has been established, the commutator generated by this one and a BMO function is then bounded on $L^p(\mathbb{R}^n)$. For example, let $B^{\frac{n-1}{2}}$ be the Bochner-Riesz operator at the critical index. In 1992, Shi and Sun [ShS1] established the L^p_ω boundedness of $B^{\frac{n-1}{2}}$, where $\omega \in A_p$. From the judge rule above, the commutator generated by a BMO function and $B^{\frac{n-1}{2}}$ is bounded on $L^p(\mathbb{R}^n)$.

On the other hand, it should be noticed that there are lots of important operators which do not satisfy the condition of the above judge rule. For example, the Bochner-Riesz operator below the critical index is one of those operators.

Let us consider the case $b \in \mathrm{BMO}(\mathbb{R}^n)$. It is natural to expect that the commutator $[b, B_r^\alpha]$ enjoy the same L^p convergence as B_r^α as $r \to \infty$, or $[b, B^\alpha]$ enjoy the same L^p-boundedness of B^α. In this case, the following problem on the L^p boundedness of their commutators is raised naturally.

Question 1 Let $b \in \mathrm{BMO}(\mathbb{R}^n)$. The commutator $[b, B^\alpha]$ is bounded on $L^p(\mathbb{R}^n)$ if and only if

$$\frac{2n}{n+1+2\alpha} < p < \frac{2n}{n-1-2\alpha}.$$

For the necessity in Question 1, Lu and Xia [LX1] gave affirmative answers.

Theorem 2.11.1 Let B^α be the Bochner-Riesz operator, $b \in \mathrm{BMO}(\mathbb{R}^n)$, $0 < \alpha < \frac{n-1}{2}$ and $p > 1$. If the commutator $[b, B^\alpha]$ is bounded on $L^p(\mathbb{R}^n)$, then we have

$$\frac{2n}{n+1+2\alpha} < p < \frac{2n}{n-1-2\alpha}.$$

Proof. If we set $b(x) = \log|x|$, then $b \in \mathrm{BMO}(\mathbb{R}^n)$.

Choose $f \in C_c^\infty(\mathbb{R}^n)$ such that $0 \leq f \leq 1$ and

$$f(x) = \begin{cases} 1, & |x| \leq \frac{1}{2}, \\ 0, & |x| \geq 1. \end{cases}$$

Then we have $f \in L^p(\mathbb{R}^n)$.

Assume that $[b, T^\alpha]$ is bounded on $L^p(\mathbb{R}^n)$ for $1 < p \leq 2$. For the sake of simplicity, we only prove the case

$$p > \frac{2n}{n+1+2\alpha}. \tag{2.11.1}$$

Fix $\psi \in C_c^\infty(\mathbb{R}^n)$, $0 \leq \psi(x) \leq 1$ such that

$$\psi(x) = \begin{cases} 1, & |x| \leq 1, \\ 0, & |x| \geq 2. \end{cases}$$

It follows immediately that

$$\psi(x) + \sum_{j=1}^{\infty} \left(\psi\left(2^{-j}x\right) - \psi\left(2^{-j+1}x\right) \right) = 1,$$

2.11 Commutator of Bochner-Riesz operator

for $x \in \mathbb{R}^n$.

Write $\varphi_0(x) = \psi(x)$,

$$\varphi_j(x) = \psi\left(2^{-j}x\right) - \psi\left(2^{-j+1}x\right),$$

for $j \in \mathbb{N}$, and

$$B^\alpha(x) = C_\alpha \frac{J_{\frac{n}{2}+\alpha}(|x|)}{|x|^{\frac{n}{2}+\alpha}},$$

where $J_\gamma(t)$ denotes the Bessel function of order γ.

Thus we have

$$B^\alpha f(x) = \sum_{j=0}^{\infty} (B^\alpha \varphi_j) * f(x) = \sum_{j=0}^{\infty} T_j f(x).$$

Noting

$$\mathrm{supp}\varphi_0 \subset \{x : |x| \leq 2\}$$

and

$$\mathrm{supp}\varphi_j \in \{x : 2^{j-1} \leq |x| \leq 2^{j+1}\}$$

for $j \geq 1$, we have

$$\mathrm{supp}[b, T_j]f \subset \{x : 2^{j-1} - 1 \leq |x| \leq 2^{j+1} + 1\}$$

and

$$\mathrm{supp}[b, T_0]f \subset \{x : |x| \leq 3\}.$$

Since

$$\|[b, B^\alpha]f\|_{L^p(\mathbb{R}^n)} \leq C < +\infty,$$

we conclude that

$$\|[b, B^\alpha]f\|^p_{L^p(\mathbb{R}^n)} = \int_{\mathbb{R}^n} \left|\sum_{j=0}^{\infty} [b, T_j]f(x)\right|^p dx$$

$$= \int_{|x|<4} \left|\sum_{j=0}^{\infty} [b, T_j]f(x)\right|^p dx$$

$$+ \sum_{i=2}^{\infty} \int_{2^i \leq |x| < 2^{i+1}} \left|\sum_{j=0}^{\infty} [b, T_j]f(x)\right|^p dx$$

$$= \int_{|x|<4} \left|\sum_{j=0}^{3}[b,T_j]f(x)\right|^p dx$$

$$+ \sum_{i=2}^{\infty}\int_{2^i\leq|x|<2^{i+1}} \left|\sum_{j=i-1}^{i+2}[b,T_j]f(x)\right|^p dx.$$

It follows that

$$\sum_{i\geq 6}\int_{2^i\leq|x|<2^{i+1}} \left|\sum_{j=i-1}^{i+2}[b,T_j]f(x)\right|^p dx \leq C < +\infty. \qquad (2.11.2)$$

Set

$$\phi_i(x) = \sum_{j=i-1}^{i+2} \varphi_j(x).$$

We have

$$\operatorname{supp}\phi_i \subset \left\{x : 2^{i-2} \leq |x| \leq 2^{i+3}\right\}.$$

Noting that

$$J_\gamma(t) = \left(\frac{2}{t\pi}\right)^{\frac{1}{2}} \cos\left(t - \frac{\pi\gamma}{2} - \frac{\pi}{4}\right) + r(t), \qquad (2.11.3)$$

as $t \to +\infty$, and

$$r(t) = O(t^{-\frac{3}{2}}), \qquad (2.11.4)$$

as $t \to +\infty$, respectively, we conclude that

$$\sum_{j=i-1}^{i+2}[b,T_j]f(x) = \left(\frac{2}{\pi}\right)^{\frac{1}{2}} C_\alpha \int_{|x-y|\geq 1} \cos\left(|x-y| - \frac{\pi(n+1+2\alpha)}{4}\right)$$

$$\times \frac{\phi_i(x-y)}{|x-y|^{\frac{n+1}{2}+\alpha}}[b(x)-b(y)]f(y)dy$$

$$+ C_\alpha \int_{|x-y|\geq 1} \frac{r(|x-y|)\phi_i(x-y)}{|x-y|^{\frac{n}{2}+\alpha}}[b(x)-b(y)]f(y)dy$$

$$+ C_\alpha \int_{|x-y|\leq 1} \frac{J_{\frac{n}{2}+\alpha}(|x-y|)\phi_i(x-y)}{|x-y|^{\frac{n}{2}+\alpha}}[b(x)-b(y)]f(y)dy$$

$$=: P_i f(x) + Q_i f(x) + R_i f(x).$$

For each $i \geq 6$, we have $R_i f(x) = 0$.

2.11 Commutator of Bochner-Riesz operator

Set
$$m_i(b) = C2^{-(i+1)n} \int_{|x|\leq 2^{i+1}} b(x)dx$$

and
$$m(b) = C \int_{|x|\leq 1} b(x)dx.$$

Noting that supp $f \subset \{x : |x| \leq 1\}$, it follows that

$$|Q_i f(x)| \leq C_\alpha \int_{|x-y|\geq 1} |x-y|^{-(\frac{n+3}{2}+\alpha)} |b(x)-b(y)||\phi_i(x-y)f(y)|dy$$

$$\leq 2C_\alpha |b(x)-m_i(b)|$$

$$\times \int_{2^{i-2}\leq|x-y|\leq 2^{i+3}} |x-y|^{-(\frac{n+3}{2}+\alpha)}|m_i(b)-b(y)|f(y)dy$$

$$\leq 2C_\alpha 2^{-(i-2)(\frac{n+3}{2}+\alpha)}|b(x)-m_i(b)| \int_{|y|\leq 1} |m_i(b)-b(y)|dy$$

$$\leq C_\alpha 2^{-i(\frac{n+3}{2}+\alpha)}|b(x)-m_i(b)||m_i(b)-m(b)|$$

$$+ C_\alpha 2^{-i(\frac{n+3}{2}+\alpha)}|b(x)-m_i(b)| \int_{|y|\leq 1} |m(b)-b(y)|dy$$

$$\leq C_\alpha 2^{-i(\frac{n+3}{2}+\alpha)} 2^n i \|b\|_{\mathrm{BMO}} |b(x)-m_i(b)|$$

$$\leq C_{\alpha,n} 2^{-i(\frac{n+3}{2}+\alpha)} 2^{\frac{i}{2}} \|b\|_{\mathrm{BMO}} |b(x)-m_i(b)|.$$

Therefore, we obtain

$$\int_{2^i \leq |x| < 2^{i+1}} |Q_i f(x) + R_i f(x)|^p dx$$

$$\leq C_{\alpha,n}^p 2^{-ip(\frac{n+2}{2}+\alpha)} \|b\|_{\mathrm{BMO}}^p \int_{|x|<2^{i+1}} |b(x)-m_i(b)|^p dx$$

$$\leq C_1 2^{-i[(\frac{n+2}{2}+\alpha)p-n]}. \tag{2.11.5}$$

Let
$$I_k = \left[2k\pi - \frac{\pi}{3} + \frac{\pi(n+1+2\alpha)}{4} + 1, 2k\pi + \frac{\pi}{3} + \frac{\pi(n+1+2\alpha)}{4} - 1\right]$$

and
$$A_k = \left[2k\pi + \frac{\pi}{3} + \frac{\pi(n+1+2\alpha)}{4} - 1, 2(k+1)\pi - \frac{\pi}{3} + \frac{\pi(n+1+2\alpha)}{4} + 1\right].$$

Since
$$|I_k| + |A_k| = 2\pi < \frac{2^i - 2}{3},$$
for $i \geq 6$, we have
$$M_i = \{k \in \mathbb{N} : I_k \subset [2^i + 1, 2^{i+1} - 1]\} \neq \emptyset,$$
for $i \geq 6$.

From the fact $|I_k| = C|A_k|$ with the constant C being independent of k, it follows that

$$\int_{|x| \in A_k} dx = C \int_{A_k} r^{n-1} dr$$
$$\leq C \left(2(k+1)\pi - \frac{\pi}{3} + \frac{\pi(n+1+2\alpha)}{4} + 1 \right)^{n-1} |A_k|$$
$$\leq C 2^{n-1} \left(2k\pi - \frac{\pi}{3} + \frac{\pi(n+1+2\alpha)}{4} + 1 \right)^{n-1} |I_k|$$
$$\leq C \int_{I_k} r^{n-1} dr$$
$$= C \int_{|x| \in I_k} dx. \tag{2.11.6}$$

Similar to the estimate above, we have

$$\int_{|x| \in I_{k-1}} dx \leq \int_{|x| \in I_k} dx \leq C \int_{|x| \in I_{k-1}} dx, \tag{2.11.7}$$

where C is independent of k.

For each $|x| \in I_k$ for $k \in M_i$, $|x - y| \leq 1$ easily implies
$$2^i \leq |x| - 1 \leq |y| \leq |x| + 1 \leq 2^{i+1},$$
$$2k\pi - \frac{\pi}{3} \leq |y| - \frac{\pi(n+1+2\alpha)}{4} \leq 2k\pi + \frac{\pi}{3},$$
and
$$\log |x| - \log |x - y| = \log \frac{|x|}{|x - y|} \geq 1.$$

Furthermore, it follows from the definition ϕ_i that
$$\phi_i(y) = 1$$
for $2^i \leq |y| \leq 2^{i+1}$.

2.11 Commutator of Bochner-Riesz operator

Hence, for $|x| \in I_k$, and noting that $\mathrm{supp} f \subset \{y : |y| \leq 1\}$, we conclude that

$$P_i f(x) = \left(\frac{2}{\pi}\right)^{\frac{1}{2}} C_\alpha \int_{|x-y|\geq 1} \cos\left(|x-y| - \frac{\pi(n+1+2\alpha)}{4}\right)$$
$$\times \frac{\phi_i(x-y)}{|x-y|^{\frac{n+1}{2}+\alpha}} [b(x) - b(y)] f(y) dy$$

$$= \left(\frac{2}{\pi}\right)^{\frac{1}{2}} C_\alpha \int_{|y|\leq 1} \cos\left(|x-y| - \frac{\pi(n+1+2\alpha)}{4}\right)$$
$$\times \frac{\phi_i(x-y)}{|x-y|^{\frac{n+1}{2}+\alpha}} [b(x) - b(y)] f(y) dy$$

$$= \left(\frac{2}{\pi}\right)^{\frac{1}{2}} C_\alpha \int_{|x-y|\leq 1} \cos\left(|y| - \frac{\pi(n+1+2\alpha)}{4}\right)$$
$$\times \frac{\phi_i(y)}{|y|^{\frac{n+1}{2}+\alpha}} [\log|x| - \log|x-y|] f(x-y) dy$$

$$\geq \left(\frac{1}{2\pi}\right)^{\frac{1}{2}} C_\alpha 2^{-i(\frac{n+1}{2}+\alpha)} \int_{|x-y|\leq 1} (\log|x| - \log|x-y|) f(x-y) dy$$

$$\geq 2^{-2} \left(\frac{2}{\pi}\right)^{\frac{1}{2}} C_\alpha 2^{-i(\frac{n+1}{2}+\alpha)} \int_{|x-y|\leq 1} f(x-y) dy$$

$$\geq C_2 2^{-i(\frac{n+1}{2}+\alpha)}.$$

Therefore, using (2.11.6) and (2.11.7), we have that

$$\int_{2^i \leq |x| < 2^{i+1}} |P_i f(x)|^p dx \geq \sum_{k \in M_i} C_2 2^{-pi(\frac{n+1}{2}+\alpha)} \int_{|x| \in I_k} dx$$
$$\geq C_2 2^{-pi(\frac{n+1}{2}+\alpha)} \sum_{k \in M_i} \int_{|x| \in I_k \cup A_k} dx$$
$$\geq C_2 2^{-pi(\frac{n+1}{2}+\alpha)} \int_{2^i \leq |x| \leq 2^{i+1}} dx$$
$$\geq C_2 2^{-i((\frac{n+1}{2}+\alpha)p - n)}.$$

Using (2.11.5) and (2.11.8), for $i \geq 6$, we have

$$\left(\int_{2^i \leq |x| < 2^{i+1}} \left|\sum_{j=i-1}^{i+2} [b, T_j]f(x)\right|^p dx\right)^{1/p}$$

$$= \left(\int_{2^i \leq |x| < 2^{i+1}} |P_i f(x) + Q_i f(x) + R_i f(x)|^p dx\right)^{1/p}$$

$$\geq \left(\int_{2^i \leq |x| < 2^{i+1}} |P_i f(x)|^p dx\right)^{1/p} - \left(\int_{2^i \leq |x| < 2^{i+1}} |Q_i f(x) + R_i f(x)|^p dx\right)^{1/p}$$

$$\geq \left(C_2 2^{-i((\frac{n+1}{2}+\alpha)p-n)}\right)^{1/p} - \left(C_1 2^{-i((\frac{n+2}{2}+\alpha)p-n)}\right)^{1/p}$$

$$= 2^{-i((\frac{n+1}{2}+\alpha)p-n)/p} \left(C_2^{1/p} - C_1^{1/p} 2^{-i/2}\right),$$

where C_1, C_2 are two constants independent of f and i.

Consequently, if (2.11.1) holds, we must have

$$\left(\frac{n+1}{2}+\alpha\right)p - n > 0,$$

which implies (2.11.1). ∎

For the sufficiency in Question 1, Hu and Lu [HL1] gave an affirmative result when $n = 2$.

Theorem 2.11.2 *Let $0 < \alpha < 1/2$ and $b \in \mathrm{BMO}(\mathbb{R}^2)$. If $4/(3+2\alpha) < p < 4/(1-2\alpha)$, then the commutator $[b, B^\alpha]$ is bounded on $L^p(\mathbb{R}^2)$ with the norm being smaller than $C(p)\|b\|_{\mathrm{BMO}}$.*

Proof. Write

$$[b, B^\alpha]f(x) = \int_{\mathbb{R}^2} (b(x) - b(y)) B^\alpha(x-y) f(y) dy,$$

where

$$B^\alpha(x) = C_\alpha \frac{J_{1+\alpha}(|x|)}{|x|^{1+\alpha}},$$

and $J_\beta(t)$ denotes the Bessel function of order β.

Noting that

$$J_\beta(t) = Ct^{-\frac{1}{2}} \cos\left(t - \frac{\pi\beta}{2} - \frac{\pi}{4}\right) + r(t),$$

2.11 Commutator of Bochner-Riesz operator

and
$$r(t) = O\left(t^{-\frac{3}{2}}\right),$$
as $t \to \infty$, we hence have that

$$\begin{aligned}
[b, B^\alpha]f(x) \\
= C\int_{|x-y|\geq 1} \cos\left(|x-y| - \frac{\pi(3+2\alpha)}{4}\right)(b(x) - b(y))\frac{f(y)}{|x-y|^{\frac{3}{2}+\alpha}}dy \\
+ C\int_{|x-y|\geq 1} r(|x-y|)(b(x) - b(y))\frac{f(y)}{|x-y|^{1+\alpha}}dy \\
+ C\int_{|x-y|\leq 1} \frac{J_{1+\alpha}(|x-y|)}{|x-y|^{1+\alpha}}(b(x) - b(y))f(y)dy \\
= Pf(x) + Qf(x) + Rf(x). \quad (2.11.8)
\end{aligned}$$

Since
$$|J_\beta(t)| \leq C_\beta|t|^\beta,$$
as $t \to 0$, it follows that

$$\left|\int_{|x-y|\leq 1} \frac{J_{1+\alpha}(|x-y|)}{|x-y|^{1+\alpha}} f(y)dy\right| \leq CMf(x),$$

where Mf denotes the Hardy-Littlewood maximal function of f. Thus by the weighted estimate for M (see[GR1]), we have

$$\|Rf\|_p \leq C\|b\|_{\text{BMO}}\|f\|_p,$$

for $1 < p < \infty$.

For $\alpha > 0$, we have

$$\left|\int_{|x-y|\geq 1} r(|x-y|)\frac{f(y)}{|x-y|^{1+\alpha}}dy\right| \leq C\sum_{k=1}^\infty \int_{2^{k-1}\leq|x-y|<2^k} \frac{|f(y)|}{|x-y|^{5/2+\alpha}}dy$$
$$\leq CMf(x).$$

This implies that
$$\|Qf\|_p \leq C\|b\|_{\text{BMO}}\|f\|_p,$$

for $1 < p < \infty$.

Obviously, the L^p norm of the operator P defined by (2.11.8) can be controlled by that of the operator $\widetilde{P}$ defined by

$$\widetilde{P}f(x) = \int_{|x-y|\geq 1} e^{i|x-y|}(b(x) - b(y))\frac{f(y)}{|x-y|^{\frac{3}{2}+\alpha}}dy.$$

Furthermore, since $\alpha < \frac{1}{2}$ and

$$\left| \int_{|x-y|<1} e^{i|x-y|} \frac{f(y)}{|x-y|^{\frac{3}{2}+\alpha}} dy \right| \leq CMf(x),$$

We have already known that

$$\left\| \int_{|x-y|<1} e^{i|x-y|} (b(x)-b(y)) \frac{f(y)}{|x-y|^{\frac{3}{2}+\alpha}} dy \right\|_p \leq C\|b\|_{\mathrm{BMO}} \|f\|_p$$

for $1 < p < \infty$.

Actually we may thus view the operator P as

$$Pf(x) = \int_{R^2} e^{i|x-y|} (b(x)-b(y)) \frac{f(y)}{|x-y|^{\frac{3}{2}+\alpha}} dy. \qquad (2.11.9)$$

By Stein's interpolation theorem (see [SW1]), to prove Theorem 2.11.2, it is enough to show that for any $0 < \alpha < \frac{1}{2}$, the operator P defined by (2.11.9) is bounded on $L^4(\mathbb{R}^2)$.

Denote $I = [0,1], I^2 = I \times I$, and $F(I^2) = [-1.5, 2.5]^2 \setminus [-0.5, 1.5]^2$. For fixed $\lambda > 0$, define

$$P^\lambda f(x) = \int_{I^2} e^{i\lambda |x-y|} \frac{f(y)}{|x-y|^{\frac{3}{2}+\alpha}} dy$$

and the corresponding commutator

$$P_b^\lambda f(x) = \int_{I^2} e^{i\lambda |x-y|} (b(x)-b(y)) \frac{f(y)}{|x-y|^{\frac{3}{2}+\alpha}} dy.$$

Set

$$S^\lambda f(x) = \lambda^{\frac{1}{2}-\alpha} P^\lambda f(x)$$

and

$$S_b^\lambda f(x) = \lambda^{\frac{1}{2}-\alpha} P_b^\lambda f(x).$$

Note that if $b \in \mathrm{BMO}(\mathbb{R}^n)$, then we can easily have

$$b(t(\cdot)) \in \mathrm{BMO}(\mathbb{R}^n)$$

and

$$\|b(t(\cdot))\|_{\mathrm{BMO}} = \|b\|_{\mathrm{BMO}}$$

for any $t > 0$.

2.11 Commutator of Bochner-Riesz operator

By the same argument as in Section 2.6, we can conclude that the proof of Theorem 2.11.2 can be reduced to the following inequality

$$\left\|S_b^\lambda f\right\|_{L^4(F(I^2))} \le C\lambda^{-\delta}\|b\|_{\text{BMO}}\|f\|_{L^4(I^2)} \qquad (2.11.10)$$

for some positive constant $\delta > 0$.

Now we prove (2.11.10). Let s be a small positive constant which will be determined later. Set $0 < r < \frac{1}{2}$ and $\sigma > 0$ such that

$$\frac{1}{4+\sigma} = \frac{1}{4} - \frac{r}{2}.$$

Observe that if $x \in F(I^2)$, then we have

$$\left|P^\lambda f(x)\right| \le C \int_{I^2} |f(y)|dy$$
$$\le C_r \int_{R^2} \frac{1}{|x-y|^{2-r}} |f(y)\chi_{I^2}(y)|dy$$
$$= C_r I_r(f\chi_{I^2})(x),$$

where χ_{I^2} is the characteristic function of I^2, and I_r is the usual fractional integral operator of order r. By the Hardy-Littlewood-Sobolev theorem, it follows that

$$\left\|S^\lambda f\right\|_{L^{4+\sigma}(F(I^2))} \le C\lambda^{\frac{1}{2}-\alpha}\left\|P^\lambda f\right\|_{L^{4+\sigma}(R^2)} \le C\lambda^{\frac{1}{2}-\alpha}\|f\|_{L^4(I^2)}. \qquad (2.11.11)$$

Similarly, if σ is small enough, we have

$$\left\|S^\lambda f\right\|_{L^4(F(I^2))} \le C\lambda^{\frac{1}{2}-\alpha}\|f\|_{L^{4-\sigma}(I^2)}. \qquad (2.11.12)$$

By the key estimate used in [CS1], we have

$$\left\|S^\lambda f\right\|_{L^4(F(I^2))} \le C\lambda^{-\varepsilon}\|f\|_{L^4(I^2)}, \qquad (2.11.13)$$

where $\varepsilon > 0$. An interpolation between the inequalities (2.11.11) and (2.11.13) yields

$$\left\|S^\lambda f\right\|_{L^{4+s\sigma}(F(I^2))} \le C\lambda^{-\varepsilon+(1/2-\alpha+\varepsilon)s}\|f\|_{L^4(I^2)} \qquad (2.11.14)$$

with $0 < s < 1$.

On the other hand, interpolation between the inequalities (2.11.12) and (2.11.13) gives

$$\left\|S^\lambda f\right\|_{L^4(F(I^2))} \leq C\lambda^{-\varepsilon+(1/2-\alpha+\varepsilon)s}\|f\|_{L^{4-s\sigma}(I^2)}. \quad (2.11.15)$$

We can also get by the inequalities (2.11.14) and (2.11.15) that

$$\left\|S^\lambda f\right\|_{L^{4+s^2\sigma}(F(I^2))} \leq C\lambda^{-\varepsilon+(1/2-\alpha+\varepsilon)s}\|f\|_{L^{4-s^2\sigma}(I^2)}. \quad (2.11.16)$$

Let $\phi(x) \in C_0^\infty(\mathbb{R}^2)$ such that $\phi(x) = 1$ if $|x| \leq 50$ and

$$\operatorname{supp}\phi \subset \{x : |x| \leq 100\}.$$

Denote
$$\widetilde{b}(y) = [b(y) - m_{10I^2}(b)]\phi(y),$$

where $m_{10I^2}(b)$ denotes the mean value of b on $10I^2$. Obviously, if $x \in F(I^2)$, then

$$S_b^\lambda f(x) = \widetilde{b}(x)S^\lambda f(x) + S^\lambda\left(\widetilde{b}f\right)(x) = \mathrm{I}(x) + \mathrm{II}(x).$$

For the first term, we have

$$\|\mathrm{I}\|_{L^4(F(I^2))} \leq \left\|\widetilde{b}\right\|_{L^q(\mathbb{R}^2)}\left\|S^\lambda f\right\|_{L^{4+s\sigma}(F(I^2))}$$
$$\leq C(\sigma,s)\|b\|_{\mathrm{BMO}}\lambda^{-\varepsilon+(1/2-\alpha+\varepsilon)s}\|f\|_{L^4(I^2)},$$

where $1/q = 1/4 - 1/(4+s\sigma)$, and the second inequality follows from the inequality (2.11.14) and the fact

$$\left\|\widetilde{b}\right\|_{L^q(\mathbb{R}^2)} \leq C(s,\sigma)\|b\|_{\mathrm{BMO}}.$$

The estimate for the second term follows from the inequality (2.11.15) by

$$\|\mathrm{II}\|_{L^4(F(I^2))} \leq C\lambda^{-\varepsilon+(1/2-\alpha+\varepsilon)s}\|\widetilde{b}f\|_{L^{4-s\sigma}(I^2)}$$
$$\leq C\lambda^{-\varepsilon+(1/2-\alpha+\varepsilon)s}\|b\|_{\mathrm{BMO}}\|f\|_{L^4(I^2)}.$$

Choose s so small that $\delta = \varepsilon - (1/2 - \alpha + \varepsilon)s > 0$. Combining the estimates above we get

$$\left\|S_b^\lambda f\right\|_{L^4(F(I^2))} \leq C\lambda^{-\delta}\|b\|_{\mathrm{BMO}}\|f\|_{L^4(I^2)}.$$

This concludes the proof of the lemma. ∎

Chapter 3

BOCHNER-RIESZ MEANS OF MULTIPLE FOURIER SERIES

3.1 The case of being over the critical index

3.1.1 Bochner formula

For a locally integrable function f, we define

$$f_x(t) = \frac{1}{\omega_{n-1}} \int_{\mathbb{S}^{n-1}} f(x - t\xi) d\sigma(\xi) \qquad (3.1.1)$$

for $t \geq 0$, where

$$\omega_{n-1} = \frac{2\pi^{\frac{n}{2}}}{\Gamma\left(\frac{n}{2}\right)}.$$

Theorem 3.1.1 (Bochner) *If $f \in L(Q)$ and $\mathrm{Re}\,\alpha > \frac{n-1}{2}$, then we have*

$$S_R^\alpha(f, x) = \frac{2^{\alpha+1-\frac{n}{2}}\Gamma(\alpha+1)}{\Gamma(\frac{n}{2})} R^{\frac{n}{2}-\alpha} \int_0^\infty f_x(t) \frac{J_{\frac{n}{2}+\alpha}(Rt)}{t^{\alpha-\frac{n}{2}+1}} dt. \qquad (3.1.2)$$

Proof. Denote

$$\Phi^\alpha(x) = \begin{cases} (1-|x|^2)^\alpha & |x| < 1, \\ 0 & |x| \geq 1, \end{cases} \qquad (3.1.3)$$

$$B^\alpha(x) = \frac{2^\alpha \Gamma(\alpha+1)}{(2\pi)^{\frac{n}{2}}} \frac{J_{\frac{n}{2}+\alpha}(|x|)}{|x|^{\frac{n}{2}+\alpha}},$$

and
$$\varphi^\alpha(x) = (2\pi)^n B^\alpha(x) = 2^{\alpha+\frac{n}{2}} \pi^{\frac{n}{2}} \Gamma(\alpha+1) \frac{J_{\frac{n}{2}+\alpha}(|x|)}{|x|^{\frac{n}{2}+\alpha}}. \quad (3.1.4)$$

At the end of Section 1.4, we have mentioned that, if $\operatorname{Re}\alpha > \frac{n-1}{2}$, then we have
$$\widehat{\varphi^\alpha} = \Phi,$$
which satisfies the conditions of Theorem 1.4.1 in Chapter 1. Hence, by (1.4.3) and (1.4.4) in Chapter 1, we obtain that

$$\begin{aligned} S_R^\alpha(f;x) &= \sum_{m\in\mathbb{Z}^n} \Phi^\alpha\left(\frac{m}{R}\right) C_m(f) e^{im\cdot x} \\ &= f * \left(\sum \varphi^\alpha_{\frac{1}{R}}(y+2\pi m)\right)(x) \\ &= \frac{1}{(2\pi)^n} \int_Q f(x-y) \sum_{m\in\mathbb{Z}^n} \varphi^\alpha_{\frac{1}{R}}(y+2\pi m) dy \\ &= \frac{1}{(2\pi)^n} \sum_{m\in\mathbb{Z}^n} \int_{Q+2\pi m} f(x-y)\varphi^\alpha_{\frac{1}{R}}(y) dy \\ &= \frac{1}{(2\pi)^n} \int_{\mathbb{R}^n} f(x-y)\varphi^\alpha_{\frac{1}{R}}(y) dy \\ &= \frac{1}{(2\pi)^n} \int_0^\infty \left[\int_{\mathbb{S}^{n-1}} f(x-t\xi)\varphi^\alpha_{\frac{1}{R}}(t\xi) d\sigma(\xi)\right] t^{n-1} dt. \end{aligned}$$

Substitute (3.1.4) into $\varphi^\alpha_{\frac{1}{R}}(u) = R^n \varphi^\alpha(Ru)$, and then it follows

$$\begin{aligned} S_R^\alpha(f;x) &= \frac{1}{(2\pi)^n} 2^{\alpha+\frac{n}{2}} \pi^{\frac{n}{2}} \Gamma(\alpha+1)\omega_n \int_0^\infty f_x(t) R^n t^{n-1} \frac{J_{\frac{n}{2}+\alpha}(Rt)}{(Rt)^{\frac{n}{2}+\alpha}} dt \\ &= 2^{\alpha+1-\frac{n}{2}} \frac{\Gamma(\alpha+1)}{\Gamma(\frac{n}{2})} \int_0^\infty f_x(t) \frac{J_{\frac{n}{2}+\alpha}(Rt)}{t^{\alpha-\frac{n}{2}+1}} dt R^{\frac{n}{2}-\alpha}. \end{aligned}$$

■

3.1.2 The localization theorem

Now we formulate the localization theorem.

Theorem 3.1.2 *Suppose that $f \in L(Q)$, and vanishes at the ball $B(0,\varepsilon) = \{x \in \mathbb{R}^n : |x| < \varepsilon\}$ for some $\varepsilon > 0$. If $\operatorname{Re}\alpha > \frac{n-1}{2}$, then we have*

$$S_R^\alpha(f;0) = O\left(R^{\frac{n-1}{2}-\operatorname{Re}\alpha}\right) \quad (3.1.5)$$

3.1 The case of being over the critical index

Proof. By (3.1.2), we have

$$S_R^\alpha(f;0) = CR^{\frac{n}{2}-\alpha} \int_\varepsilon^\infty f_0(t) \frac{J_{\frac{n}{2}+\alpha}(Rt)}{t^{\alpha-\frac{n}{2}+1}} dt.$$

Using the asymptotic formula

$$\left|J_{\frac{n}{2}+\alpha}(Rt)\right| \le \frac{C}{\sqrt{Rt}},$$

for $t > \frac{1}{R}$, when $R > \frac{1}{\varepsilon}$, we can obtain

$$|S_R^\alpha(f;0)| \le CR^{\frac{n}{2}-\mathrm{Re}\alpha} \int_\varepsilon^\infty \frac{|f_0(t)|}{\sqrt{R}\, t^{\mathrm{Re}\alpha-\frac{n-1}{2}+1}} dt.$$

Denote

$$\beta = \mathrm{Re}\alpha - \frac{n-1}{2} > 0$$

and

$$|f_0(t)|t^{n-1} = g(t).$$

We have

$$|S_R^\alpha(f;0)| \le CR^{-\beta} \int_\varepsilon^\infty \frac{g(t)}{t^{\beta+n}} dt. \tag{3.1.6}$$

Since it follows that

$$\int_\varepsilon^\infty \frac{g(t)}{t^{\beta+n}} dt = \lim_{A \to +\infty} \int_\varepsilon^A \frac{g(t)}{t^{\beta+n}} dt$$

$$= \lim_{A \to +\infty} \left\{ \frac{1}{t^{\beta+n}} \int_\varepsilon^t g(\tau)d\tau \bigg|_\varepsilon^A + (\beta+n) \int_\varepsilon^A \frac{\int_\varepsilon^t g(\tau)d\tau}{t^{\beta+n+1}} dt \right\}$$

$$= (\beta+n) \int_\varepsilon^\infty \frac{O(t^n)}{t^{\beta+n+1}} dt$$

$$= O(1),$$

we have

$$S_R^\alpha(f;0) = O(R^{-\beta})$$

as $R \to \infty$, which completes the proof of Theorem 3.1.2. ∎

Remark 3.1.1 *Here, the symbol 'O' depends on β. As $\beta \to 0^+$, 'O' increases as $\frac{1}{\beta}$.*

3.1.3 The maximal operator S_*^α

Definition 3.1.1 *Let $f \in L(Q)$. Define*

$$S_*^\alpha(f)(x) = \sup_{R>0} |S_R^\alpha(f;x)|. \qquad (3.1.7)$$

We denote by Mf the Hardy-Littlewood maximal operator of a locally integrable function f as

$$Mf(x) = \sup_{r>0} \frac{1}{r^n} \int_{|y|<r} |f(x-y)|dy. \qquad (3.1.8)$$

Theorem 3.1.3 *If $\operatorname{Re}\alpha > \frac{n-1}{2}$ and $f \in L(Q)$, the following inequality*

$$S_*^\alpha(f)(x) \leq C\left(1 + \frac{1}{\operatorname{Re}\alpha - \frac{n-1}{2}}\right) Mf(x) \qquad (3.1.9)$$

holds, where the positive number $C \leq |2^\alpha \Gamma(\alpha+1)| e^{2\pi |\operatorname{Im}\alpha|} C_n$.

Proof. Denote again $\operatorname{Re}\alpha - \frac{n-1}{2} = \beta > 0$, $|f_x(t)|t^{n-1} = g(t)$, then

$$\int_0^r g(t)dt \leq C_n \int_{|y|<r} |f(x-y)|dy \leq C_n r^n Mf(x).$$

By the Bochner formula, we have

$$S_R^\alpha(f;x) = C_n 2^\alpha \Gamma(\alpha+1) R^{\frac{n}{2}-\alpha} \left(\int_0^{R^{-1}} + \int_{R^{-1}}^\infty\right) f_x(t) t^{n-1} \frac{J_{\frac{n}{2}+\alpha}(Rt)}{t^{\alpha+\frac{n}{2}}} dt$$

$$= I_1 + I_2.$$

Next we will estimate I_1 and I_2, respectively.
For I_1, it follows that

$$|I_1| \leq C_n |2^\alpha \Gamma(\alpha+1)| R^n \int_0^{R^{-1}} g(t)dt \leq C_n |2^\alpha \Gamma(\alpha+1)| Mf(x). \qquad (3.1.10)$$

3.1 The case of being over the critical index

On the other hand, we have

$$\left|\frac{J_{\frac{n}{2}+\alpha}(Rt)}{t^{\frac{n}{2}+\alpha}}\right| \leq \frac{1}{\sqrt{R}\, t^{\frac{n+1}{2}+\operatorname{Re}\alpha}} C_n e^{\frac{3\pi}{2}|\operatorname{Im}\alpha|},$$

for $t > R^{-1}$.

About the appearance of the constant $e^{\frac{3\pi}{2}|\operatorname{Im}\alpha|}$, which is concerned about the imaginary part of the order in the asymptotic equation of the multi-ordered Bessel function, we omit the details here. One can refer to the related contents in Watson [Wat1]. Hence

$$\begin{aligned}|I_2| &\leq C_n e^{2\pi|\operatorname{Im}\alpha|} R^{-\beta} \int_{R^{-1}}^{\infty} g(t)\frac{dt}{t^{n+\beta}} \\ &\leq C_n e^{2\pi|\operatorname{Im}\alpha|} Mf(x) \int_{R^{-1}}^{\infty} \frac{dt(n+\beta)}{t^{1+\beta} R^{\beta}} \\ &\leq C_n (1+\frac{1}{\beta}) e^{2\pi|\operatorname{Im}\alpha|} Mf(x). \end{aligned} \qquad (3.1.11)$$

Combining (3.1.10) with (3.1.11), we immediately obtain (3.1.9). The proof of Theorem 3.1.3 has been completed. ∎

Corollary 3.1.1 *If* $\operatorname{Re}\alpha > \frac{n-1}{2}$, *then the operator* S_*^α *is of type* (p,p) *for* $1 < p \leq \infty$, *and is of weak type* $(1,1)$. *More precisely, we have*

$$\|S_*^\alpha(f)\|_p \leq C_n e^{2\pi|\operatorname{Im}\alpha|} \left(\frac{1}{\operatorname{Re}\alpha - \frac{n-1}{2}}+1\right) \frac{p}{p-1}\|f\|_p, \qquad (3.1.12)$$

for $1 < p \leq \infty$, *and*

$$|\{x \in Q : S_*^\alpha(f)(x) > \lambda\}| \leq \frac{C_n}{\lambda}\|f\|_1, \qquad (3.1.13)$$

for any $\lambda > 0$.

We point out that (3.1.12) and (3.1.13) are direct consequences of Theorem 3.1.3 and the estimate of the Hardy-Littlewood maximal operator. One can refer to Section 2.3 in Stein and Weiss [SW1]. Although they discussed functions defined on $\mathbb{R}^n$, the conclusions are completely the same as that of Q.

3.2 The case of the critical index (general discussion)

3.2.1 Localization problems

For the Bochner-Riesz means at the critical index, the localization principle does not hold on $L(Q^n)$ with $n > 1$ any more. It was proven by Bochner [Bo1].

Theorem 3.2.1 (Bochner) *If $n > 1$, then there exists $f \in L(Q^n)$ which vanishes on $B(0, \delta)$ for some $\delta > 0$ such that*

$$\limsup_{R \to +\infty} S_R^{\frac{n-1}{2}}(f; 0) = +\infty.$$

Proof. By Lemma 1.3.5 in Chapter 1, for almost every y, $D_R^{\frac{n-1}{2}}(y)$ is unbounded about R. Define

$$L_\delta = \{f \in L(Q^n) : f \text{ vanishes at } B(0, \delta)\}.$$

Let $\delta < 1$. Obviously L_δ is a closed subspace of $L(Q^n)$. Define a linear functional F_R $(R > 0)$ on L_δ,

$$F_R(f) = S_R^{\frac{n-1}{2}}(f; 0)$$

for $f \in L_\delta$. It follows that

$$\|F_R\| = \sup_{f \in L_\delta, \|f\|_1 = 1} |F_R(f)|$$

$$= \sup_{f \in L_\delta, \|f\|_1 = 1} \left| \frac{1}{(2\pi)^n} \int_{Q^n} f(y) D_R^{\frac{n-1}{2}}(-y) dy \right|$$

$$= \frac{1}{(2\pi)^n} \sup_{y \in Q^n \setminus B(0, \delta)} \left| D_R^{\frac{n-1}{2}}(-y) \right|.$$

Since $\{\|F_R\| : R > 0\}$ is unbounded, it implies from the uniform boundedness theorem that there exists a $f \in L_\delta$ such that $\{F_R(f) : R > 0\}$ is unbounded. This finishes the proof of Theorem 3.2.1. ∎

3.2 The case of the critical index (general discussion)

3.2.2 An example of being divergent almost everywhere

Theorem 3.2.2 (Stein) *There exists $f \in L(Q^n)$ $(n > 1)$, such that*

$$\limsup_{R \to +\infty} \left| S_R^{\frac{n-1}{2}} (f;x) \right| = +\infty$$

for a.e. $x \in Q^n$, and f can be constructed so that it is supported in an arbitrary small given neighborhood of the origin. It should be pointed out that the support discussed here is restricted in Q^n.

Proof. Denote $B(0,1) = \{x \in \mathbb{R}^n : |x| \leq 1\}$. Choose a function ψ, satisfying

$$\operatorname{supp}\psi \subset B, \ \psi \in C^\infty(\mathbb{R}^n), \ \frac{1}{(2\pi)^n} \int_{\mathbb{R}^n} \psi(x) dx = 1. \qquad (3.2.1)$$

By (3.2.1), it follows from the properties of the the Fourier transform of ψ that

(i) $\hat{\psi}(0) = 1$;

(ii) $\left|\hat{\psi}(y)\right| \leq |y|^{-2k} \left\|\Delta^k \psi\right\|_{L(\mathbb{R}^n)}$ for all $k \in \mathbb{N}$ and $y \neq 0$,

where Δ is the Laplace operator and $\Delta^k = \Delta(\Delta^{k-1})$.

Choose $\varepsilon \in (0,1)$. Define

$$\psi_\varepsilon(x) = \varepsilon^{-n} \psi(\varepsilon^{-1} x).$$

The periodization of ψ_ε is

$$\varphi_\varepsilon(x) := \sum_{m \in \mathbb{Z}^n} \psi_\varepsilon(x + 2\pi m).$$

Since $\operatorname{supp}\psi_\varepsilon = \varepsilon B \subset Q^n$, then φ_ε is exactly the 2π-periodic extension of ψ_ε when restricted on Q^n. It follows that

$$\|\varphi_\varepsilon\|_C = \|\psi_\varepsilon\|_C = \varepsilon^{-n} \|\psi\|_C.$$

By the Poisson summation formula (see Section 1.2),

$$\sigma(\varphi_\varepsilon)(x) \sim \sum \widehat{\psi_\varepsilon}(m) e^{im \cdot x} = \sum \hat{\psi}(\varepsilon m) e^{im \cdot x}. \qquad (3.2.2)$$

Since we have

$$\sum |\widehat{\psi}(\varepsilon m)| = \left(\sum_{|m|<\varepsilon^{-1}} + \sum_{|m|\geq \varepsilon^{-1}}\right) |\widehat{\psi}(\varepsilon m)|$$

$$\leq \|\psi\|_{L(\mathbb{R}^n)} \sum_{|m|<\varepsilon^{-1}} 1 + \|\Delta^n \psi\|_{L(\mathbb{R}^n)} \sum_{|m|\geq \varepsilon^{-1}} \frac{1}{|\varepsilon m|^{2n}}$$

$$\leq C\varepsilon^{-n},$$

Hence, it follows that

$$\left\| S_R^{\frac{n-1}{2}} (\varphi_\varepsilon) \right\|_C \leq C\varepsilon^{-n}. \tag{3.2.3}$$

And it is obvious that

$$\lim_{\varepsilon \to 0^+} S_R^{\frac{n-1}{2}} (\varphi_\varepsilon; x) = D_R^{\frac{n-1}{2}} (x) \tag{3.2.4}$$

is valid for (x, R) on $Q \times (0, R_0)$ uniformly for any $R_0 > 0$.

By Lemma 1.3.5 in Chapter 1, as long as $x^0 \in \mathscr{S}$, we have

$$\limsup_{R \to \infty} \left| D_R^{\frac{n-1}{2}} (x^0) \right| = +\infty.$$

The condition of $x^0 \in S$ actually implies that $\{|x^0 + 2\pi m| : m \in \mathbb{Z}^n\}$ is a linearly independent set (see Definitions 1.3.1 and 1.3.2 in Chapter 1), while S is a full measurable set, when $n > 1$.

For $k, l \in \mathbb{N}$, define

$$E_{k,l} = \left\{ x \in S \cap Q : \sup_{0<R<k} \left| D_R^{\frac{n-1}{2}} (x) \right| > l \right\}.$$

Then it is evident that $E_{k,l} \subset E_{k+1,l}$ and $\bigcup_{k=1}^\infty E_{k,l} \subset S \cap Q$. Thus

$$\lim_{k \to \infty} |E_{k,l}| = \left| \bigcup_{k=1}^\infty E_{k,l} \right| = |S \cap Q| = (2\pi)^n,$$

for all $l \in \mathbb{N}$.

According to the above argument, for any $A > 0$, $m \in \mathbb{N}$, choose $l \geq A+1$ and $k = k(A, m, l)$ large enough such that

$$|E_{k,l}| > \left(1 - \frac{1}{m+1}\right) |Q|.$$

3.2 The case of the critical index (general discussion)

Then, if $x \in E_{k,l}$, we have

$$\sup_{0<R<k} \left| D_R^{\frac{n-1}{2}}(x) \right| > A + 1.$$

Thus, according to (3.2.4), there exists $\varepsilon_0 = \varepsilon_0(k) > 0$, such that if $x \in E_{k,l}$ and $\varepsilon \in (0, \varepsilon_0]$,

$$\sup_{0<R<k} \left| S_R^{\frac{n-1}{2}}(\varphi_\varepsilon; x) \right| > A. \tag{3.2.5}$$

For any $\varepsilon, \delta : 0 < \varepsilon < \delta < 1$, we have

$$|\widehat{\psi}(\varepsilon m) - \widehat{\psi}(\delta m)| \leq \frac{1}{(2\pi)^n} \int_{\mathbb{R}^n} |\psi(x)||e^{-im\varepsilon x} - e^{-im\delta x}|dx$$

$$\leq \frac{1}{|Q|}|\varepsilon - \delta||m| \int_{\mathbb{R}^n} |\psi(x)||x|dx$$

$$\leq C|m|(\delta - \varepsilon) \left\| S_R^{\frac{n-1}{2}}(\varphi_\varepsilon - \varphi_\delta) \right\|_C$$

$$\leq \sum_{|m|<R} |\widehat{\psi}(\varepsilon m) - \widehat{\psi}(\delta m)| \leq C\delta R^{n+1}.$$

Consequently, for arbitrary $R_0 > 0$, there exists $\delta = \delta(R_0) \in (0,1)$ such that if $\varepsilon', \varepsilon'' \in (0, \delta]$, then

$$\sup_{0<R<R_0} \left\| S_R^{\frac{n-1}{2}}(\varphi_\varepsilon - \varphi_\delta) \right\|_C < 1. \tag{3.2.6}$$

Finally,

$$\|\varphi_\varepsilon\|_{L(Q)} = \left\| \sum \psi_\varepsilon(x + 2\pi m) \right\|_{L(Q)} = \|\psi_\varepsilon\|_{L(\mathbb{R}^n)}$$

$$= \|\psi\|_{L(\mathbb{R}^n)} < +\infty;$$

thus for any sequence $\{\varepsilon_j\}_{j=0}^\infty$ ($\varepsilon_j > 0$),

$$\sum_{j=0}^\infty \frac{1}{2^j} \varphi_{\varepsilon_j}(x) \tag{3.2.7}$$

absolutely converges in $L(Q)$-norm.

Now we take any $\varepsilon_0 \in (0,1)$, $R_0 > 0$ and let $\delta_0 = \varepsilon_0$. By (3.2.6), there exists $\delta_1 < \varepsilon_0 = \delta_0$ such that $\forall \varepsilon \in (0, \delta_1]$,

$$\sup_{0<R<R_0} \left\| S_R^{\frac{n-1}{2}}(\varphi_\varepsilon - \varphi_{\delta_1}) \right\|_C < 1.$$

According to (3.2.3), we define

$$A_1 = \sup_{R>0} \left\| 2^{-1} S_R^{\frac{n-1}{2}}(\varphi_{\delta_1}) \right\|_C, \quad A_1 < +\infty.$$

By (3.2.5), there exists a set $E_1 \subset S \bigcap Q$, $R_1 > R_0$, $\varepsilon_1 \in (0, \delta_1)$ such that

$$\begin{cases} |E_1| > \left(1 - \dfrac{1}{2}\right)|Q|, \\ \forall\, x \in E_1,\ \sup\limits_{0<R<R_1} 2^{-1} \left| S_R^{\frac{n-1}{2}}(\varphi_{\varepsilon_1}; x) \right| > A_1 + 2. \end{cases}$$

Repeat the above steps.

By (3.2.6), there exists $\delta_2 < \varepsilon_1 < \delta_1$ such that for each $\varepsilon \in (0, \delta_2]$,

$$\sup_{0<R<R_1} \left\| S_R^{\frac{n-1}{2}}(\varphi_\varepsilon - \varphi_{\delta_2}) \right\|_C < 1.$$

By (3.2.3), set

$$A_2 = \sup_{R>0} \left\| S_R^{\frac{n-1}{2}}\left(\sum_{j=0}^{1} 2^{-j}(\varphi_{\varepsilon_j} - \varphi_{\delta_j}) - 2^{-2}\varphi_{\delta_2} \right) \right\|_C < +\infty.$$

By (3.2.5), there exists a set $E_2 \subset S \bigcap Q$, $R_2 > R_1$ and $\varepsilon_2 \in (0, \delta_2)$ such that

$$\begin{cases} |E_2| > \left(1 - \dfrac{1}{2+1}\right) |Q|, \\ \forall\, x \in E_2,\ \sup\limits_{0<R<R_2} \left| 2^{-2} S_R^{\frac{n-1}{2}}(\varphi_{\varepsilon_2}; x) \right| > A_2 + 3. \end{cases}$$

Repeating these steps many times, we have

$$\varepsilon_0 \geq \varepsilon_1 > \varepsilon_2 > \cdots > \varepsilon_j > 0,$$
$$\delta_0 \geq \delta_1 > \delta_2 > \cdots > \delta_j > 0,$$
$$0 < R_0 < R_1 < R_2 < \cdots < R_j < \cdots,$$

for $j \in \mathbb{N}$, and

$$A_k = \sup_{R>0} \left\| S_R^{\frac{n-1}{2}}\left(\sum_{j=0}^{k-1} 2^{-j}(\varphi_{\varepsilon_j} - \varphi_{\delta_j}) - 2^{-k}\varphi_{\delta_k} \right) \right\|_C < +\infty,$$

$$E_k \subset S \bigcap Q,$$

3.2 The case of the critical index (general discussion)

for $k = 1, 2, 3, \ldots$, such that

$$|E_k| \geq \left(1 - \frac{1}{k+1}\right)|Q|$$

and

$$\sup_{0<R<R_{k-1}} \left\| S_R^{\frac{n-1}{2}} (\varphi_{\varepsilon_k} - \varphi_{\delta_k}) \right\|_C < 1$$

for $k \in \mathbb{N}$, and for any $x \in E_k$ for $k \in \mathbb{N}$,

$$\sup_{0<R<R_k} \left| 2^{-k} S_R^{\frac{n-1}{2}} (\varphi_{\varepsilon_k}; x) \right| > A_k + k + 1.$$

By (3.2.7), define

$$f = \sum_{j=1}^{\infty} \frac{1}{2^j} (\varphi_{\varepsilon_j} - \varphi_{\delta_j})$$

in the sense of $L(Q)$ norm.

Restricted to Q, we can see

$$\operatorname{supp} f \subset \varepsilon_0 B,$$

and for any $k \in \mathbb{N}$, and

$$S_R^{\frac{n-1}{2}}(f) = S_R^{\frac{n-1}{2}} \left(\sum_{j=0}^{k-1} 2^{-j} (\varphi_{\varepsilon_j} - \varphi_{\delta_j}) - 2^{-k} \varphi_{\delta_k} \right)$$

$$+ 2^{-k} S_R^{\frac{n-1}{2}} (\varphi_{\varepsilon_k}) + \sum_{j=k+1}^{\infty} 2^{-j} S_R^{\frac{m-1}{2}} (\varphi_{\varepsilon_j} - \varphi_{\delta_j}).$$

When $x \in E_k$, we have

$$\sup_{0<R<R_k} \left| S_R^{\frac{n-1}{2}}(f; x) \right| \geq \sup_{R \in (0, R_k)} \left| 2^{-k} S_R^{\frac{n-1}{2}} (\varphi_{\varepsilon_k}; x) \right| - A_k$$

$$- \sum_{j=k+1}^{\infty} 2^{-j} \sup_{0<R<R_k \leq R_{j-1}} \left\| S_R^{\frac{n-1}{2}} (\varphi_{\varepsilon_j} - \varphi_{\delta_j}) \right\|_C$$

$$> k.$$

It immediately follows that

$$\sup_{R>0} \left| S_R^{\frac{n-1}{2}} (f; x) \right| = +\infty,$$

for
$$x \in \bigcap_{l=1}^{\infty}\bigcup_{k=l}^{\infty} E_k.$$

However, we have
$$\left|\bigcap_{l=1}^{\infty}\bigcup_{k=l}^{\infty} E_k\right| = \lim_{l\to\infty}\left|\bigcup_{k=l}^{\infty} E_k\right| = |Q|.$$

This completes the proof of Theorem 3.2.2. ∎

Theorem 3.2.2 can be viewed as the multi-dimensional analogy of the classic Kolmogorov Theorem in the case of one variable where we draw an analogy between $S_R^{\frac{n-1}{2}}$ and partial Fourier sum of one variable. Many facts show that such an analogy is quite appropriate.

We have seen that the conclusion of Theorem 3.2.2 is stronger than that of Theorem 3.2.1, and the former gave such a function which is integrable on Q and vanishes at some neighborhood of a point, where $S_R^{\frac{n-1}{2}}$ is unbounded almost everywhere.

3.2.3 The relation between the series and the integral

Let $f \in L(Q^n)$. Define $\tilde{f} = f\chi_Q$, then $\tilde{f}$ defined on $\mathbb{R}^n$ is supported in Q^n, $\tilde{f} \in L(\mathbb{R}^n)$. Denote $B_R^\alpha(\tilde{f};x)$ as the Bochner-Riesz means of the Fourier integral of $\tilde{f}$. Stein [St2] established the following important theorem, which joined $S_R^{\frac{n-1}{2}}(f)$ with the convergence property of $B_R^{\frac{n-1}{2}}(f;x)$ which gave a new way in the research of $S_R^{\frac{n-1}{2}}(f)$. Compared with the case of the series, lots of problems are much easily handled with after they are transfered into the Fourier integral.

Theorem 3.2.3 (Stein) *Suppose that $f \in L\log^+ L(Q^n)$ with $n > 1$, and G is a closed set in Q^n. Then we have*
$$\lim_{R\to\infty}\max_{x\in G}\left|S_R^{\frac{n-1}{2}}(f;x) - B_R^{\frac{n-1}{2}}(\tilde{f};x)\right| = 0.$$

To prove Theorem 3.2.3, some lemmas are needed. Firstly, set
$$\mathscr{D} = \{\sigma + i\tau : 0 \leq \sigma \leq 1,\ -\infty < \tau < +\infty\}.$$

3.2 The case of the critical index (general discussion)

For $z \in \mathscr{D}$ and $R > 0$, define

$$D_R^{\frac{n-1}{2}+z}(u) = \sum_{|m|<R} \left(1 - \frac{|m|^2}{R^2}\right)^{\frac{n-1}{2}+z} e^{imu},$$

$$H_R^{\frac{n-1}{2}+z}(u) = \int_{|\xi|<R} \left(1 - \frac{|\xi|^2}{R^2}\right)^{\frac{n-1}{2}+z} e^{i\xi u} d\xi,$$

and

$$\triangle_R^{\frac{n-1}{2}+z}(u) = D_R^{\frac{n-1}{2}+z}(u) - H_R^{\frac{n-1}{2}+z}(u),$$

for $u \in Q^n$.

Obviously, we have that

$$\left| D_R^{\frac{n-1}{2}+z} \right| \leq \sum_{|m|<R} \left(1 - \frac{|m|^2}{R^2}\right)^{\frac{n-1}{2}+\sigma} \leq CR^n, \qquad (3.2.8)$$

and

$$\left| H_R^{\frac{n-1}{2}+z} \right| \leq CR^n. \qquad (3.2.9)$$

In the following, we set $R > 1$ and $q > 2$. We define the linear operator

$$T_z : L(Q^n) \to C(Q^n)$$

as

$$T_z(f)(x) = R^{\frac{1}{2}(z-1+\frac{2}{q})} \int_Q f(x-y) \triangle_R^{\frac{n-1}{2}+\frac{1}{2}(z-1+\frac{2}{q})}(y) dy.$$

For any $f, g \in L(Q)$, the function

$$F(z) = \int_Q T_z(f)(x) g(x) dx$$

is analytic on $\mathscr{D}$, and

$$|T_z(f)(x)| \leq R^{\frac{1}{q}} \int_Q |f(x-y)| CR^n dy \leq CR^{n+\frac{1}{2}} \|f\|_{L(Q)},$$

$$|F(z)| \leq CR^{n+\frac{1}{2}} \|f\|_{L(Q)} \|g\|_{L(Q)} \text{ (independent of } z\text{)}.$$

We can see that $\{T_z\}_{z \in \mathscr{D}}$ is an allowed family (see [SW1]).

As the operator from $L^r(Q)$ with $1 \le r < \infty$ to $L^\infty(Q)$, the norm of T_z is

$$\|T_z\|_{r,\infty} = \sup_{\|f\|_r=1} \|T_z(f)\|_\infty$$

$$= \sup_{\|f\|_r=1} \sup_{x \in Q} \left| \int_Q f(x-y) \cdot \triangle_R^{\frac{n-1}{2}+\frac{1}{2}(z-1+\frac{2}{q})}(y) dy \right| \left| R^{\frac{1}{2}(z-1+\frac{2}{q})} \right|$$

$$= R^{\frac{1}{2}(\sigma-1+\frac{2}{q})} \left\| \triangle_R^{\frac{n-1}{2}+\frac{1}{2}(z-1+\frac{2}{q})} \right\|_{r'},$$

where $\sigma = \operatorname{Re} z \in [0,1]$, $r' = \frac{r}{r-1}$, and $r' = +\infty$, if $r=1$. Then we have

$$\|T_{i\tau}\|_{2,\infty} = R^{\frac{1}{2}(-1+\frac{2}{q})} \left\| \triangle_R^{\frac{n-1}{2}+\frac{1}{2}(-1+\frac{2}{q})+i\frac{\tau}{2}} \right\|_2, \qquad (3.2.10)$$

and

$$\|T_{1+i\tau}\|_{1,\infty} = R^{\frac{1}{q}} \left\| \triangle_R^{\frac{n-1}{2}+\frac{1}{q}+i\frac{\tau}{2}} \right\|_\infty. \qquad (3.2.11)$$

With proper estimates of the two quantities of (3.2.10) and (3.2.11) as well as the complex interpolation theorem of Stein (see Stein and Weiss [SW1]), we can interpolate at $t = 1 - \frac{2}{q}$ to get an appropriate estimate of the quantity

$$\|T_t\|_{q',\infty} \le \left\| \triangle_R^{\frac{n-1}{2}} \right\|_q. \qquad (3.2.12)$$

Our purpose is to get the following inequality

$$\left\| \triangle_R^{\frac{n-1}{2}} \right\|_q \le Mq, \qquad (3.2.13)$$

for $q \ge 1$ and $R > 1$.

The following assertion holds.

Lemma 3.2.1 *If (3.2.13) holds, then Theorem 3.2.3 is valid.*

Proof. Suppose that (3.2.13) holds. Choose $a = \frac{1}{2eM}$ for $M > 0$.

$$e^{a\left| \triangle_R^{\frac{n-1}{2}}(u) \right|} = \sum_{k=0}^\infty \frac{a^k}{k!} \left| \triangle_R^{\frac{n-1}{2}}(u) \right|^k.$$

By Levi's theorem, we have

3.2 The case of the critical index (general discussion)

$$\int_Q e^{a\left|\triangle_R^{\frac{n-1}{2}}(u)\right|} du = \sum_{k=0}^{\infty} \frac{a^k}{k!} \int_Q \left|\triangle_R^{\frac{n-1}{2}}(u)\right|^k du$$

$$\leq |Q| + \sum_{k=1}^{\infty} \frac{a^k}{k!}(Mk)^k$$

$$= |Q| + \sum_{k=1}^{\infty} \frac{1}{k!}\left(\frac{k}{2e}\right)^k.$$

By the Stirling formula, we conclude that the right side of the above inequality is less than $1 + |Q|$.

On the other hand, if we write

$$\Phi(\mu) = e^\mu - \mu - 1$$

and

$$\Psi(\nu) = (\nu + 1)\log(\nu + 1)$$

for $\mu, \nu \geq 0$, then Φ, Ψ is a pair of conjugate functions in the sense of Young (see Zygmund [Zy1]). Then

$$\mu\nu \leq \Phi(\mu) + \Psi(\nu)$$

holds for every $\mu, \nu \geq 0$.

It follows that

$$\left|a\triangle_R^{\frac{n-1}{2}}(y)f(x-y)\right| \leq \Phi\left(\left|a\triangle_R^{\frac{n-1}{2}}(y)\right|\right) + \Psi(|f(x-y)|)$$

$$\leq e^{a\left|\triangle_R^{\frac{n-1}{2}}(y)\right|} + (|f(x-y)| + 1)\log(|f(x-y)| + 1).$$

Hence, we have

$$\int_Q \left|\triangle_R^{\frac{n-1}{2}}(y)f(x-y)\right| dy \leq C\left(\int_Q |f|\log^+|f|dx + 1\right). \qquad (3.2.14)$$

By the definition, we conclude that

$$S_R^{\frac{n-1}{2}}(f;x) - B_R^{\frac{n-1}{2}}(\tilde{f};x) = \frac{1}{(2\pi)^n}\int_Q f(y)D_R^{\frac{n-1}{2}}(x-y)dy$$
$$+ \frac{1}{(2\pi)^n}\int_Q f(y)H_R^{\frac{n-1}{2}}(x-y)dy$$
$$= \frac{1}{(2\pi)^n}\int_Q f(x-y)\triangle_R^{\frac{n-1}{2}}(y)dy$$
$$+ \frac{1}{(2\pi)^n}\int_Q f(x-y)H_R^{\frac{n-1}{2}}(y)dy$$
$$- \frac{1}{(2\pi)^n}\int_{\mathbb{R}^n}(f\chi_Q)(x-y)H_R^{\frac{n-1}{2}}(y)dy.$$

Let $\delta = \text{dis}(G, Q^c) > 0$. When $x \in G$, if $|y| < \delta$, then $x - y \in Q$. Thus we have

$$S_R^{\frac{n-1}{2}}(f;x) - B_R^{\frac{n-1}{2}}(\tilde{f};x) = \frac{1}{(2\pi)^n}\int_Q f(x-y)\triangle_R^{\frac{n-1}{2}}(y)dy$$
$$+ \frac{1}{(2\pi)^n}\int_{y\in Q, |y|>\delta} f(x-y)H_R^{\frac{n-1}{2}}(y)dy$$
$$- \frac{1}{(2\pi)^n}\int_{|y|>\delta}(f\chi_Q)(x-y)H_R^{\frac{n-1}{2}}(y)dy$$
$$= I_1 + I_2 + I_3. \qquad (3.2.15)$$

From Chapter 2, we have

$$H_R^\alpha(y) = (2\pi)^{\frac{n}{2}} 2^\alpha \Gamma(\alpha+1)\frac{J_{\frac{n}{2}+\alpha}(R|y|)}{(R|y|)^{\frac{n}{2}+\alpha}} \cdot R^n$$

for $\text{Re}\,\alpha > -1$.

Put $\alpha = \frac{n-1}{2}$. By the asymptotic equation of Bessel function, we have that

$$\left|H_R^{\frac{n-1}{2}}(y)\right| \le C_n|y|^{-n} \le \frac{C_n}{\delta^n},$$

for $|y| > \delta$, $R \ge 1$.

It follows that

$$|I_2| \le C\frac{1}{\delta^n}\|f\|_{L(Q)},$$

and

$$|I_3| \le C\frac{1}{\delta^n}\|f\chi_Q\|_{L(\mathbb{R}^n)} = C\delta^{-n}\|f\|_{L(Q)}.$$

3.2 The case of the critical index (general discussion)

Besides, the estimate of I_1 is contained in (3.2.14). Thus we have

$$\max_{x \in G} \left| S_{R^2}^{\frac{n-1}{2}}(f;x) - B_{R^2}^{\frac{n-1}{2}}(\tilde{f};x) \right| \leq C_\delta \left(\int_Q |f| \log^+ |f| dx + 1 \right). \quad (3.2.16)$$

Combining the above results and the fact that trigonometric polynomials are dense in $L(Q)$, by a standard argument, the proof of Theorem 3.2.3 is concluded. ∎

It suffices to prove (3.2.13). We shall first get the estimate of (3.2.11).

Lemma 3.2.2 *Let $\sigma \in (0,1)$ and $\tau \in (-\infty, \infty)$. We have*

$$R^\sigma \left\| \Delta_{R^2}^{\frac{n-1}{2}+\sigma+i\tau} \right\|_\infty \leq \frac{C}{\sigma} e^{\pi|\tau|}, \quad (3.2.17)$$

for $R > 1$.

Proof. By the expression

$$H_{R^2}^{\frac{n-1}{2}+z}(u) = (2\pi)^{\frac{n}{2}} 2^{\frac{n-1}{2}+z} \Gamma\left(\frac{n+1}{2}+z\right) \frac{J_{n-\frac{1}{2}+z}(R|u|)}{(R|u|)^{n-\frac{1}{2}+z}} R^n, \quad (3.2.18)$$

for $\operatorname{Re} z > -\frac{n+1}{2}$, and the formula (see Watson [Wat1])

$$J_\nu(t) = \sqrt{\frac{2}{\pi t}} \cos\left(t - \frac{1}{2}\nu\pi - \frac{\pi}{4}\right) + O\left(e^{\pi|\operatorname{Im}\nu|} \frac{1}{t^{\frac{3}{2}}}\right),$$

$t \geq 1$, where the real part of ν takes values within a bounded set, we have

$$|J_\nu(t)| \leq C e^{\pi|\operatorname{Im}\nu|} t^{-\frac{1}{2}} \quad (3.2.19)$$

and

$$\left| H_{R^2}^{\frac{n-1}{2}+z}(u) \right| \leq C e^{\pi|\tau|}/(R^\sigma |u|^{n+\sigma}), \quad (3.2.20)$$

for $\sigma > 0$, if $\operatorname{Re} z \in (0,1]$.

By Theorem 1.2.2, we have

$$D_{R^2}^{\frac{n-1}{2}+z}(u) = \sum_{m \in \mathbb{Z}^n} \widehat{H_{R^2}^{\frac{n-1}{2}+z}}(m) e^{im \cdot x}$$

$$= \sum_{m \in \mathbb{Z}^n} H_{R^2}^{\frac{n-1}{2}+z}(u + 2\pi m),$$

for $\operatorname{Re} z = \sigma > 0$.

It follows that

$$\triangle_R^{\frac{n-1}{2}+z}(u) = \sum_{m \neq 0} H_R^{\frac{n-1}{2}+z}(u+2\pi m), \qquad (3.2.21)$$

for $\operatorname{Re} z = \sigma > 0$.

Substituting the above equation into (3.2.20), when $u \in Q^n$, we can obtain that

$$\left|\triangle_R^{\frac{n-1}{2}+z}(u)\right| \leq C e^{\pi|\tau|} R^{-\sigma} \cdot \sum_{m \neq 0} \frac{1}{|u+2\pi m|^{n+\sigma}}$$

$$\leq \frac{C}{\sigma} e^{\pi|\tau|} R^{-\sigma},$$

for $0 < \sigma \leq 1$, $z = \sigma + i\tau$.

This completes the proof of Lemma 3.2.2. ∎

Let $\sigma = \frac{1}{q}$ with $q > 2$. By Lemma 3.2.2 and (3.2.11), we have that

$$\|T_{1+i\tau}\|_{1,\infty} \leq C e^{\pi|\tau|} q := M_1(\tau). \qquad (3.2.22)$$

In the following, we shall estimate (3.2.10).

Choose a function $\psi(t) \in C^1[0,1]$ such that

$$\int_0^1 \psi(t) dt = 1$$

and

$$\int_0^1 t^k \psi(t) dt = 0,$$

for $k = 1, 2, \ldots, n-1$.

Define

$$\widetilde{D_R}^{\frac{n-1}{2}+z}(x) = \int_0^1 \left(1+\frac{t}{R}\right)^{n-1+2z} \psi(t) D_{R+t}^{\frac{n-1}{2}+z}(x) dt,$$

$$\widetilde{H_R}^{\frac{n-1}{2}+z}(x) = \int_0^1 \left(1+\frac{t}{R}\right)^{n-1+2z} \psi(t) H_{R+t}^{\frac{n-1}{2}+z}(x) dt$$

and

$$\widetilde{\triangle_R}^{\frac{n-1}{2}+z}(x) = \widetilde{D_R}^{\frac{n-1}{2}+z}(x) - \widetilde{H_R}^{\frac{n-1}{2}+z}(x).$$

3.2 The case of the critical index (general discussion)

Lemma 3.2.3 *If* $-1 < \sigma \leq 0$, *then*

$$\left\| \widetilde{\Delta_R}^{-\frac{n-1}{2}+\sigma+i\tau} \right\|_\infty \leq \frac{C}{1+\sigma} e^{\pi|\tau|} R^{-\sigma} \qquad (3.2.23)$$

holds.

Proof. Let us define a function of z

$$F(z) = \sum_{m \neq 0} \int_0^1 \left(1 + \frac{t}{R}\right)^{n-1+2z} \psi(t) H_{R+t}^{\frac{n-1}{2}+z}(x + 2\pi m) dt. \qquad (3.2.24)$$

When $\operatorname{Re} z > 0$, by (3.2.20) and (3.2.21), we have

$$F(z) = \widetilde{\Delta_R}^{-\frac{n-1}{2}+z}(x). \qquad (3.2.25)$$

It is easy to notice that both $\widetilde{\Delta_R}^{-\frac{n-1}{2}+z}(x)$ and $\Delta_R^{-\frac{n-1}{2}+z}(x)$ are analytic in the region $\operatorname{Re} z > -\frac{n+1}{2}$. We shall now prove that $F(z)$ is analytic, when $\operatorname{Re} z = \sigma > -1$.

Denote

$$C(z) = (2\pi)^{\frac{n}{2}} 2^{\frac{n-1}{2}+z} \Gamma\left(\frac{n+1}{2} + z\right).$$

By (3.2.18), we have

$$\widetilde{H_R}^{\frac{n-1}{2}+z}(y) = C(z) \int_0^1 \left(1 + \frac{t}{R}\right)^{n-1+2z} \psi(t) \frac{J_{n-\frac{1}{2}+z}(R+t)|y|}{(R+t)^{-\frac{1}{2}+z}|y|^{n-\frac{1}{2}+z}} dt,$$

which implies that it is analytic when $\operatorname{Re} z > -\frac{n+1}{2}$.

Using the formula

$$-\frac{d}{dt} \frac{J_{\nu-1}(t)}{t^{\nu-1}} = \frac{J_\nu(t)}{t^{\nu-1}}$$

and integration by parts, we have that

$$\widetilde{H_R}^{\frac{n-1}{2}+z}(y) = \frac{C(z)}{R^{n-1+2z}} \left\{ \left. \frac{-(R+t)^{2n-2+2z} J_{n-\frac{3}{2}+z}((R+t)|y|)}{|y|^2 [(R+t)|y|]^{n-3/2+z}} \right|_{t=0}^{t=1} \right.$$

$$+ \frac{1}{|y|^2} \int_0^1 \left[(2n - 2 + 2z)(R+t)^{2n-3+2z} \psi(t) \right.$$

$$\left. \left. + (R+t)^{2n-2+2z} \psi'(t) \right] \frac{J_{n-\frac{3}{2}+z}((R+t)|y|)}{[(R+t)y|^{n-\frac{3}{2}+z}} dt \right\},$$

and

$$\left|\widetilde{H_R}^{\frac{n-1}{2}+z}(y)\right| \leq \frac{|C(z)|}{R^{n-1+2\sigma}}\left\{\frac{R^{n-1+\sigma}}{|y|^2}\cdot\frac{Ce^{\pi|\tau|}}{|y|^{n-1+\sigma}}\right.$$
$$\left.+\frac{1}{|y|^2}\int_0^1 \frac{Ce^{\pi|\tau|}}{|Ry|^{n-1+\sigma}}(R^{2n-3+2\sigma}|\psi(t)|+R^{2n-2+2\sigma}|\psi'(t)|)dt\right\}$$
$$\leq C_\sigma e^{\pi|\tau|}R^{-\sigma}/|y|^{n+1+\sigma}. \tag{3.2.26}$$

Therefore, for $x \in Q$ and $R \geq 1$, $\sum_{m\neq 0}\widetilde{H_R}^{\frac{n-1}{2}+z}(x+2\pi m)$ is uniformly convergent in every compact subset of the half plane $\text{Re} z = \sigma > -1$. Thus $F(z)$ is analytic in the half plane.

Consequently, as analytic functions defined on $\text{Re} z > -1$, both sides of (3.2.25) coincide in $\text{Re} z > -1$. Invoking (3.2.26), we can get

$$|F(z)| = \left|\widetilde{\Delta_R}^{\frac{n-1}{2}+z}(x)\right| \leq \sum_{m\neq 0}\left|\widetilde{H_R}^{\frac{n-1}{2}+z}(x+2\pi m)\right|$$
$$\leq Ce^{\pi|\tau|}R^{-\sigma}\sum_{m\neq 0}\frac{1}{|x+2\pi m|^{n+1+\sigma}}$$
$$\leq C\frac{1}{1+\sigma}e^{\pi|\tau|}R^{-\sigma},$$

for $-1 < \sigma \leq 0$, $x \in Q$ and $R \geq 1$.

This completes the proof of Lemma 3.2.3. ∎

Lemma 3.2.4 *If $-1 < \sigma \leq 0$, then*

$$\sup_{x\in Q}\left|\widetilde{H_R}^{\sigma+i\tau+\frac{n-1}{2}}(x)-H_R^{\sigma+i\tau+\frac{n-1}{2}}(x)\right| \leq C_n e^{2\pi|\tau|}R^{-\sigma}. \tag{3.2.27}$$

Proof. We conclude that

$$\widetilde{H_R}^{\frac{n-1}{2}+z}(x)-H_R^{\frac{n-1}{2}+z}(x) = \int_0^1 \left[\left(1+\frac{t}{R}\right)^{n-1+2z}H_{R+t}^{\frac{n-1}{2}+z}(x)\right.$$
$$\left.-H_R^{\frac{n-1}{2}+z}(x)\right]\psi(t)dt$$
$$= \frac{1}{R^{n-1+2z}}\int_0^1\left[(R+u)^{n-1+2z}H_{R+t}^{\frac{n-1}{2}+z}(x)\right.$$
$$\left.-R^{n-1+2z}H_R^{\frac{n-1}{2}+z}(x)\right]\psi(u)du.$$

3.2 The case of the critical index (general discussion)

Write $\varphi(u) = (R+u)^{n-1+2z} H_{R+u}^{\frac{n-1}{2}}(x)$. Then, we have

$$\varphi(u) = C(z)(R+u)^{2n-1+2z} \frac{J_{n-\frac{1}{2}+z}((R+u)|x|)}{[\,(R+u)|x|\,]^{n-\frac{1}{2}+z}}.$$

Denote $\nu = n - \frac{1}{2} + z$, $y = (R+u)|x|$, $0 < u < 1$. We have

$$\frac{d}{du}\frac{J_\nu(y)}{y^\nu} = -\frac{J_{\nu+1}(y)}{y^\nu}|x| = -\frac{J_{\nu+1}(y)}{y^{\nu+1}}y|x|$$

and

$$\frac{d^2}{du^2}\frac{J_\nu(y)}{y^\nu} = \frac{J_{\nu+2}(y)}{y^{\nu+1}}y|x|^2 - \frac{J_{\nu+1}(y)}{y^{\nu+1}}|x|^2$$

$$= \left(\frac{J_{\nu+2}(y)}{y^\nu} - \frac{J_{\nu+1}(y)}{y^{\nu+1}}\right)|x|^2.$$

By an induction argument, it is easy to show that

$$\frac{d^k}{du^k}\left(\frac{J_\nu(y)}{y^\nu}\right) = \left(a_0^{(k)}\frac{J_{\nu+k}(y)}{y^\nu} + a_1^{(k)}\frac{J_{\nu+k-1}(y)}{y^{\nu+1}}\right. \qquad (3.2.28)$$

$$\left. + \cdots + a_{k-1}^{(k)}\frac{J_{\nu+1}(y)}{y^{\nu+k-1}}\right)|x|^k,$$

for $k \in \mathbb{N}$, and

$$\left|\frac{d^k}{du^k}\left(\frac{J_\nu(y)}{y^\nu}\right)\right| \leq \frac{C_k e^{\pi|\tau|}}{|y|^{n+\sigma}}|x|^k \leq \frac{C_k e^{\pi|\tau|}|x|^k}{(R|x|)^{n+\sigma}}. \qquad (3.2.29)$$

By Taylor's formula, we obtain that

$$\varphi(u) = \varphi(0) + \varphi'(0)u + \cdots + \frac{1}{(n-1)!}\varphi^{(n-1)}(0)u^{n-1} + \frac{1}{n!}\varphi^{(n)}(\theta_u)u^n, \qquad (3.2.30)$$

for some $0 < \theta < u$.

In addition, we use the orthogonal condition of $\psi(u)$ to obtain

$$\widetilde{H_R}^{\frac{n-1}{2}+z}(x) - H_R^{\frac{n-1}{2}+z}(x) = \frac{1}{R^{n-1+2z}}\int_0^1 [\varphi(u) - \varphi(0)]\psi(u)du$$

$$= \frac{1}{R^{n-1+2z}}\int_0^1 \frac{1}{n!}\varphi^{(n)}(\theta_u)u^n\psi(u)du.$$

Hence we obtain

$$\left|\widetilde{H_R^{\frac{n-1}{2}+z}}(x) - H_R^{\frac{n-1}{2}+z}(x)\right| \leq \frac{1}{R^{n-1+2\sigma}}\|\varphi^{(n)}\|_{C[0,1]}. \tag{3.2.31}$$

By Leibniz' formula, we obtain that

$$\varphi^{(n)}(\theta) = C(z)\sum_{k=0}^{n} C_n^k \left\{\frac{d^k}{du^k}\frac{J_\nu(y)}{y^\nu} \cdot \frac{d^{n-k}}{du^{n-k}}(R+u)^{2n-1+2z}\right\}\bigg|_{u=\theta}.$$

Since

$$\left|\frac{d^{n-k}}{du^{n-k}}(R+u)^{2n-1+2z}\right| \leq C_n(1+|\tau|^n)R^{n-1+2\sigma+k}, \quad u \in [0,1], \tag{3.2.32}$$

if we substitute this into the (3.2.29), then we get that $|x| > R^{-1}$,

$$|\varphi^{(n)}(\theta)| \leq C_n(1+|\tau|^n)e^{\pi|\tau|}\sum_{k=0}^{n}\frac{|x|^k R^{n-1+2\sigma+k}}{(R|x|)^{n+\sigma}}$$

$$\leq C_n e^{2\pi|\tau|}R^{n-1+\sigma}. \tag{3.2.33}$$

If $|x| \leq R^{-1}$, it follows immediately from (3.2.28) that

$$\left|\frac{d^k}{du^k}\frac{J_\nu(y)}{y^\nu}\right| \leq C_n e^{\pi|\tau|}|x|^k \leq \frac{C_n}{R^k}e^{\pi|\tau|}. \tag{3.2.34}$$

Invoking this estimate and (3.2.32), we obtain that

$$|\varphi^{(n)}(\theta)| \leq C_n e^{\pi|\tau|}\sum_{k=0}^{n} C_n^k \cdot \frac{R^{n-1+2\sigma+k}}{R^k}$$

$$\leq C_n e^{\pi|\tau|}R^{n-1+2\sigma}, \tag{3.2.35}$$

for $|x| \leq R^{-1}$. Observe that $\sigma \in (-1,0]$, we have that (3.2.33) holds for every $x \in Q^n$.

By (3.2.31) and (3.2.33), the proof of Lemma 3.2.4 is complete. ∎

Lemma 3.2.5 *If $-\frac{n-1}{2} \leq \sigma \leq 0$, then we have*

$$\left\|D_R^{\frac{n-1}{2}+\sigma+i\tau} - \widetilde{D_R^{\frac{n-1}{2}+\sigma+i\tau}}\right\|_2 \leq C_n e^{\pi|\tau|}R^{-\sigma}, \tag{3.2.36}$$

where $R \geq 1$.

3.2 The case of the critical index (general discussion)

Proof. We conclude that

$$D_R^{\frac{n-1}{2}+\sigma+i\tau}(x) - \widetilde{D_R}^{\frac{n-1}{2}+\sigma+i\tau}(x)$$
$$= \int_0^1 \Big\{ \Big(1+\frac{u}{R}\Big)^{n-1+2z} \sum_{|m|<R+u} (1-\frac{|m|^2}{(R+u)^2})^{\frac{n-1}{2}+z} e^{im\cdot x}$$
$$- \sum_{|m|<R} \Big(1-\frac{|m|^2}{R^2}\Big)^{\frac{n-1}{2}+z} e^{im\cdot x} \Big\} \psi(u) du$$
$$= \frac{1}{R^{n-1+2z}} \int_0^1 \Big\{ \sum_{|m|<R+u} ((R+u)^2 - |m|^2)^{\frac{n-1}{2}+z} e^{im\cdot x}$$
$$- \sum_{|m|<R} (R^2 - |m|^2)^{\frac{n-1}{2}+z} e^{im\cdot x} \Big\} \psi(u) du$$
$$= \frac{1}{R^{n-1+2z}} \int_0^1 \sum_{|m|<R} \Big\{ [(R+u)^2 - |m|^2]^{\frac{n-1}{2}+z}$$
$$- (R^2 - |m|^2)^{\frac{n-1}{2}+z} \Big\} e^{im\cdot x} \psi(u) du$$
$$+ \frac{1}{R^{n-1+2z}} \int_0^1 \sum_{R\leq |m|<R+1} ((R+u)^2 - |m|^2)^{\frac{n-1}{2}+z}$$
$$\times e^{im\cdot x} \chi_{(|m|-R,1]}(u) \psi(u) du.$$

Define

$$\lambda_m = \begin{cases} \int_0^1 \Big\{ ((R+u)^2 - |m|^2)^{\frac{n-1}{2}+z} - (R^2 - |m|^2)^{\frac{n-1}{2}+z} \Big\} \\ \qquad \times \psi(u) du, \qquad |m| < R, \\ \int_0^1 ((R+u)^2 - |m|^2)^{\frac{n-1}{2}+z} \chi_{(|m|-R,1]}(u) \psi(u) du, \\ \qquad R \leq |m| < R+1, \\ 0, \qquad |m| \geq R+1. \end{cases}$$

Then

$$\widetilde{D_R}^{\frac{n-1}{2}+z}(x) - D_R^{\frac{n-1}{2}+z}(x) = \frac{1}{R^{n-1+2z}} \sum_{m\in\mathbb{Z}^n} \lambda_m e^{im\cdot x},$$

and

$$\Big\|\widetilde{D_R}^{\frac{n-1}{2}+z} - D_R^{\frac{n-1}{2}+z}\Big\|_2 = \frac{1}{R^{n-1+2\sigma}} \frac{1}{\sqrt{|Q|}} \Big(\sum |\lambda_m|^2\Big)^{\frac{1}{2}}. \qquad (3.2.37)$$

When $R-1 \leq |m| < R+1$, for $\sigma \geq -\frac{n-1}{2}$, it is obvious that

$$|\lambda_m| \leq C[(R+1)^2 - (R-1)^2]^{\frac{n-1}{2}+\sigma} \leq CR^{\frac{n-1}{2}+\sigma}.$$

Hence

$$\sum_{R-1 \leq |m| < +\infty} |\lambda_m|^2 = \sum_{R-1 \leq |m| < R+1} |\lambda_m|^2 \leq C_n R^{2(n-1+\sigma)}. \tag{3.2.38}$$

When $|m| < R-1$, with the Taylor formula and the orthogonal property of $\psi(u)$ and $\{u, \ldots, u^{n-1}\}$, we can get that

$$\lambda_m = \int_0^1 \frac{1}{n!} \frac{d^n}{dt^n}[(R+t)^2 - |m|^2]^{\frac{n-1}{2}+z}\Big|_{t=\theta_u} u^n \psi(u) du \quad (0 < \theta_u < u \leq 1).$$

However

$$\left|\frac{d^k}{dt^k}[(R+t)^2 - |m|^2]^{\frac{n-1}{2}+z}\right| \leq C_n(1+|z|)^k R^k (R^2 - |m|^2)^{\frac{n-1}{2}+\sigma-k},$$

$$k = 1, 2, \ldots, n, \ 0 < t < 1.$$

And it follows that

$$\sum_{|m|<R-1} |\lambda_m|^2 \leq C_n e^{\pi|\tau|} R^{2n} \sum_{|m|<R-1} \frac{1}{(R^2 - |m|^2)^{n+1-2\sigma}}$$

$$\leq C_n e^{\pi|\tau|} R^{2n} \frac{1}{R^{n+1-2\sigma}} \int_0^{R-1} \frac{t^{n-1} dt}{(R-t)^{n+1-2\sigma}}$$

$$\leq C_n e^{\pi|\tau|} R^{n-1+2\sigma} \int_1^\infty \frac{R^{n-1} dt}{t^{n+1-2\sigma}}$$

$$\leq C_n e^{\pi|\tau|} R^{2(n-1+\sigma)}. \tag{3.2.39}$$

Combining (3.2.38), (3.2.39) and then substituting the above two into (3.2.37), we can obtain (3.2.36).

This finishes the proof of Lemma 3.2.5. ∎

Remark 3.2.1 *Considering the above case, if $n > 1$, then $\frac{n-1}{2} \geq \frac{1}{2}$. Therefore, Lemma 3.2.5 is still valid in the case $-\frac{1}{2} \leq \sigma \leq 0$.*

3.2 The case of the critical index (general discussion)

Lemma 3.2.6 *If* $-\frac{1}{2} \le \sigma \le 0$, *then we have*

$$\left\| \triangle_R^{\frac{n-1}{2}+\sigma+i\tau} \right\|_2 \le C_n e^{\pi|\tau|} R^{-\sigma}, \qquad (3.2.40)$$

for $R \ge 1$ *and* $n > 1$.

Proof. By the definition, we have

$$\triangle_R^{\frac{n-1}{2}+z} = \left(D_R^{\frac{n-1}{2}+z} - \widetilde{D_R}^{\frac{n-1}{2}+z} \right) + \left(\widetilde{H_R}^{\frac{n-1}{2}+z} - H_R^{\frac{n-1}{2}+z} \right) + \widetilde{\triangle_R}^{\frac{n-1}{2}+z}.$$

Therefore, Lemma 3.2.6 is a direct consequence of Lemmas 3.2.3, 3.2.4 and 3.2.5. ∎

Lemma 3.2.7 *If* $n > 1$, *then (3.2.13) is valid.*

Proof. Take $q > 2$, $R > 1$ and write (3.2.22) as

$$\|T_{1+i\tau}\|_{1,\infty} \le M_1(\tau) = C_n e^{\pi|\tau|} q.$$

Substituting $\sigma = \frac{1}{2}\left(-1 + \frac{2}{q} \right) \in \left(-\frac{1}{2}, 0 \right)$ into (3.2.40), and invoking (3.2.10), we can get that

$$\|T_{i\tau}\|_{2,\infty} = R^{\sigma} \left\| \triangle_R^{\frac{n-1}{2}+\sigma+i\frac{\tau}{2}} \right\|_2 \le C_n e^{\frac{\pi}{2}|\tau|} := M_0(\tau). \qquad (3.2.41)$$

By the interpolation theorem of Stein, we interpolate at $t = 1 - \frac{2}{q}$ and then get

$$\|T_{1-\frac{2}{q}}\|_{q,\infty} = \left\| \triangle_R^{\frac{n-1}{2}} \right\|_q$$

$$\le \exp\left\{ \frac{1}{2} \sin \pi t \int_{-\infty}^{\infty} \left(\frac{\log M_0(\tau)}{\operatorname{ch}\pi\tau - \cos \pi t} + \frac{\log M_1(\tau)}{\operatorname{ch}\pi\tau + \cos \pi t} \right) d\tau \right\}$$

$$\le C_n q.$$

This completes the proof of Lemma 3.2.7, and then that of Theorem 3.2.3. ∎

From Theorem 3.2.3, we can directly acquire a corollary about the localization problem.

Corollary 3.2.1 *Let $f \in L\log^+ L(Q^n)$, $n > 1$. If f vanishes at the ball $B(x,\delta)$ for some $\delta > 0$, then*

$$\lim_{R\to\infty} S_R^{\frac{n-1}{2}}(f;x) = 0$$

holds.

Proof. Because the convergence of the Bochner-Riesz means at the critical index of Fourier integral is a local property, in other words, the localization principle is valid (see Chapter 2), then by Theorem 3.2.3 we can transfer the problem into the integral and then obtain the conclusion immediately. This finishes the proof. ∎

3.2.4 The order of Lebesgue constant

In the above, we have mentioned that it is reasonable to draw an analogy between $S_R^{\frac{n-1}{2}}$ and the Fourier partial sum with $n=1$; and between the kernel of $S_R^{\frac{n-1}{2}}$ defined as

$$D_R^{\frac{n-1}{2}}(x) = \sum_{|m|<R} \left(1 - \frac{|m|^2}{R^2}\right)^{\frac{n-1}{2}} e^{im\cdot x}$$

and the Dirichlet kernel with unitary variable. Then we call $\frac{1}{(2\pi)^n}\|D_R^{\frac{n-1}{2}}\|_{L(Q^n)}$ as Lebesgue constant. It is meaningful to figure out the order which shows how it increases as $R \to \infty$.

By Lemma 3.2.7, it follows that

$$\left\|\triangle_R^{\frac{n-1}{2}}\right\|_1 \leq M,$$

so we have that

$$\left\|D_R^{\frac{n-1}{2}}\right\|_1 \leq \left\|H_R^{\frac{n-1}{2}}\right\|_1 + M.$$

3.3 The convergence at fixed point

A simple computation gives

$$\left\|H_R^{\frac{n-1}{2}}\right\|_1 = C_n \int_Q \frac{\left|J_{n-\frac{1}{2}}(R|y|)\right|}{(R|y|)^{n-\frac{1}{2}}} R^n dy$$

$$\leq C_n \int_{1/R}^{\sqrt{n}\pi} \frac{|\cos(Rt - \frac{n\pi}{2})|}{t} dt \qquad (3.2.42)$$

$$+ O\left(\int_{|y|<1/R} R^n dy\right) + O\left(\int_{1/R}^{\sqrt{n}\pi} \frac{dt}{Rt^2}\right)$$

$$= O(\log R)$$

for $R > 2$.

Therefore, we have

$$\left\|D_R^{\frac{n-1}{2}}\right\|_1 = O(\log R), \qquad (3.2.43)$$

as $R \to \infty$. On the other hand, since

$$\left\|D_R^{\frac{n-1}{2}}\right\|_1 \geq \left\|H_R^{\frac{n-1}{2}}\right\|_1 - \left\|\triangle_R^{\frac{n-1}{2}}\right\|_1 \geq \left\|H_R^{\frac{n-1}{2}}\right\|_1 - M,$$

it is easy to see that (3.2.42) is accurate, so is (3.2.43) in the sense of order.

Remark 3.2.2 *It is well known that the estimate of Lebesgue constant is very important in the study of convergence and approximation of linear means of Fourier series. Further reference on the estimates of Lebesgue constant can see Trigub [Tr1].*

3.3 The convergence at fixed point

Theorem 3.3.1 *Let $f \in L\log^+ L(Q^n)$, $n \geq 2$, satisfies*

$$\int_0^t u^{n-1}(f_x(u) - f(x)) du = o(t^n) \quad (t \to 0), \qquad (3.3.1)$$

and

$$\int_t^\eta \frac{|f_x(u+t) - f_x u|}{u} du = o(1) \quad (t \to 0, \eta > 0). \qquad (3.3.2)$$

Then we have

$$\lim_{R\to\infty} S_R^{\frac{n-1}{2}}(f;x) = f(x). \qquad (3.3.3)$$

Proof. We might as well take it for granted that $x = 0$ and $f(0) = 0$. Let $g(y) = f(y)\chi_Q(y)$, $y \in \mathbb{R}^n$. By Theorem 3.2.3

$$\lim_{R\to\infty}\left(S_R^{\frac{n-1}{2}}(f;0) - B_R^{\frac{n-1}{2}}(g;0)\right) = 0,$$

it suffices to show that

$$\lim_{R\to\infty} B_R^{\frac{n-1}{2}}(g;0) = 0. \qquad (3.3.4)$$

It is well known that

$$B_R^{\frac{n-1}{2}}(g;0) = \frac{\Gamma(\frac{n+1}{2})}{\sqrt{2}\pi^{n/2}} \int_Q g(y) \frac{J_{n-\frac{1}{2}}(R|y|)}{(R|y|)^{n-\frac{1}{2}}} R^n dy$$

$$= \frac{\sqrt{2}\Gamma(\frac{n+1}{2})}{\Gamma(n/2)} \int_0^\infty g_0(t) t^{n-1} \frac{J_{n-\frac{1}{2}}(Rt)}{(Rt)^{n-\frac{1}{2}}} R^n dt$$

$$= c_n \left(\int_0^{\frac{\pi}{R}} + \int_{\frac{\pi}{R}}^\infty\right).$$

Denote

$$\varphi(t) = \int_0^t g_0(\tau)\tau^{n-1} d\tau.$$

By the condition of (3.3.1), we have

$$\varphi(t) = o(t^n), \qquad (3.3.5)$$

as $t \to 0$.

Denote $V_\nu(t) = \frac{J_\nu(t)}{t^\nu}$, then by integration by parts we can get that

$$\int_0^{\frac{\pi}{R}} g_0(t) t^{n-1} V_{n-\frac{1}{2}}(Rt) R^n dt$$

$$= \varphi(t) V_{n-\frac{1}{2}}(Rt) R^n \Big|_0^{\frac{\pi}{R}} - \int_0^{\frac{\pi}{R}} \varphi(t) V'_{n-\frac{1}{2}}(Rt) R^{n+1} dt$$

$$= o(1) + \int_0^{\pi/R} \varphi(t) \frac{J_{n-\frac{1}{2}}(Rt)}{(Rt)^{n-\frac{1}{2}}} R^{n+1} dt$$

$$= o(1) + \int_0^{\pi/R} o(t^n) O(Rt) R^{n+1} dt = o(1) \ (R \to \infty).$$

It follows that

$$B_R^{\frac{n-1}{2}}(g;0) = o(1) + C_n \int_{\frac{\pi}{R}}^\infty g_0(t) t^{n-1} \frac{J_{n-\frac{1}{2}}(Rt)}{(Rt)^{n-\frac{1}{2}}} R^n dt. \qquad (3.3.6)$$

3.3 The convergence at fixed point

Let

$$K_n(u) = \frac{1}{u^{n-\frac{1}{2}}}\left\{J_{n-\frac{1}{2}}(u) - \sqrt{\frac{2}{\pi u}}\left[\cos\left(u - \frac{n\pi}{2}\right)\right.\right.$$
$$\left.\left. - \frac{n(n-1)}{2u}\sin\left(u - \frac{n\pi}{2}\right)\right]\right\}$$

and

$$I_1 = \int_{\frac{\pi}{R}}^{\infty} g_0(t) t^{n-1} \cdot \sqrt{\frac{2}{\pi}} \frac{\cos(Rt - \frac{n\pi}{2})}{(Rt)^n} R^n dt,$$

$$I_2 = \int_{\frac{\pi}{R}}^{\infty} g_0(t) t^{n-1} \left(-\sqrt{\frac{2}{\pi}}\right) \frac{n(n-1)}{2} \frac{\sin(Rt - \frac{n\pi}{2})}{(Rt)^{n+1}} R^n dt,$$

$$I_3 = \int_{\frac{\pi}{R}}^{\infty} g_0(t) t^{n-1} \cdot K_n(Rt) \cdot R^n dt.$$

It follows from (3.3.6) that

$$B_R^{\frac{n-1}{2}}(g;0) = C_n(I_1 + I_2 + I_3) + o(1) \quad (R \to \infty). \tag{3.3.7}$$

$$I_1 = \sqrt{\frac{2}{\pi}} \int_{\frac{\pi}{R}}^{\infty} g_0(t) \frac{\cos(Rt - \frac{n\pi}{2})}{t} dt$$

$$= \sqrt{\frac{2}{\pi}} \int_{\frac{\pi}{R}}^{\eta} g_0(t) \frac{\cos(Rt - \frac{n\pi}{2})}{t} dt + \sqrt{\frac{2}{\pi}} \int_{\eta}^{\infty} g_0(t) \frac{\cos(Rt - \frac{n\pi}{2})}{t} dt,$$

where the second part is infinitely small, since $g_0(t) t^{-1} \in L((\eta, \infty))$, by the Riemann-Lebesgue lemma. It follows that

$$I_1 = \sqrt{\frac{2}{\pi}} \int_{\frac{\pi}{R}}^{\eta} g_0(t) \cdot \frac{\cos(Rt - \frac{n\pi}{2})}{t} dt + o(1) \ (R \to \infty).$$

Obviously,

$$\int_{\frac{\pi}{R}}^{\eta} g_0(t) \cdot \frac{\cos(Rt - \frac{n\pi}{2})}{t} dt$$

$$= \frac{1}{2} \int_{\frac{\pi}{R}}^{\eta} \left[\frac{g_0(t)}{t} - \frac{g_0(t + \frac{\pi}{R})}{t + \frac{\pi}{R}}\right] \cos(Rt - \frac{\pi}{2}n) dt$$

$$+ \frac{1}{2} \int_{\eta - \frac{\pi}{R}}^{\eta} \frac{g_0(t + \frac{\pi}{R})}{t + \frac{\pi}{R}} \cos(Rt - \frac{n\pi}{2}) dt$$

$$- \frac{1}{2} \int_{0}^{\frac{\pi}{R}} \frac{g_0(t + \frac{\pi}{R})}{t + \frac{\pi}{R}} \cos(Rt - \frac{\pi}{2}n) dt.$$

By integration by parts and (3.3.5), we have

$$-\int_0^{\frac{\pi}{R}} \frac{g_0(t+\frac{\pi}{R})}{t+\frac{\pi}{R}} \cos\left(Rt - \frac{\pi}{2}n\right) dt$$

$$= \varphi(t) \frac{\cos(Rt - \frac{\pi}{2}n)}{t^n} \Big|_{\frac{\pi}{R}}^{\frac{2\pi}{R}}$$

$$+ \int_{\frac{\pi}{R}}^{\frac{2\pi}{R}} \varphi(t) \left(\frac{R\sin(Rt - \frac{\pi}{2}n)}{t^n} + \frac{n\cos(Rt - \frac{\pi}{2}n)}{t^{n+1}} \right) dt$$

$$= o(1),$$

as $R \to \infty$.

By the absolute continuity of integration, it follows that

$$\left| \int_{\eta - \frac{\pi}{R}}^{\eta} \frac{g_0(t+\frac{\pi}{R})}{t+\frac{\pi}{R}} \cos(Rt - \frac{n\pi}{2}) dt \right| \leq \frac{1}{\eta^n} \int_{\eta < |y| < \eta + \frac{\pi}{R}} |g(y)| dy = o(1).$$

Besides, we have

$$\int_{\frac{\pi}{R}}^{\eta} \left[\frac{g_0(t)}{t} - \frac{g_0(t+\frac{\pi}{R})}{t+\frac{\pi}{R}} \right] \cos(Rt - \frac{\pi}{2}n) dt$$

$$= \int_{\frac{\pi}{R}}^{\eta} \frac{g_0(t) - g_0(t+\frac{\pi}{R})}{t} \cos(Rt - \frac{\pi}{2}n) dt$$

$$+ \int_{\frac{\pi}{R}}^{\eta} g\left(t+\frac{\pi}{R}\right) \left(\frac{1}{t} - \frac{1}{t+\frac{\pi}{R}} \right) \cos(Rt - \frac{\pi}{2}n) dt.$$

Making integration by parts to the second term on the right side of the above equality and (3.3.5), we can obtain that

$$-\int_{\frac{2\pi}{R}}^{\eta+\frac{\pi}{R}} g(t) \frac{1}{(t-\frac{\pi}{R})t} \frac{\pi}{R} \cdot \cos(Rt - \frac{\pi}{2}n) dt$$

$$= -\varphi(t) \frac{\pi}{R} \frac{\cos(Rt - \frac{\pi}{2}n)}{t^n(t-\frac{\pi}{R})} \Big|_{\frac{2\pi}{R}}^{\eta+\frac{\pi}{R}}$$

$$+ \int_{\frac{2\pi}{R}}^{\eta+\frac{\pi}{R}} \varphi(t) \left[\frac{\cos(Rt - \frac{\pi}{2}n)}{t^n(t-\frac{\pi}{R})} \right]' dt \frac{\pi}{R}$$

$$= o(1).$$

From (3.3.2), it follows that

$$\int_{\frac{\pi}{R}}^{\eta} \frac{g_0(t) - g_0(t+\frac{\pi}{R})}{t} \cos(Rt - \frac{\pi}{2}n) dt = o(1) \quad (R \to \infty).$$

3.3 The convergence at fixed point

In another words, we get $I_1 = o(1)$ as $R \to \infty$.

Similarly, we can have $I_2 = o(1)$.

In order to estimate I_3, we shall now consider

$$K'_n(u) = \left(\frac{J_{n-\frac{1}{2}}(u)}{u^{n-\frac{1}{2}}}\right)' + \sqrt{\frac{2}{\pi}} \left[\sin\left(u - \frac{n\pi}{2}\right) + \frac{n(n-1)}{2u}\cos\left(u - \frac{n\pi}{2}\right)\right]\frac{1}{u^n}$$

$$+\sqrt{\frac{2}{\pi}}\left\{n\frac{\cos\left(u - \frac{n\pi}{2}\right)}{u^{n+1}} - \frac{(n+1)n(n-1)}{2u^{n+2}}\sin\left(u - \frac{n\pi}{2}\right)\right\}$$

$$= -\frac{J_{n+\frac{1}{2}}(u)}{u^{n-\frac{1}{2}}} + \sqrt{\frac{2}{\pi}}\left(\frac{\sin\left(u - \frac{n\pi}{2}\right)}{u^n} + \frac{n(n+1)}{2}\frac{\cos\left(u - \frac{n\pi}{2}\right)}{u^{n+1}}\right)$$

$$+O\left(\frac{1}{u^{n+2}}\right). \tag{3.3.8}$$

From the asymptotic formula (see Watson [Wat1]), it follows that

$$J_\nu(u) = \sqrt{\frac{2}{\pi}}\frac{\cos(u - \frac{\pi}{2}\nu - \frac{\pi}{4})}{\sqrt{u}} - \sqrt{\frac{2}{\pi}}\frac{(\nu + \frac{1}{2})(\nu - \frac{1}{2})}{2u^{3/2}}\sin\left(u - \frac{\pi}{2}\nu - \frac{\pi}{4}\right)$$

$$+O\left(\frac{1}{u^{5/2}}\right), \tag{3.3.9}$$

for $u \geq 1$. We thus have that

$$K'_n(u) = O\left(\frac{1}{u^{n+2}}\right). \tag{3.3.10}$$

Now we proceed to compute I_3. In the method of integration by parts, with $(3.3.1)'$ and $(3.3.10)$, we get that

$$I_3 = \varphi(t)K_n(Rt)R^n\Big|_{\frac{\pi}{R}}^{\infty} - \int_{\frac{\pi}{R}}^{\infty}\varphi(t)K'_n(Rt)R^{n+1}dt$$

$$= o(1) - \int_{\frac{\pi}{R}}^{\infty} o(t^n)O\left(\frac{1}{(Rt)^{n+2}}\right)R^{n+1}dt$$

$$= o(1).$$

Thus (3.3.4) holds and this finishes the proof. ∎

The above Theorem 3.3.1 was published in the essay by Lu [Lu1]. At the same time, Lu also proved it.

Theorem 3.3.2 Let $f \in L\log^+ L(Q^n)$, $n \geq 2$. If f satisfies the following conditions at x_0,

(i) $\int_0^t u^{n-1}(f_{x_0}(u) - f(x_0))du = o(t^n)$, as $t \to 0$,

(ii) The function $Q_{x_0}(t) := t(f_{x_0}(t) - f(x_0))$ has bounded variation on $[0, \eta]$ with $\eta > 0$, and

$$\bigvee_0^t (Q_{x_0}) = O(t), \quad \text{as } t \to 0^+,$$

then (3.3.3) holds at x_0.

Proof. In fact, condition (i) is (3.3.1).

Make the same assumptions as that of 3.3.1. We can see that it only suffices to show that

$$\lim_{R \to \infty} \int_{\frac{\pi}{R}}^{\eta} g_0(t) \frac{\cos(Rt - \frac{\pi n}{2})}{t} dt = 0. \qquad (3.3.11)$$

By (3.3.5), for every $\varepsilon > 0$, there exists $\delta > 0$, such that if $0 < t \leq \delta$,

$$|\varphi(t) t^{-n}| < \varepsilon. \qquad (3.3.12)$$

Of course we can take it for granted that $\delta < \eta$. Now let $R > \frac{\pi}{\sqrt{\varepsilon \delta}}$ and we denote $\tau = \frac{\pi}{\sqrt{\varepsilon R}}$.

$$\int_{\frac{\pi}{R}}^{\tau} g_0(t) \frac{\cos(Rt - \frac{\pi n}{2}) dt}{t} = \varphi(t) \frac{\cos(Rt - \frac{\pi n}{2})}{t^n} \bigg|_{\frac{\pi}{R}}^{\tau}$$
$$+ \int_{\frac{\pi}{R}}^{\tau} \varphi(t) \left(\frac{R \sin(Rt - \frac{n\pi}{2})}{t^n} + \frac{n \cos(Rt - \frac{n\pi}{2})}{t^{n+1}} \right) dt.$$

Since $0 < \tau < \delta$, we substitute (3.3.12) into the above inequality and get

$$\left| \int_{\frac{\pi}{R}}^{\tau} g_0(t) \frac{\cos(Rt - \frac{\pi n}{2})}{t} dt \right| \leq \varepsilon + \int_{\frac{\pi}{R}}^{\tau} \left(\varepsilon R + n \frac{\varepsilon}{t} \right) dt$$
$$\leq \varepsilon + \left(1 + \frac{n}{\pi}\right) \varepsilon \tau R \qquad (3.3.13)$$
$$\leq C_n \sqrt{\varepsilon}.$$

3.3 The convergence at fixed point

Besides, with the formula

$$d\left(\frac{g_0(t)}{t}\right) = \frac{1}{t^2}d\theta_0(t) - 2\frac{\theta_0(t)}{t^3}dt,$$

and $(\theta_0(t) = tg_0(t))$, we denote the total variation as

$$\bigvee_0^t(\theta_0) = h(t).$$

Then, we have

$$\int_\tau^\eta g_0(t)\frac{\cos(Rt - \frac{n\pi}{2})}{t}dt = \frac{1}{R}\sin\left(Rt - \frac{\pi n}{2}\right) \cdot \frac{g_0(t)}{t}\bigg|_\tau^\eta$$
$$- \int_\tau^\eta \frac{\sin(Rt - \frac{\pi n}{2})}{Rt^2}d\theta_0(t)$$
$$+ 2\int_\tau^\eta \frac{\sin(Rt - \frac{\pi n}{2})}{Rt^3}\theta_0(t)dt.$$

Therefore,

$$\left|\int_\tau^\eta g_0(t)\frac{\cos(Rt - \frac{n\pi}{2})}{t}dt\right| \le \frac{1}{\eta}\frac{g_0(\eta)}{R} + \frac{1}{R}\frac{h(\tau)}{\tau^2} + \int_\tau^\eta \frac{1}{Rt^2}dh(t) + 2\int_\tau^\eta \frac{h(t)}{Rt^3}dt$$
$$= O\left(\frac{1}{R}\right) + O\left(\frac{1}{R\tau}\right) + O\left(\int_\tau^\eta \frac{dt}{Rt^2}\right)$$
$$\le M\sqrt{\varepsilon},$$

(3.3.14)

where M is independent of ε, R with $R > \frac{\pi}{\sqrt{\varepsilon\delta}}$.

Combining (3.3.13) with (3.3.14), we can obtain (3.3.11), which finishes the proof. ∎

In the following, we will prove a sufficient condition which guarantees (3.3.2) to be valid which was proven by Lu [Lu5]. Let us first introduce some concepts.

Definition 3.3.1 *Define a finitely valued function $f(t)$ on a finite interval $[a,b]$. If for any open interval of $[a,b]$, that is, $I_n = [a_n, b_n]$, $n = 1, 2, \ldots$, which does not intersect with each other, the following holds:*

$$HV(f;[a,b]) := \sup_{\{I_n\}} \sum_{n=1}^\infty \frac{|f(b_n) - f(a_n)|}{n} < \infty,$$

then we say f is of harmonic bounded variation on $[a,b]$, denoted by $f \in$ $\mathrm{HBV}_{[a,b]}$.

From the definition, we can see that if $f \in \mathrm{HBV}_{[a,b]}$, then $f(x_0+0)$ exists and is finite at every point x_0 on $(a,b]$. In fact, if

$$\liminf_{x \to x_0^+} f(x) = \alpha < \limsup_{x \to x_0^+} f(x) = \beta,$$

then we choose the point sequence

$$b_1 > a_1 > b_2 > a_2 > \cdots, \ a_n \to x_0^+,$$

such that $f(b_n) > \frac{3}{4}\beta + \frac{1}{4}\alpha$, $f(a_n) < \frac{1}{4}\beta + \frac{3}{4}\alpha$. Therefore,

$$\sum_{n=1}^{\infty} \frac{f(b_n) - f(a_n)}{n} \geq \sum_{n=1}^{\infty} \frac{1}{n}\frac{1}{2}(\beta - \alpha) = +\infty,$$

which contradicts the fact $f \in \mathrm{HBV}_{[a,b]}$. Similarly, $f(x_0 - 0)$ exists and is finite $(x_0 \in [a,b))$.

Besides, the following relation is quite evident,

$$\mathrm{BV}_{[a,b]} \subset \mathrm{HBV}_{[a,b]} \subset B_{[a,b]},$$

where BV represents for the class of usual bounded variation functions, and B means the class of bounded functions.

Lemma 3.3.1 *Let $f \in L(Q^n)$ with $n \geq 2$. If $f_x(t)$ is of bounded variation at x on $[0,\eta]$ $(\eta > 0)$, then (3.3.2) holds.*

Proof. Firstly, we will prove that, for any two positive numbers η_1, η_2 with $\eta_1 < \eta_2$,

$$\limsup_{h \to 0} \int_h^{\eta_1} \frac{|f_x(t+h) - f_x(t)|}{t} dt = \limsup_{h \to 0} \int_h^{\eta_2} \frac{|f_x(t+h) - f_x(t)|}{t} dt. \tag{3.3.15}$$

In fact,

$$\int_h^{\eta_1} \frac{|f_x(t+h) - f_x(t)|}{t} dt \leq \int_h^{\eta_2} \frac{|f_x(t+h) - f_x(t)|}{t} dt$$

$$\leq \int_h^{\eta_1} \frac{|f_x(t+h) - f_x(t)|}{t} dt \tag{3.3.16}$$

$$+ \frac{1}{\eta_1} \int_{\eta_1}^{\eta_2} |f_x(t+h) - f_x(t)| dt,$$

3.3 The convergence at fixed point

for $h < \eta_1$.

Since $f_x(t) \in L(\eta_1, \eta_1 + \eta_2)$, by the property of the norm of the integration, the second term on the right side of the above inequality is a finitely small quantity ($h \to 0^+$). Hence we can get (3.3.15).

If (3.3.2) is not valid, then according to (3.3.15), there exists a positive number ε, such that for every $\delta > 0$,

$$\limsup_{h \to 0^+} \int_h^\delta \frac{|f_x(t+h) - f_x(t)|}{t} dt > \varepsilon. \tag{3.3.17}$$

Therefore, we could choose such a sequence by induction:

$$\eta \geq \delta_1 > h_1 > \delta_2 > h_2 > \cdots$$

such that, for $n \in \mathbb{N}$,

$$\delta_n = \lambda_n h_n, \tag{3.3.18}$$

for $\lambda_n \in \mathbb{N}$, $\lambda_n \geq 2$,

$$h_n \geq 2\delta_{n+1}, \tag{3.3.19}$$

$$\int_{h_n}^{\delta_n} \frac{|f_x(t+h_n) - f_x(t)|}{t} dt > \varepsilon, \tag{3.3.20}$$

and

$$\sum_{j=1}^{\lambda_n - 1} \frac{1}{j} \sup_{0 < t \leq \delta_n,\ 0 < h \leq h_n} |f_x(t+h) - f_x(t)| < \frac{1}{2}\varepsilon, \tag{3.3.21}$$

for $n = 2, 3, \ldots$.

By (3.3.16), (3.3.17) and the existence of $f_x(0+0)$, there exists such a sequence. Thus we have

$$\varepsilon < \int_{h_n}^{\delta_n} \frac{|f_x(t+h_n) - f_x(t)|}{t} dt$$

$$= \sum_{j=1}^{\lambda_n - 1} \int_0^{h_n} \frac{|f_x(t+(j+1)h_n) - f_x(t+jh_n)|}{t + jh_n} dt$$

$$\leq \frac{1}{h_n} \int_0^{h_n} \left(\sum_{j=1}^{\lambda_n - 1} \frac{1}{j} |f_x(t+(j+1)h_n) - f_x(t+jh_n)| \right) dt.$$

Since $f_x(t)$ is a bounded function, there exists $\theta_n \in (0, h_n)$, such that

$$\varepsilon \leq \sum_{j=1}^{\lambda_n-1} \frac{1}{j} |f_x(\theta_n + (j+1)h_n) - f_x(\theta_n + jh_n)|$$

$$= \sum_{j=\lambda_{n-1}-1}^{\lambda_n-1} \frac{1}{j} |f_x(\theta_n + (j+1)h_n) - f_x(\theta_n + jh_n)|$$

$$+ \sum_{j=1}^{\lambda_{n-1}-1} \frac{1}{j} |f_x(\theta_n + (j+1)h_n) - f_x(\theta_n + jh_n)|.$$

It follows that

$$\frac{1}{2}\varepsilon < \sum_{j=\lambda_{n-1}}^{\lambda_n-1} \frac{1}{j} |f_x(\theta_n + (j+1)h_n) - f_x(\theta_n + jh_n)|,$$

for $n = 2, 3, \ldots$.

And thus

$$\sum_{n=2}^{\infty} \sum_{j=\lambda_{n-1}}^{\lambda_n-1} \frac{1}{j} |f_x(\theta_n + (j+1)h_n) - f_x(\theta_n + jh_n)| = +\infty,$$

which contradicts the fact $f_x(t) \in \text{HBV}_{[0,\eta]}$, and this finishes the proof. ∎

From Theorem 3.3.1 and Lemma 3.3.1, we can deduce the main result of Lu [Lu5] immediately.

Theorem 3.3.3 *Let $f \in L \log^+ L(Q^n)$, for $n \geq 2$. If $f_x(t)$ is of harmonic bounded variation on some interval $[0, \eta]$ with $\eta > 0$, and*

$$\lim_{t \to 0^+} f_x(t) = f(x), \qquad (3.3.22)$$

then (3.3.3) is valid.

Obviously, if we remove the assumption (3.3.22), we can obtain

$$\lim_{R \to \infty} S_R^{\frac{n-1}{2}}(f; x) = f_x(0+0),$$

which is just the original form in Lu's theorem [Lu5].

At the end of this section, we will point out that the proof methods of several sufficient conditions to guarantee convergence at fixed points are similar. They all transfer the problems into the Fourier integral under the condition that $L\log L$ is integrable, where Theorem 3.2.3 weighs a lot. At the same time, we notice that these conclusions are undoubtedly applicable for the Bochner-Riesz means of Fourier integral at the critical index, and in the case of integral, as the condition of convergence, we do not need the local $L\log L$-integrability any more.

As far as the convergence problem of the Bochner-Riesz means of the multiple Fourier series is concerned, Wang [Wan1] and Chang [Chan1] also dedicated in the study while Shi [Sh1] investigated the Fourier series of functions of the $\bigwedge$BMV class.

3.4 L^p approximation

When $f \in L^p(Q^n)$, $n > 1$, we shall consider the convergence (L^p-convergence and a.e.-convergence) of the Bochner-Riesz means of the order $\alpha > \alpha_p := \frac{n-1}{2}\left|\frac{2}{p} - 1\right|$. We shall see that for functions of L^p class, when $1 < p < \infty$, the index $\frac{n-1}{2}$ of the Bochner-Riesz means all lose critical meanings in problems of L^p-convergence and a.e.-convergence. We think that for L^p- functions and when $1 < p \leq 2$, under the conclusion in Theorem 3.10.3, the critical order of the Bochner-Riesz means might be regarded as $\alpha_p = \frac{n-1}{2}\left(\frac{2}{p} - 1\right)$. At this time, it makes sense to consider the properties of the α_p order of the Bochner-Riesz means, however, it is not an easy task.

Remark 3.4.1 *The study on approximation of Bochner-Riesz means is usually based on their convergence. To study approximation of Bochner-Riesz means in L^p with $1 < p < \infty$, we have to consider Theorem 3.10.3 as our starting point. In this sense, we temporarily regard the critical order of the Bochner-Riesz means in L^p approximation with $1 < p < \infty$ as $\alpha_p = \frac{n-1}{2}\left(\frac{2}{p} - 1\right)$. In fact, according to the conjecture in Section 2.2, the critical order of the Bochner-Riesz means in L^p convergence or approximation should be $\alpha_p = n\left|\frac{1}{p} - \frac{1}{2}\right| - \frac{1}{2}$ (see Sogge [S1], Stein [St5] or Davis-Chang [DC1]).*

3.4.1 The estimate of the maximal operator

Let $f \in L(Q^n)$ with $n > 1$. Let $\alpha \in \mathbb{C}$ and $\text{Re}\,\alpha > -1$. $S^\alpha_*(f)$ is the maximal operator given by Definition 3.1.1. The estimate of S^α_* above the critical index $\frac{n-1}{2}$ has been figured out by Theorem 3.1.3. Now we assume

$f \in L^p$, $1 < p < 2$, and we will discuss in the case of $\operatorname{Re}\alpha \in \left(\alpha_p, \frac{n-1}{2}\right]$. Generally, we define

$$G^\alpha(f)(x) = \left\{\int_0^\infty |S_R^{\alpha+1}(f;x) - S_R^\alpha(f;x)|^2 \frac{dR}{R}\right\}^{\frac{1}{2}}$$

and

$$M^\alpha(f)(x) = \sup_{R>0} \left\{\frac{1}{R}\int_0^R |S_u^\alpha(f;x)|^2 du\right\}^{\frac{1}{2}}.$$

Lemma 3.4.1 If $-\frac{1}{2} < \operatorname{Re}\alpha < n$ and $\tau = \operatorname{Im}\alpha$, then we have

$$\|M^\alpha(f)\|_2 \leq A_n e^{2\pi|\tau|}\left(1 + \frac{1}{\sqrt{\operatorname{Re}\alpha + \frac{1}{2}}}\right)\|f\|_2, \qquad (3.4.1)$$

and

$$\|G^\alpha(f)\|_2 \leq A_n\left(1 + \frac{1}{\sqrt{\operatorname{Re}\alpha + \frac{1}{2}}}\right)\|f\|_2. \qquad (3.4.2)$$

Proof. We conclude that

$$\|G^\alpha(f)\|_2^2 = \int_0^\infty \int_Q |S_R^{\alpha+1}(f;x) - S_R^\alpha(f;x)|^2 dx \frac{dR}{R}$$

$$= \int_0^\infty \sum_{|m|<R} \left|\left(1 - \frac{|m|^2}{R^2}\right)^{\alpha+1} - \left(1 - \frac{|m|^2}{R^2}\right)^\alpha\right|^2 |C_m(f)|^2 |Q| \frac{dR}{R}$$

$$= |Q|\int_0^\infty \frac{1}{R} \sum_{|m|<R} \left(1 - \frac{|m|^2}{R^2}\right)^{2\operatorname{Re}\alpha} \cdot \frac{|m|^4}{R^4} |C_m(f)|^2 dR$$

$$= |Q|\sum_m \left\{\int_0^\infty \frac{1}{R^5}\left(1 - \frac{|m|^2}{R^2}\right)^{2\operatorname{Re}\alpha} \chi_{[0,R)}(|m|) dR\right\} |m|^4 |C_m|^2$$

$$= |Q|\sum_{m\neq 0} |C_m|^2 \cdot |m|^4 \cdot \int_{|m|}^\infty \frac{1}{R^5}\left(1 - \frac{|m|^2}{R^2}\right)^{2\operatorname{Re}\alpha} dR$$

$$= \frac{1}{2}|Q|\int_0^1 u(1-u)^{2\operatorname{Re}\alpha} du \sum_{m\neq 0}|C_m|^2$$

$$\leq A\left(1 + \frac{1}{\operatorname{Re}\alpha + \frac{1}{2}}\right)\|f\|_2^2.$$

3.4 L^p approximation

Consequently, we can get (3.4.2).

In order to prove (3.4.1), we choose
$$k = \left[\frac{n+1}{2}\right] + 1 > \frac{n+1}{2},$$
and we have
$$\operatorname{Re}\alpha + k > \frac{n}{2}.$$

Since
$$\left\{\frac{1}{R}\int_0^R |S_u^\alpha(f)|^2 du\right\}^{\frac{1}{2}} \leq \sum_{j=1}^k \left(\frac{1}{R}\int_0^R |S_u^{\alpha+j-1}(f) - S_u^{\alpha+j}(f)|^2 du\right)^{\frac{1}{2}}$$
$$+ \left\{\frac{1}{R}\int_0^R |S_u^{\alpha+k}(f)|^2 du\right\}^{\frac{1}{2}}$$
$$\leq \sum_{j=1}^k G^{\alpha+j-1}(f) + S_*^{\alpha+k}(f),$$

it follows from (3.4.2) and (3.1.12) that (3.4.1) holds, which completes the proof. ∎

Lemma 3.4.2 *Let $\beta \in \mathbb{C}$, $\operatorname{Re}\beta > 0$, $\delta > -1$, then*

$$S_R^{\beta+\delta}(f;x) = \frac{2}{B(\beta, \delta+1)} \cdot \frac{1}{R^{2(\beta+\delta)}} \int_0^R (R^2 - r^2)^{\beta-1} r^{2\delta+1} S_r^\delta(f;x) dr. \quad (3.4.3)$$

Proof. Set $s = \frac{x}{t}$,
$$B(\beta, \delta+1) = \int_0^1 (1-s)^{\beta-1} s^\delta ds$$
$$= \int_0^t (1 - \frac{x}{t})^{\beta-1} \frac{x^\delta}{t^{\delta+1}} dx$$
$$= \int_0^t \frac{1}{t^{\beta+\delta}} (t-x)^{\beta-1} x^\delta dx$$

for every $t > 0$.

For any $r > 0$, let $x = u^2 - r^2$ ($u \geq r$). We have
$$B(\beta, \delta+1) = \frac{1}{t^{\beta+\delta}} \int_r^{\sqrt{r^2+t}} (t+r^2-u^2)^{\beta-1} (u^2-r^2)^\delta 2u du.$$

Substituting $R = \sqrt{r^2 + t}$ into the above equation, we have that

$$B(\beta, \delta+1) = \frac{2}{t^{\beta+\delta}} \int_r^R (R^2 - u^2)^{\beta-1} \left(1 - \frac{r^2}{u^2}\right)^\delta u^{2\delta+1} du,$$

$$t^{\beta+\delta} = (R^2 - r^2)^{\beta+\delta} = \frac{2}{B(\beta, \delta+1)} \int_r^R (R^2 - u^2)^{\beta-1} u^{2\delta+1} \left(1 - \frac{r^2}{u^2}\right)^\delta du$$

and

$$\left(1 - \frac{r^2}{R^2}\right)^{\beta+\delta} = \frac{2}{B(\beta, \delta+1)} \frac{1}{R^{2(\beta+\delta)}} \int_r^R (R^2 - u^2)^{\beta-1} u^{2\delta+1} \left(1 - \frac{r^2}{u^2}\right)^\delta du.$$

It follows that

$$S_R^{\beta+\delta}(f; x) = \sum_{|m|<R} \left\{ \frac{2}{B(\beta, \delta+1)} \frac{1}{R^{2(\beta+\delta)}} \right.$$

$$\left. \times \int_{|m|}^R (R^2 - u^2)^{\beta-1} u^{2\delta+1} \left(1 - \frac{|m|^2}{u^2}\right)^\delta du \right\} C_m(f) e^{im \cdot x}$$

$$= \frac{2}{B(\beta, \delta+1) R^{2(\beta+\delta)}} \int_0^R (R^2 - u^2)^{\beta-1} u^{2\delta+1} \cdot$$

$$\left(\sum_{|m|<R} \left(1 - \frac{|m|^2}{u^2}\right)^\delta \chi_{(|m|,R)}(u) C_m(f) e^{im \cdot x} \right) du$$

$$= \frac{2}{B(\beta, \delta+1) R^{2(\beta+\delta)}} \int_0^R (R^2 - u^2)^{\beta-1} u^{2\delta+1} S_u^\delta(f; x) du.$$

This completes the proof. ∎

Lemma 3.4.3 *Let* $\operatorname{Re}\beta > 0$, $\delta > -1$. *Then we have*

$$S_*^{\beta+\delta}(f)(x) \leq \frac{B(\sigma, \delta+1)}{|B(\beta, \delta+1)|} S_*^\delta(f)(x) \tag{3.4.4}$$

for $\sigma = \operatorname{Re}\beta$.

3.4 L^p approximation

Proof. By (3.4.3) and the following estimate

$$\frac{1}{|R^{2(\beta+\delta)}|}\int_0^R |R^2-u^2|^{\beta-1}u^{2\delta+1}du$$

$$=\frac{1}{R^{2(\sigma+\delta)}}\int_0^R (R^2-u^2)^{\sigma-1}u^{2\delta+1}du$$

$$=\frac{1}{R^{2(\sigma+\delta)}}\cdot\int_0^1 R^{2(\sigma+\delta)}\cdot(1-t^2)^{\sigma-1}t^{2\delta+1}dt$$

$$=\frac{1}{2}B(\sigma,\delta+1),$$

we can get (3.4.4), which finishes the proof. ∎

Lemma 3.4.4 *Let* $\mathrm{Re}\beta > \frac{1}{2}$ *and* $\delta > -\frac{3}{4}$. *Then*

$$S_*^{\beta+\delta}(f)(x) \le \frac{\sqrt{2B(2\sigma-1, 2\delta+3/2)}}{|B(\beta,\delta+1)|} M^\delta(f)(x), \qquad (3.4.5)$$

where $\sigma = \mathrm{Re}\beta$.

Proof. By (3.4.3) and Hölder's inequality, we can get that

$$\left|S_R^{\beta+\delta}(f)(x)\right| \le \frac{2}{|B(\beta,\delta+1)|}\frac{1}{R^{2(\sigma+\delta)}}\left\{\int_0^R (R^2-u^2)^{2(\sigma-1)}u^{4\delta+2}du\right\}^{\frac{1}{2}}$$

$$\times \left\{\int_0^R |S_u^\delta(f;x)|^2 du\right\}^{\frac{1}{2}}$$

$$\le \frac{2}{|B(\beta,\delta+1)|}\cdot\frac{1}{R^{2(\sigma+\delta)-\frac{1}{2}}}\left\{\frac{B(2\sigma-1,2\delta+3/2)}{2}R^{4(\sigma+\delta)-1}\right\}^{\frac{1}{2}}$$

$$\times M^\delta(f)(x).$$

And thus we can obtain (3.4.5), which completes the proof. ∎

Corollary 3.4.1 *Let* $f \in L^2(Q^n)$ *with* $n > 1$ *and* $0 < \mathrm{Re}\alpha < n+1$. *Then we ahve*

$$\|S_*^\alpha(f)\|_2 \le A_n e^{A_n|\tau|}\frac{1}{\sigma}\|f\|_2, \qquad (3.4.6)$$

for $\sigma = \mathrm{Re}\alpha$ *and* $\tau = \mathrm{Im}\alpha$.

Proof. Choose $\beta = \frac{\sigma+1}{2} + i\tau$, $\delta = \frac{\sigma-1}{2}$. By (3.4.5), we have

$$S_*^\alpha(f)(x) \leq \frac{\sqrt{2B(\sigma, \sigma + \frac{1}{2})}}{\left|B\left(\frac{\sigma+1}{2} + i\tau, \frac{\sigma+1}{2}\right)\right|} M^{\frac{\sigma-1}{2}}(f)(x).$$

Now, we quote a formula on Gamma functions (see Bateman [Ba1]):

$$\Gamma(x + iy) = \Gamma(x)e^{-i\gamma y} x(x+iy)^{-1} \prod_{k=1}^\infty \frac{e^{i\frac{y}{k}}}{1 + \frac{iy}{k+x}}, \qquad (3.4.7)$$

where $x > 0$, $y \in \mathbb{R}$, γ is the Euler number.

We have that

$$\frac{1}{\left|B\left(\frac{\sigma+1}{2} + i\tau, \frac{\sigma+1}{2}\right)\right|} = \frac{|\Gamma(\sigma + 1 + i\tau)|}{\left|\Gamma\left(\frac{\sigma+1}{2} + i\tau\right)\right| \left|\Gamma\left(\frac{\sigma+1}{2}\right)\right|}$$

$$= \frac{\Gamma(\sigma+1)(\sigma+1)\left|(\sigma+1+i\tau)^{-1}\right|}{\left|\Gamma\left(\frac{\sigma+1}{2}\right)\right|^2 \cdot \frac{\sigma+1}{2} \left|\left(\frac{\sigma+1}{2} + i\tau\right)^{-1}\right|} \prod_{k=1}^\infty \frac{\left|1 + \frac{i\tau}{k + \frac{\sigma+1}{2}}\right|}{\left|1 + \frac{i\tau}{k+\sigma+1}\right|}$$

$$= O\left(\left\{\prod_{k=1}^\infty \frac{1 + \frac{\tau^2}{(k+\frac{\sigma+1}{2})^2}}{1 + \frac{\tau^2}{(k+\sigma+1)^2}}\right\}^{\frac{1}{2}}\right)$$

$$= O\left(\left\{\prod_{k=1}^\infty \frac{(k+\sigma+1)^2}{(k+\frac{\sigma+1}{2})^2} \frac{(k+\frac{\sigma+1}{2})^2 + \tau^2}{(k+\sigma+1)^2 + \tau^2}\right\}^{\frac{1}{2}}\right)$$

$$= O\left(\sqrt{\prod_{k=1}^\infty \left(1 + \frac{4n\tau^2}{k^3 + k\tau^2}\right)}\right)$$

$$= O\left(\sqrt{\prod_{k=1}^\infty \left(1 + \frac{2n|\tau|}{k^2}\right)}\right).$$

Using the following fact

$$1 + \frac{2n\tau}{k^2} = \left(1 + \frac{2n\tau}{k^2}\right)^{\frac{k^2}{2n\tau} \frac{2n\tau}{k^2}} \leq e^{\frac{2n\tau}{k^2}},$$

we have that

$$\frac{1}{\left|B\left(\frac{\sigma+1}{2} + i\tau, \frac{\sigma+1}{2}\right)\right|} \leq A_n e^{A_n|\tau|} \qquad (3.4.8)$$

3.4 L^p approximation

for $0 < \sigma < n+1$.

Hence
$$S_*^\alpha(f)(x) \leq A_n \frac{1}{\sqrt{\sigma}} e^{A_n|\tau|} M^{\frac{\sigma-1}{2}}(f)(x). \tag{3.4.9}$$

Then, by (3.4.1) we can get (3.4.6).

This completes the proof. ∎

Theorem 3.4.1 (Stein) *If* $1 < p \leq 2$, $n > 1$, $\frac{n-1}{2} \geq \alpha > \alpha_p := \frac{n-1}{2}\left(\frac{2}{p} - 1\right)$, *then we have*
$$\|S_*^\alpha(f)\|_p \leq A_{n,\alpha,p}\|f\|_p.$$

The proof of the above theorem needs the interpolation theorem of the analytic family of operators (see Stein and Weiss [SW1]). The steps of the proof are as follows: Firstly, by Corollary 3.4.1, S_*^α of positive order (Re$\alpha > 0$) is of type $(2,2)$; secondly, by Corollary 3.1.1, S_*^α of the order higher than $\frac{n-1}{2}$ (Re$\alpha > \frac{n-1}{2}$) is of type (q,q) $(1 < q \leq \infty)$. Therefore, for any given $p \in (1,2)$ and $0 < \alpha \leq \frac{n-1}{2}$, we assume that μ_0 is taken close to zero, $0 < \mu_0 < \alpha$, while we choose $\mu_1 > \frac{n-1}{2}$. Make an analytic family $S_*^{\delta(z)}$, $\delta(z) = \mu_0(1-z) + \mu_1 z$ (which is not a family of linear operators, for further treatments needed), then, by Re$\delta(i\tau) = \mu_0 > 0$ and Re$\delta(1+i\tau) = \mu_1 > \frac{n-1}{2}$, we can get an estimate of $S_*^{\delta(i\tau)}$ in type $(2,2)$ as well as that of $S^{\delta(Hi\delta)}$ in type (p_1, p_1), where p_1 is some chosen positive number, $1 < p_1 < p$. Then we interpolate at t, which is decided by
$$\frac{1-t}{2} + \frac{t}{p_1} = \frac{1}{p},$$

that is,
$$t = \frac{\frac{1}{p} - \frac{1}{2}}{\frac{1}{p_1} - \frac{1}{2}} \in (0,1). \tag{3.4.10}$$

μ_0, μ_1 must satisfy the condition that
$$\delta(t) = \mu_0(1-t) + \mu_1 t = \alpha. \tag{3.4.11}$$

Hence we can induce that S_*^α is of type (p,p). Actually, we can select $p_1 \in (1,p)$ firstly, and then give the option of t to (3.4.10). Substituting the value of t into (3.4.11), then we can choose proper μ_0, μ_1 by (3.4.11).

Here we have to point out that: if α is not given ahead, then by (3.4.11), since $\mu_0 > 0$, $\mu_1 > \frac{n-1}{2}$, we have

$$\alpha > \frac{n-1}{2} t$$

$$= \frac{n-1}{2} \frac{\frac{1}{p} - \frac{1}{2}}{\frac{1}{p_1} - \frac{1}{2}}$$

$$\geq \frac{n-1}{2} \left(\frac{2}{p} - 1 \right)$$

$$:= \alpha_p.$$

That is to say that the index α, which is applicable for the method of interpolation, must be greater than α_p.

Proof of Theorem 3.4.1. When $p = 2$, the conclusion of Theorem 3.4.1 is contained in Corollary 3.4.1. If we assume $1 < p < 2$ and choose

$$p_1 = 1 + \frac{\alpha - \alpha_p}{3\alpha + \alpha_p}(p-1) \in (1, p),$$

then, we have

$$t = \frac{\frac{1}{p} - \frac{1}{2}}{\frac{1}{p_1} - \frac{1}{2}}.$$

Choose

$$\mu_1 = \frac{1}{\frac{2}{p} - 1} \frac{\alpha + \alpha_p}{2} > \frac{n-1}{2}$$

and

$$\mu_1 < \frac{n-1}{2} \frac{1}{2-p}.$$

It follows that

$$\mu_0 = \frac{\frac{1}{p_1} - \frac{1}{2}}{\frac{1}{p_1} - \frac{1}{p}} \left(\alpha - \frac{1}{\frac{2}{p_1} - 1} \frac{\alpha + \alpha_p}{2} \right)$$

$$= \frac{p(2-p)}{2(p-1)}(\alpha - \alpha_p)\frac{3\alpha + \alpha_p}{4(\alpha + \alpha_p)}$$

$$< \frac{n-1}{2} \cdot \frac{3}{4},$$

and $\mu_0 > 0$.

3.4 L^p approximation

Let $\delta(z) = \mu_0(1-z) + \mu_1 z$, $0 \leq \operatorname{Re} z = \sigma \leq 1$, $\operatorname{Im} z = \tau$. By Corollary 3.4.1, we have that

$$|\operatorname{Im}\delta(i\tau)| = (\mu_1 - \mu_0)|\tau| < \frac{n-1}{2}\frac{1}{2-p}|\tau|,$$

and

$$\left\|S_*^{\delta(i\tau)}(f)\right\|_2 \leq A_n e^{A_n \frac{1}{2-p}|\tau|} \frac{1}{\mu_0}\|f\|_2. \tag{3.4.12}$$

Similarly, from Corollary 3.1.1, we have

$$|\operatorname{Im}\delta(1+i\tau)| = (\mu_1 - \mu_0)|\tau| < \frac{n-1}{2} \times \frac{1}{2-p} \cdot |\tau|,$$

and therefore

$$\left\|S_*^{\delta(1+i\tau)}(f)\right\|_{p_1} \leq A_n e^{A_n \frac{1}{2-p}|\tau|} \left(\frac{1}{\mu_1 - \frac{n-1}{2}} + 1\right) \frac{p_1}{p_1 - 1}\|f\|_{p_1}. \tag{3.4.13}$$

Denote by $\mathscr{F}$ the collection of all non-negative bounded measurable functions defined on Q. Choose any $R = R(x) \in \mathscr{F}$.

The mapping

$$T_z = T_z^R : f(x) \to S_{R(x)}^{\delta(z)}(f;x),$$

for $f \in L(Q))$, is linear. The image of the mapping $S_{R(x)}^{\delta(z)}(f;x)$ is a bounded measurable function defined on Q, and the family $\{T_z : 0 \leq \operatorname{Re} z \leq 1\}$ is admissible.

Obviously, by (3.4.12) and (3.4.13), we can deduce that

$$|T_z(f)(x)| \leq S_R^{\delta(z)}(f;x). \quad (z = \sigma + i\tau),$$

$$\|T_{i\tau}(f)\|_2 \leq M_0(\tau)\|f\|_2, \tag{3.4.14}$$

$$\|T_{1+i\tau}(f)\|_{p_1} \leq M_1(\tau)\|f\|_{p_1}, \tag{3.4.15}$$

$$M_0(\tau) = A_n \frac{1}{2-p}\frac{p-1}{\alpha - \alpha_p} \cdot e^{A_n \frac{1}{2-p}|\tau|}, \tag{3.4.16}$$

$$M_1(\tau) = A_n \left(\frac{1}{\mu_1 - \alpha_1} + 1\right) \frac{\alpha + \alpha_p}{(\alpha - \alpha_p)(p-1)} \cdot e^{A_n \frac{1}{2-p}|\tau|}. \tag{3.4.17}$$

By Stein's interpolation theorem, for $t = \frac{\frac{1}{p} - \frac{1}{2}}{\frac{1}{p_1} - \frac{1}{2}}$, we get that

$$\|T_t(f)\|_p \leq M_t \|f\|_p, \tag{3.4.18}$$

$$M_t = \exp\left\{\frac{1}{2}\sin t\pi \int_{-\infty}^{\infty} \left(\frac{\log M_0(\tau)}{\operatorname{ch}\pi\tau - \cos\pi t} + \frac{\log M_1(\tau)}{\operatorname{ch}\pi\tau + \cos\pi\tau}\right) d\tau\right\}. \tag{3.4.19}$$

And (3.4.18) gives that

$$\|S_{R(x)}^\alpha(f;x)\|_p \leq A_{n,\alpha,p}\|f\|_p, \quad \forall R \in \mathscr{F}. \tag{3.4.20}$$

Choose $R_0 > 1$ arbitrarily. Line all the rational numbers in $[0, R_0)$ as a sequence of $\{R_j\}_{j=1}^\infty$. Obviously,

$$\sup_{0 < R < R_0} |S_R^\alpha(f;x)| = \sup_{j \in \mathbb{N}}\{|S_{R_j}^\alpha(f;x)|\}.$$

Let

$$E_j = \left\{x \in Q^n : \sup_{0 < R < R_0} |S_R^\alpha(f;x)| \leq |S_{R_j}^\alpha(f;x)| + \frac{1}{R_0}\right\}, \quad j \in \mathbb{N}.$$

Define

$$F_1 = E_1, \quad F_{j+1} = E_{j+1} \setminus \bigcup_{k=1}^j E_k, \quad j \in \mathbb{N},$$

and

$$R_0(x) = R_j, \quad \text{when } x \in F_j, \; j \in \mathbb{N}.$$

Then $R_0(x) \in \mathscr{F}$ and

$$\sup_{0 < R < R_0} |S_R^\alpha(f;x)| \leq |S_{R_0(x)}^\alpha(f;x)| + \frac{1}{R_0}.$$

By the Fatou lemma, we can deduce that

$$\|S_*^\alpha(f)\|_p \leq \liminf_{R_0 \to +\infty} \|S_{R_0(x)}^\alpha(f;x)\|_p.$$

Thus we have transfered from (3.4.20) to

$$\|S_*^\alpha(f)\|_p \leq A_{n,\alpha,p}\|f\|_p \tag{3.4.21}$$

for all $f \in L^p(Q^n)$, which completes the proof of Theorem 3.4.1. ∎

3.4 L^p approximation

If we restrict $p \in (1, 3/2]$, $n > 1$, $\alpha = \frac{n-1}{2}$, and in the proof of Theorem 3.4.1, we choose
$$p_1 = \frac{p+1}{2}, \quad \mu_0 = \frac{n-1}{4}$$
and
$$\mu_1 = \frac{n-1}{2} \frac{1+2p-p^2}{2+p-p^2},$$
then by the method of interpolation, we can obtain that:

Theorem 3.4.2 *If $1 < p \leq 3/2$ and $n > 1$, then we have*
$$\left\| S_*^{\frac{n-1}{2}}(f) \right\|_p \leq A_n \frac{1}{(p-1)^2} \|f\|_p. \tag{3.4.22}$$

Proof. From the proof of Theorem 3.4.1, we can see that: the constant in (3.4.21) actually can be chosen as M_t in (3.4.19). When $1 < p \leq 3/2$, and $\alpha = \alpha_1 = \frac{n-1}{2}$, with the above interpolation method, we can compute the corresponding $M_0(\tau)$ and $M_1(\tau)$ as
$$M_0(\tau) = A_n e^{A_n |\tau|}, \quad M_1(\tau) = A_n \frac{1}{(p-1)^2} e^{A_n |\tau|}.$$

With the help of
$$\frac{1}{2} \sin \pi t \int_{-\infty}^{\infty} \left(\frac{1}{\operatorname{ch} \pi \tau - \cos \pi t} + \frac{1}{\operatorname{ch} \pi \tau + \cos \pi t} \right) d\tau = 1$$
and
$$\frac{1}{2} \sin \pi t \int_{-\infty}^{\infty} \left(\frac{|\tau|}{\operatorname{ch} \pi \tau - \cos \pi t} + \frac{|\tau|}{\operatorname{ch} \pi \tau + \cos \pi t} \right) d\tau \leq A < +\infty,$$
we can get that
$$M_t = e^{\log \frac{A_n}{(p-1)^2} + A_n} \leq A_n \frac{1}{(p-1)^2}.$$
It follows (3.4.22), which completes the proof. ∎

Similar to the proof of Theorem 3.4.1, if we interpolate between 2 and $+\infty$, then we can get the conclusion of $p > 2$. At this time, the critical index for L^p is $\alpha_p = \frac{n-1}{2}\left(1 - \frac{2}{p}\right)$. Detailed steps are much simpler. And t is decided by the interpolation formula
$$\frac{1-t}{2} + \frac{t}{\infty} = \frac{1}{p},$$

that is, $t = 1 - \frac{2}{p}$. And then we choose $\mu_0 > 0$, $\mu_1 > \alpha_1 = \frac{n-1}{2}$, satisfying

$$\mu_0(1-t) + \mu_1 t = \alpha.$$

By the above equation, as long as $\alpha > \alpha_p = \frac{n-1}{2}(1 - \frac{2}{p})$, we can make it, and that is where the meaning of the critical property of α_p lies. We can take

$$\mu_1 = \frac{\alpha + \alpha_p}{2t} = \frac{\alpha + \alpha_p}{2(p-2)}p, \quad \mu_0 = \frac{\alpha - \alpha_p}{4}p,$$

where

$$\alpha_p < \alpha \leq \alpha_1 = \frac{n-1}{2}.$$

Making an estimate in type $(2,2)$ with (3.4.6), we can get

$$\left\| S_*^{\mu_0 + i\tau(\mu_1 - \mu_0)}(f) \right\|_2 \leq A_n e^{A_n(\mu_1 - \mu_0)|\tau|} \frac{1}{\mu_0} \|f\|_2. \tag{3.4.23}$$

Make an estimate in type (∞, ∞) with Corollary (3.1.1) (see (3.1.12))

$$\left\| S_*^{\mu_1 + i\tau(\mu_1 - \mu_0)}(f) \right\|_\infty \leq A_n e^{A_n(\mu_1 - \mu_0)|\tau|} \left(\frac{1}{\mu_1 - \alpha_1} + 1 \right) \|f\|_\infty. \tag{3.4.24}$$

Therefore, we can easily obtain the following theorem.

Theorem 3.4.3 Let $p > 2$, $n > 1$, $\frac{n-1}{2} \geq \alpha > \alpha_p := \frac{n-1}{2}\left(1 - \frac{2}{p}\right)$. The inequality (3.4.21) still holds.

In detail, we substitute with $\alpha = \alpha_1 = \frac{n-1}{2}$, and we can see from (3.4.23), (3.4.24) the corresponding quantities: $\mu_1 = \frac{n-1}{2}\frac{p-1}{p-2}$, $\mu_0 = \frac{n-1}{4}$, $\mu_1 - \mu_0 = \frac{n-1}{4}\frac{p}{p-2}$, and

$$M_0(\tau) = A_n e^{A_n \frac{p}{p-2}|\tau|},$$
$$M_1(\tau) = A_n e^{A_n \frac{p}{p-2}|\tau|} p.$$

And thus we get the following statement.

Theorem 3.4.4 Let $p > 3$, $n > 1$, then

$$\left\| S_*^{\frac{n-1}{2}}(f) \right\|_p \leq A_n p \|f\|_p. \tag{3.4.25}$$

Combining (3.4.22) with (3.4.25), we have

$$\left\| S_*^{\frac{n-1}{2}}(f) \right\|_p \leq A_n \frac{p^3}{(p-1)^2} \|f\|_p \tag{3.4.26}$$

for $1 < p < \infty$.

3.4 L^p approximation

3.4.2 The L^p approximation

With the estimate of the maximal operator S_*^α in type (p,p), $1 < p < \infty$, the L^p-convergence of $S_R^\alpha(f)$ is a natural corollary. Now let us compute it more accurately and then we can acquire the approximation order characterized by the modulus of continuity.

Let $f \in L^p(Q^n)$. Define

$$\omega(f;t)_p = \sup_{|h|\leq t} \|f(\cdot + h) - f(\cdot)\|_p \quad (t \geq 0)$$

as the modulus of continuity of f in L^p.

Denote by $E_R(f)_p$ the best approximation of trigonometric polynomial f with order below R (spherical order) in the norm of L^p. If $g_R(x)$ is the best L^p approximation trigonometric polynomial with order R, then by the well-known theorem about the relation between trigonometric polynomial and the norm of its derivative, we can get

$$\|\triangle g_R\|_p \leq C_{n,p} R^2 \omega_2\left(f; \frac{1}{R}\right)_p, \qquad (3.4.27)$$

where ω_2 represents for 2-ordered the modulus of continuity, and its definition is as follows

$$\omega_2(f;t)_p = \sup_{|h|<t} \|f(\cdot + h) + f(\cdot - h) - 2f(\cdot)\|_p$$

and $\triangle$ is the well-known Laplace operator.

Having got all the above preparations, it is time for us to prove the following theorem on the L^p approximation:

Theorem 3.4.5 *If $n > 1$, $1 < p < \infty$, $\alpha > \alpha_p := \frac{n-1}{2}\left|\frac{2}{p} - 1\right|$, then we have*

$$\|S_R^\alpha(f) - f\|_p \leq C_{n,\alpha,p} \cdot \omega_2\left(f; \frac{1}{R}\right)_p. \qquad (3.4.28)$$

Proof. Suppose that $R > 0$, $g = g_R$ is the best R-ordered L^p approximation trigonometric polynomial. Then,

$$\|S_R^\alpha(f) - f\|_p \leq \|S_R^\alpha(f) - S_R^\alpha(g)\|_p + \sum_{j=1}^n \|S^{\alpha+j-1}(g) - S_R^{\alpha+j}(g)\|_p$$

$$+ \|S_R^{\alpha+n}(g) - g\|_p + \|g - f\|_p. \qquad (3.4.29)$$

By Theorem 3.4.1 (and (3.4.3)), we can get that
$$\|S_R^\alpha(f) - S_R^\alpha(g)\|_p \le \|S_*^\alpha(f-g)\|_p \le C_{n,\alpha,p}\|f-g\|_p,$$
and with the Jackson inequality
$$\|f-g\|_p = E_R(f)_p \le C_{n,p}\,\omega_2\left(f;\frac{1}{R}\right)_p \qquad (3.4.30)$$
we have
$$\|S_R^\alpha(f) - S_R^\alpha(g)\|_p \le C_{n,\alpha,p}\|f-g\|_p \le C_{n,\alpha,p}\,\omega_2\left(f;\frac{1}{R}\right)_p. \qquad (3.4.31)$$

Besides
$$S_R^{\alpha+j-1}(g) - S_R^{\alpha+j}(g) = \sum_{|m|<R}\left(1-\frac{|m|^2}{R^2}\right)^{\alpha+j-1}\frac{|m|^2}{R^2}C_m(g)e^{im\cdot x}$$
$$= \frac{-1}{R^2}S_R^{\alpha+j-1}(\Delta g) \quad \forall\, j.$$

With the estimate in the strong type and (3.4.27), it follows that
$$\left\|S_R^{\alpha+j-1}(g) - S_R^{\alpha+j}(g)\right\|_p \le C_{n,\alpha,p}\frac{1}{R^2}\|\Delta g\|_p$$
$$\le C_{n,\alpha,p}\,\omega_2\left(f;\frac{1}{R}\right)_p, \qquad (3.4.32)$$
for $j = 1, 2, \ldots, n$.

Finally, it is obvious that $\alpha + n > \frac{n-1}{2}$. In the next lemma, we will prove that when $\beta > \frac{n-1}{2}$,
$$\|S_R^\beta(f) - f\|_p \le C_{n,\beta,p}\cdot\omega_2\left(f;\frac{1}{R}\right)_p. \qquad (3.4.33)$$

Then, substituting $\beta = \alpha + n$ into (3.4.33), and combining (3.4.32) with (3.4.31), we put them together into (3.4.29), and thus we get (3.4.28). This finishes the proof of the theorem. ∎

Lemma 3.4.5 *If* $\beta > \frac{n-1}{2}$, $1 < p < \infty$, *then (3.4.33) is valid.*

3.4 L^p approximation

Proof. By Bochner's formula (1.2),

$$S_R^\alpha(f;x) - f(x) = C_{n,\beta} \cdot R^{\frac{n}{2}-\beta} \int_0^\infty [f_x(t) - f(x)] \frac{J_{\frac{n}{2}+\beta}Rt}{t^{\beta+1-\frac{n}{2}}} dt. \quad (3.4.34)$$

Denote

$$\varphi_x(t) = f_x(t) - f(x) = \frac{1}{\omega_{n-1}} \int_{\mathbb{S}^{n-1}} [f(x - t\xi) - f(x)] d\sigma(\xi).$$

Divide the sphere $\mathbb{S}^{n-1}$ into two parts:

$$S^+ = \{\xi : |\xi| = 1, \; \xi_1 > 0\}$$

and

$$S^- = \{\xi : |\xi| = 1, \; \xi_1 < 0\},$$

then

$$\varphi_x(t) = \frac{1}{\omega_{n-1}} \int_{S^+} [f(x+t\xi) + f(x-t\xi) - 2f(x)] d\sigma(\xi).$$

Rewrite (3.4.34) as

$$\begin{aligned}
&S_R^\alpha(f;x) - f(x) \\
&= C_{n,\beta} \cdot \int_0^\infty \varphi_x\left(\frac{t}{R}\right) t^{n-1} \frac{J_{\frac{n}{2}+\beta}(t)}{t^{\frac{n}{2}+\beta}} dt \\
&= C_{n,\beta} \left\{ \int_0^1 \varphi_x\left(\frac{t}{R}\right) t^{n-1} \cdot \frac{J_{\frac{n}{2}+\beta}(t)}{t^{\frac{n}{2}+\beta}} dt + \int_1^\infty \varphi_x\left(\frac{t}{R}\right) t^{n-1} \frac{J_{\frac{n}{2}+\beta}(t)}{t^{\frac{n}{2}+\beta}} dt \right\} \\
&= I_1 + I_2.
\end{aligned} \quad (3.4.35)$$

$$\begin{aligned}
\|I_1\|_p &\leq C \left\{ \int_Q \left| \int_0^1 \varphi_x\left(\frac{t}{R}\right) t^{n-1} dt \right|^p \cdot dx \right\}^{1/p} \\
&\leq C \int_0^1 \left\{ \int_Q \left| \varphi_x\left(\frac{t}{R}\right) t^{n-1} \right|^p dx \right\}^{1/p} dt \\
&\leq C \int_0^1 \int_{S^+} \left\{ \int_Q \left| f\left(x + \frac{t}{R}\xi\right) + f\left(x - \frac{t}{R}\xi\right) - 2f(x) \right|^p dx \right\}^{\frac{1}{p}} d\sigma(\xi) dt \\
&\leq C_{n,\beta,p} \cdot \omega_2\left(f; \frac{1}{R}\right)_p.
\end{aligned} \quad (3.4.36)$$

In order to estimate the L^p norm of I_2, with the asymptotic formula (3.3.8) of Bessel functions, we separate I_2 into three terms. In the following, we omit the constants and write down that

$$I_2 = \int_1^\infty \varphi_x\left(\frac{t}{R}\right) \frac{\cos(t-\theta)}{t^{\beta-\frac{n}{2}+\frac{3}{2}}} dt + \int_1^\infty \varphi_x\left(\frac{t}{R}\right) \frac{\sin(t-\theta)}{t^{\beta-\frac{n}{2}+\frac{5}{2}}} dt$$
$$+ \int_1^\infty \varphi_x\left(\frac{t}{R}\right) \cdot O\left(\frac{1}{t^{\beta-\frac{n}{2}+\frac{7}{2}}}\right) dt \qquad (3.4.37)$$
$$= J_1 + J_2 + J_3.$$

Let $\delta = \beta - \frac{n-1}{2} > 0$, and we denote $\varphi_x\left(\frac{t}{R}\right) \frac{1}{t^{\beta-\frac{n}{2}+\frac{3}{2}}} = \varphi_x\left(\frac{t}{R}\right) \frac{1}{t^{1+\delta}} = h(t)$. Then we have

$$J_1 = \int_1^\infty h(t) \cos(t-\theta) dt$$
$$= \frac{1}{8} \int_1^\infty \Delta_\pi^3 h(t) \cos(t-\theta) dt - \int_{1-\pi}^1 \left[\frac{1}{8}\Delta_\pi^2 h(t+\pi)\right. \qquad (3.4.38)$$
$$\left. + \frac{1}{4}\Delta_\pi h(t+\pi) + \frac{1}{2}h(t+\pi)\right] \cdot \cos(t-\theta) dt,$$

where 'Δ_π' is a differential operation for the step of π:

$$\Delta_\pi h(t) = h(t) - h(t+\pi), \quad \Delta_\pi^2 = \Delta_\pi(\Delta_\pi),$$
$$\Delta_\pi^3 = \Delta_\pi(\Delta_\pi^2).$$

Then just the same as the estimate of $\|I_1\|_p$, we can obtain that

$$\left\{\int_Q \left|\int_{1-\pi}^1 \left[\frac{1}{8}\Delta_\pi^2 h(t+\pi) + \frac{1}{4}\Delta_\pi h(t+\pi) + \frac{1}{2}h(t+\pi)\right] \cos(t-\theta) dt\right|^p dx\right\}^{1/p}$$
$$\leq C_{n,\beta,p} \cdot \omega_2\left(f; \frac{1}{R}\right)_p. \qquad (3.4.39)$$

We substitute $u(t) = \varphi_x\left(\frac{t}{R}\right)$ and $v(t) = \frac{1}{t^{1+\delta}}$ into the formula

$$\Delta_\pi^3(u(t)v(t)) = \Delta_\pi^3 v(t) v(t) + 3\Delta_\pi^2 u(t+\pi) \Delta_\pi v(t)$$
$$+ 3\Delta_\pi u(t+2\pi) \Delta_\pi^2 v(t) + u(t+3\pi) \Delta_\pi^3 v(t),$$

and with the notice of the fact that when $t \geq 1$,

3.4 L^p approximation

$$\|\triangle_\pi^3 u(t)\|_{L^p(Q)} = O\left(\|\triangle_\pi^2 u(t)\|_{L^p(Q)}\right) = O\left(\omega_2\left(f;\frac{1}{R}\right)_p\right),$$

$$\|\triangle_\pi u(t)\|_{L^p(Q)} = O\left(\|u(t)\|_{L^p(Q)}\right) = O\left(\omega_2\left(f;\frac{t}{R}\right)_p\right)$$

$$= O\left(t^2\omega_2\left(f;\frac{1}{R}\right)_p\right),$$

$$\triangle_\pi v(t) = O\left(\frac{1}{t^{2+\delta}}\right),\ \triangle_\pi^2 v(t) = O\left(\frac{1}{t^{3+\delta}}\right),$$

$$\triangle_\pi^3 v(t) = O\left(\frac{1}{t^{4+\delta}}\right),$$

we can acquire that

$$\left\|\int_1^\infty \triangle_\pi^3 h(t)\cos(t-\theta)dt\right\|_{L^p(Q)}$$

$$= O\left\{\int_1^\infty \frac{\omega_2\left(f;\frac{1}{R}\right)_p}{t^{1+\delta}}dt + \int_1^\infty \frac{\omega_2\left(f;\frac{t}{R}\right)_p}{t^{3+\delta}}dt\right\}$$

$$= O\left(\omega_2\left(f;\frac{1}{R}\right)_p\right).$$

Consequently, we have

$$\|J_1\|_p = O\left(\omega_2\left(f;\frac{1}{R}\right)_p\right).$$

For J_2, we have the same estimate.
Finally, we conclude that

$$\|J_3\|_p = \left\|\int_1^\infty \varphi_x\left(\frac{t}{R}\right)O\left(\frac{1}{t^{3+\delta}}\right)dt\right\|_p$$

$$= O\left(\int_1^\infty \left\|\varphi_x\left(\frac{t}{R}\right)\right\|_p \frac{1}{t^{3+\delta}}dt\right)$$

$$= O\left(\omega_2\left(f;\frac{1}{R}\right)_p\right).$$

Combining all of the above results together, Lemma 3.4.5 can be proved. ∎

We notice that $\alpha_1 = \alpha_\infty = \frac{n-1}{2}$, and it is obvious that Lemma 3.4.5 is valid for $p = 1$ and $p = \infty$. It is to be noted that we have taken $L^\infty(Q)$ to represent for $C(Q)$, and thus Theorem 3.4.5 is also valid for $p = 1$ and $p = \infty$.

By the way, we would like to point out that the above approximation results in the case of $p = \infty$, that is, in the situation of $C(Q)$, were firstly given by Cheng and Chen [CheC1](in the form of unitary ordered continuous norm). For the case of L^p with $1 \le p < \infty$, Wang [Wa6] considered the uniform approximation problems with the order of $\alpha > \frac{n-3}{2}$.

We have seen that Theorem 3.4.5 only figured out the order of L^p approximation above the critical order α_p. As far as the case of being below or equaling to the order of α_p, there is no result in that aspect.

3.5 Almost everywhere convergence (the critical index)

First of all, we will consider the problem of almost everywhere convergence in the case at the critical index $(\alpha_1 = \frac{n-1}{2})$ in the context of $L(Q^n)$.

Theorem 3.5.1 (Stein) *If $n > 1$ and $f \in L(\log^+ L)^2(Q^n)$, then we have*

$$\left\|S_*^{\frac{n-1}{2}}(f)\right\|_1 \le A_n \left\{\int_Q |f|(\log^+ |f|)^2 dx + 1\right\}.$$

Proof. According to Theorem 3.4.2, we have

$$\left\|S_*^{\frac{n-1}{2}}(g)\right\|_p \le \frac{A_n}{(p-1)^2}\|g\|_p, \qquad (3.5.1)$$

for $1 < p < 2$.
  Let

$$E_0 = \{x \in Q^n : |f(x)| < 1\},$$

and

$$E_k = \left\{x \in Q^n : 2^{k-1} \le |f(x)| < 2^k\right\},$$

for $k = 1, 2, 3, \ldots$.

3.5 Almost everywhere convergence (the critical index)

Set $f_k = \chi_{E_k} f$, and make a periodical extension, then

$$f(x) = \sum_{k=0}^{\infty} f_k(x). \tag{3.5.2}$$

We first assume that f is a simple function, then the summation of (3.5.2) is actually finite. Therefore,

$$\left\| S_*^{\frac{n-1}{2}}(f) \right\|_1 \leq \sum_{k=0}^{\infty} S_*^{\frac{n-1}{2}}(f_k),$$
$$\left\| S_*^{\frac{n-1}{2}}(f) \right\|_1 \leq \sum_{k=0}^{\infty} \left\| S_*^{\frac{n-1}{2}}(f_k) \right\|_1. \tag{3.5.3}$$

We take $g = f_k$, $p = 1 + \frac{1}{k}$, $k = 1, 2, \ldots$, and then substitute them into (3.5.1)

$$\left\| S_*^{\frac{n-1}{2}}(f_k) \right\|_1 \leq A_n \left\| S_*^{\frac{n-1}{2}}(f_k) \right\|_p$$
$$\leq A_n k^2 \|f_k\|_p$$
$$\leq A_n k^2 2^k |E_k|^{\frac{k}{k+1}}.$$

By Young's inequality,

$$a \cdot b \leq \frac{1}{p} a^p + \frac{1}{q} b^q, \quad \frac{1}{p} + \frac{1}{q} = 1, \quad p = 1 + \frac{1}{k}, \quad q = k+1$$

$$4|E_k|^{\frac{k}{k+1}} \cdot \frac{1}{4} \leq \frac{1}{1+\frac{1}{k}} \left(4|E_k|^{\frac{k}{k+1}}\right)^{\frac{k+1}{k}} + \frac{1}{k+1} \frac{1}{4^{k+1}},$$

we can deduce that

$$\left\| S_*^{\frac{n-1}{2}}(f_k) \right\|_1 \leq A_n \cdot \left(k^2 \cdot 2^k |E_k| + \frac{k^2}{k+1} \frac{1}{2^k}\right),$$

for $k = 1, 2, \ldots$.

At the same time, we have

$$\left\| S_*^{\frac{n-1}{2}}(f_0) \right\|_1 \leq A_n \left\| S_*^{\frac{n-1}{2}}(f_0) \right\|_2 \leq A_n \|f_0\|_2 \leq A_n.$$

Substituting all the above into (3.5.3), we obtain

$$\left\|S_*^{\frac{n-1}{2}}(f)\right\|_1 \le A_n \left\{\sum_{k=1}^{\infty} k^2 2^k |E_k| + 1\right\}$$

$$\le A_n \left\{\sum_{k=1}^{\infty} \int_Q (\log |f_k(x)|)^2 |f_k(x)| dx + 1\right\}$$

$$= A_n \left\{\int_Q |f(x)|(\log^+ |f(x)|)^2 dx + 1\right\}, \qquad (3.5.4)$$

which is valid for every simple functions. Starting from here, it is not difficult to move onto the general situation. And that is the end of the proof. ∎

Theorem 3.5.2 (Stein) *If $f \in L(\log^+ L)^2(Q^n)$ and $n > 1$, then we have*

$$\lim_{R \to \infty} S_R^{\frac{n-1}{2}}(f; x) = f(x), \qquad (3.5.5)$$

for a.e. $x \in Q^n$.

Proof. The proof method is quite common: with the help of Theorem 3.5.1, we can choose good functions as the approximation of f in $L(Q^n)$, and then pass to the limit(ones may give a proof as an exercise). ∎

Theorem 3.2.2 in Section 2 asserts that: $S_R^{\frac{n-1}{2}}(f)$ may be divergent almost everywhere in $L(Q^n)$. Yet, Theorem 3.5.2 ensures that if $f \in L(\log^+ L)^2(Q^n)$, then $S_R^{\frac{n-1}{2}}(f)$ must be convergent almost everywhere.

It is well known that in the case of single variable, sufficient conditions that guarantee a.e.-convergence of the Fourier series are listed as follows:

(i) Marcinkiewicz's result: if

$$\int_0^h |f(x+t) - f(x)| dt = O\left(h/\log \frac{1}{h}\right) \quad (h \to 0) \qquad (3.5.6)$$

holds for every points in $[0, 2\pi]$, then $S_k(f; x) \to f(x)$ for a.e. $x \in [0, 2\pi]$;

(ii) Salem's result: if

$$\int_0^h \{f(x+t) - f(x-t)\} dt = o\left(h/\log \frac{1}{h}\right) \quad (h \to 0) \qquad (3.5.7)$$

3.5 Almost everywhere convergence (the critical index)

uniformly holds for $x \in [0, 2\pi]$, then $S_k(f;x) \to f(x)$ for a.e. $x \in [0, 2\pi]$.

In 1965, Chang [Cha2] made an extension of Marcinkiewicz-type criterion to the case of multi-variables. He proved

Theorem 3.5.3 *Let $f \in L\log^+ L(Q^n)$, $n > 1$. If f satisfies*

$$\frac{1}{h^n} \int_{|y|<h} |f(x+y) - f(x)| dy = O\left(\left(\log \frac{1}{h}\right)^{-1}\right) \quad (h \to 0), \quad (3.5.8)$$

at every point of the positive measure set $E \subset Q^n$, then $S_R^{\frac{n-1}{2}}(f;x)$ is a.e. convergent to $f(x)$ on E.

Limited by space, we omit the proof of this theorem here. In the following, we will formulate the Salem-type condition given by Lu [Lu2]. Lu defined

$$F(x,r) = \int_{B(x,r)} (f(y) - f(x)) dy, \quad (3.5.9)$$

as the spherical integral of $f(x)$ in [Lu2].

Theorem 3.5.4 *Suppose that $f \in L\log^+ L(Q^n)$ with $n > 1$ and $x \in Q^n$ is the Lebesgue point of f. At the point of x, if the condition*

$$\frac{1}{r^{n-1}}\{F(x, r+2h) + F(x,r) - 2F(x, r+h)\} = o\left(\frac{h}{\log 1/h}\right) \quad (3.5.10)$$

is uniformly valid for $r \in [h, r_0]$, as $h \to 0^+$, then we have

$$\lim_{R \to \infty} S_R^{\frac{n-1}{2}}(f;x) = f(x).$$

It is easy to see that if we let

$$F(x) = \int_0^x f(t) dt$$

then (3.5.7) can be rewritten as

$$F(x+h) + F(x-h) - 2F(x) = o\left(\frac{h}{\log 1/h}\right)$$

as $h \to 0^+$.

In this sense, we can say that (3.5.10) is the multi-dimensional analogy of (3.5.7).

Proof of Theorem 3.5.4. Since $f \in L\log^+ L(Q^n)$, then with Stein's theorem, that is, Theorem 3.2.3, we can transfer the problem of series into that of the integral.

Suppose that the point x considered is the origin which is the Lebesgue point of f. Let $g = f\chi_{Q^n}$. Then $g \in L(\mathbb{R}^n)$, and $x = 0$ is the Lebesgue point of g. It is enough to prove

$$B_R^{\frac{n-1}{2}}(g;0) \to g(0).$$

Of course, we could premise $f(0) = 0$, that is, $g(0) = 0$. We denote the spherical integral of g at the origin as

$$G(r) = \int_{B(0,r)} g(y) dy.$$

Let

$$\varphi(t) = \frac{1}{\omega_{n-1}} \int_{|\xi|=t} g(\xi) d\sigma(\xi)$$

for $t > 0$.

Then

$$G(r) = \omega_{n-1} \int_0^r \varphi(t) dt.$$

By (3.5.10), with the notice of

$$G(r+2h) + G(r) - 2G(r+h) = \omega_{n-1} \int_0^h \{\varphi(r+h+t) - \varphi(r+h-t)\} dt,$$

we can get that

$$\frac{1}{r^{n-1}} \int_0^h \{\varphi(r+h+t) - \varphi(r+h-t)\} dt = o(h/\log 1/h) \quad (3.5.11)$$

uniformly holds for $r \in [h, r_0]$, as $h \to 0^+$.

Since $x = 0$ is the Lebesgue point of g, and by the sufficient condition (see Chapter 2) about the convergence of $B_R^{\frac{n-1}{2}}(g)$ at Lebesgue points, it only suffices to prove that

$$\lim_{R \to \infty} \int_{\frac{\pi+\theta}{R}}^{r_0} \varphi(t) \frac{\cos(tR - \theta)}{t^n} dt = 0 \quad (3.5.12)$$

3.5 Almost everywhere convergence (the critical index)

holds for $\theta = \frac{n\pi}{2}$.

Let $R > \frac{2\pi + |\theta|}{r_0}$ and $m = [\frac{r_0 R - \theta}{\pi}]$. Then, we have

$$m \in \left(\frac{r_0 R - \theta}{\pi} - 1, \frac{r_0 R - \theta}{\pi}\right]$$

and $m > 2$. Divide the integral of (3.5.12) into two parts as follows,

$$\int_{\frac{\pi+\theta}{R}}^{r_0} \varphi(t) \frac{\cos(tR - \theta)}{t^n} dt = \left(\int_{\frac{\pi+\theta}{R}}^{\frac{m\pi+\theta}{R}} + \int_{\frac{m\pi+\theta}{R}}^{r_0}\right) \varphi(t) \frac{\cos(tR - \theta)}{t^n} dt.$$

Therefore, it follows from the absolute continuity of the integration that

$$\left|\int_{\frac{m\pi+\theta}{R}}^{r_0} \varphi(t) \frac{\cos(tR - \theta)}{t^n} dt\right| \le \int_{r_0 - \frac{\pi}{R}}^{r_0} |\varphi(t)| dt \frac{2^n}{r_0^n} = o(1).$$

Hence, we merely need to prove

$$I_R := \int_{\frac{\pi+\theta}{R}}^{\frac{m\pi+\theta}{R}} \varphi(t) \frac{\cos(tR - \theta)}{t^n} dt = o(1).$$

We have that

$$I_R = \sum_{j=1}^{m-1} \int_{j\pi}^{(j+1)\pi} \varphi\left(\frac{t+\theta}{R}\right) \frac{\cos t}{(t+\theta)^n} dt R^{n-1}$$

$$= \sum_{j=1}^{m-1} (-1)^{j+1} R^{n-1} \int_{-\frac{\pi}{2}}^{\frac{\pi}{2}} \varphi\left(\frac{(j+\frac{1}{2})\pi + \theta + t}{R}\right) \frac{\sin t}{((j+\frac{1}{2})\pi + \theta + t)^n} dt.$$

Denote $h_j = (j + \frac{1}{2})\pi + \theta$. It follows that

$$I_R = \sum_{j=1}^{m-1} (-1)^{j+1} R^{n+1} \int_0^{\frac{\pi}{2}} \left((h_j + t)^{-n} \varphi\left(\frac{h_j + t}{R}\right) - (h_j - t)^{-n} \varphi\left(\frac{h_j - t}{R}\right)\right)$$
$$\times \sin t \, dt.$$

Set

$$\alpha_j = R^{n-1} \int_0^{\frac{\pi}{2}} [(h_j + t)^{-n} - (h_j - t)^{-n}] \varphi\left(\frac{h_j + t}{R}\right) \sin t \, dt,$$

$$\beta_j = R^{n-1} \int_0^{\frac{\pi}{2}} [(h_j - t)^{-n} - h_j^{-n}] \left[\varphi\left(\frac{h_j - t}{R}\right) - \varphi\left(\frac{h_j - t}{R}\right)\right] \sin t \, dt$$

and
$$\gamma_j = \left(\frac{R}{h_j}\right)^n \frac{1}{R} \int_0^{\frac{\pi}{2}} \left[\varphi\left(\frac{h_j+t}{R}\right) - \varphi\left(\frac{h_j-t}{R}\right)\right] \sin t \, dt.$$

Then we have
$$I_R = \sum_{j=1}^{m-1} (-1)^{j+1}(\alpha_j + \beta_j + \gamma_j).$$

Since
$$|\alpha_j| \leq C_n \frac{R^{n-1}}{j^{n+1}} \int_0^{\frac{\pi}{2}} \left|\varphi\left(\frac{h_j+t}{R}\right)\right| dt \leq C_n \frac{R^n}{j^{n+1}} \int_{\frac{j\pi+\theta}{R}}^{\frac{(j+1)\pi+\theta}{R}} |\varphi(t)| dt,$$

we have
$$\sum_{j=1}^{m-1} |\alpha_j| \leq C_n \left(\sum_{j=1}^{m-2} \frac{R^n}{j^{n+2}} \int_0^{\frac{(j+1)\pi+\theta}{R}} |\varphi(t)| dt + \frac{R^n}{(m-1)^{n+1}} \int_0^{\frac{m\pi+\theta}{R}} |\varphi(t)| dt\right).$$

Noting that $x = 0$ is the Lebesgue point of g, and $g(0) = 0$, we have that
$$\int_0^r |\varphi(t)| dt = o(r^n),$$

as $r \to 0$.

It clearly follows that
$$\sum_{j=1}^{m-1} |\alpha_j| = o(1)$$

as $R \to \infty$.

Similarly, we also get
$$\sum_{j=1}^{m-1} |\beta_j| = o(1),$$

as $R \to \infty$.

Therefore, we immediately have
$$I_R = \sum_{j=1}^{m-1} (-1)^{j+1} \gamma_j + o(1), \tag{3.5.13}$$

as $R \to \infty$.

Denote
$$F_j(s) = \frac{1}{R} \int_0^s \left\{\varphi\left(\frac{h_j+t}{R}\right) - \varphi\left(\frac{h_j-t}{R}\right)\right\} dt$$

3.5 Almost everywhere convergence (the critical index)

for $s \in [0, \frac{\pi}{2}]$. By the condition of (3.5.11), we can get that if $s \in (0, \frac{\pi}{2}]$, then

$$\left(\frac{R}{h_j - s}\right)^{n-1} \int_0^{s/R} \left\{\varphi\left(\frac{h_j - s}{R} + \frac{s}{R} + t\right) - \varphi\left(\frac{h_j - s}{R} + \frac{s}{R} - t\right)\right\} dt$$
$$= \left(\frac{R}{h_j - s}\right)^{n-1} F_j(s)$$
$$= o\left(\frac{s}{R} \frac{1}{\log R/s}\right)$$

holds uniformly about $j = 1, 2, \ldots, m-1$, as $R \to \infty$.

From the above it follows that for $s \in (0, \frac{\pi}{2}]$,

$$F_j(s) = o\left(\frac{j^{n-1}}{R^n \log R}\right)$$

is valid uniformly about $j \in \{1, 2, \ldots, m-1\}$, as $R \to +\infty$.

Therefore, by the method of integration by parts, we can deduce that

$$\gamma_j = \left(\frac{R}{h_j}\right)^n \left\{F_j(t) \sin t \Big|_0^{\frac{\pi}{2}} - \int_0^{\frac{\pi}{2}} F_j(t) \cos t \, dt\right\} = o\left(\frac{1}{j \log R}\right)$$

uniformly about $j \in \{1, 2, \ldots, m-1\}$, as $R \to \infty$. And thus we have

$$\sum_{j=1}^{m-1} (-1)^{j+1} \gamma_j = o\left(\frac{1}{j \log R}\right) \sum_{j=1}^{m-1} \frac{1}{j} = o(1), \qquad (3.5.14)$$

as $R \to \infty$.

We substitute the formula (3.5.14) into (3.5.13), and then obtain $I_R = o(1)$.

This finishes the proof of Theorem 3.5.4. ∎

Of course, for a function belonging to $L\log^+ L(Q^n)$ with $n > 1$, if the condition of (3.5.10) is valid at every Lebesgue point, then the Bochner-Riesz means at the critical index converges almost everywhere.

Fan [Fa1] extended Theorem 3.5.4 to the case of classical groups.

Now let us turn our attention to discuss the problems of always almost everywhere convergence at the critical index $\alpha_2 = 0$ on $L^2(Q^n)$. Only a few results can be obtained in this aspect. Mathematicians have been investigating for a long time about the problem whether S_R^0 always almost everywhere converges on $L^2(Q^n)$, $n > 1$, yet no any results were acquired. In recent years, the majority of researchers are prone to negative conclusions. In the following, we will introduce a sufficient condition by Golubov [Go1].

Firstly, let us introduce some notations: Let $f \in L^2(Q^n)$, $n > 1$, and we denote

$$A_\nu = \left(\sum_{|m|^2 = \nu} |C_m(f)|^2 \right)^{\frac{1}{2}},$$

for $\nu = 0, 1, 2, \ldots$.

If $\{m \in \mathbb{Z}^n : |m|^2 = \nu\} = \emptyset$, then we take it for granted that $A_\nu = 0$.

Suppose that α is a non-negative and non-increasing function defined on $(0, 1]$, and we denote

$$\omega(\nu) = \int_{\frac{1}{\sqrt{\nu}}}^{1} \alpha(t) dt,$$

for $\nu \in \mathbb{N}$.

Theorem 3.5.5 Let $f \in L(Q^n)$ with $n > 1$. If

$$\int_0^\delta t^4 \alpha(t) dt = O\left(\delta^4 \int_\delta^1 \alpha(t) dt \right), \qquad (3.5.15)$$

as $\delta \to 0^+$, then the two conditions

$$\int_0^1 \alpha(t) \|f_x(t) - f(x)\|_2^2 dt < +\infty, \qquad (3.5.16)$$

and

$$\sum_{\nu=1}^{\infty} A_\nu^2 \omega(\nu) < \infty \qquad (3.5.17)$$

are equivalent.

Proof. Define

$$f_x(t) = \frac{1}{\omega_{n-1}} \int_{\mathbb{S}^{n-1}} f(x + t\xi) d\sigma(\xi),$$

where $\omega_{n-1} = \frac{2\pi^{n/2}}{\Gamma(n/2)}$, and for fixed t, as a function about x. Obviously we have $f_x(t) \in L^2(Q^n)$.

3.5 Almost everywhere convergence (the critical index)

The Fourier coefficients of $f_x(t)$ are

$$\frac{1}{(2\pi)^n}\int_Q f_x(t)e^{-im\cdot x}dx = \frac{1}{\omega_{n-1}}\int_{\mathbb{S}^{n-1}}\left\{\frac{1}{(2\pi)^n}\int_Q f(x+t\xi)e^{-im\cdot x}dx\right\}d\sigma(\xi)$$

$$= \frac{1}{\omega_{n-1}}\int_{\mathbb{S}^{n-1}} C_m(f)e^{it\cdot m\xi}d\sigma(\xi).$$

Let $k = \frac{n}{2} - 1$. By the formula

$$\int_{\mathbb{S}^{n-1}} e^{itm\cdot \xi}d\sigma(\xi) = (2\pi)^{\frac{n}{2}}\frac{J_{\frac{n}{2}-1}(t|m|)}{(t|m|)^{\frac{n}{2}-1}},$$

we can obtain that

$$\frac{1}{(2\pi)^n}\int_Q f_x(t)e^{-im\cdot x}dx = 2^k\Gamma(k+1)C_m(f)\frac{J_k(t|m|)}{(t|m|)^k}. \quad (3.5.18)$$

Hence,

$$\sigma(f_y(t))(x) = \sum_m 2^k\Gamma(k+1)C_m(f)\frac{J_k(t|m|)}{(t|m|)^k}e^{im\cdot x}. \quad (3.5.19)$$

By the Parseval equation and the fact that

$$\left.2^k\Gamma(k+1)\frac{J_k(u)}{u^k}\right|_{u=0} = 1,$$

we conclude that

$$\|f_x(t) - f(x)\|_2^2 = (2\pi)^n \sum_{m\neq 0} |C_m(f)|^2 \left|2^k\Gamma(k+1)\frac{J_k(t|m|)}{(t|m|)^k} - 1\right|^2 dt,$$

and

$$\int_0^1 \alpha(t)\|f_x(t) - f(x)\|_2^2 dt$$

$$= (2\pi)^n \sum_{\nu=1}^\infty \sum_{|m|^2=\nu} |C_m(f)|^2 \int_0^1 \alpha(t)\left|2^k\Gamma(k+1)\frac{J_k(t|m|)}{(t|m|)^k} - 1\right|^2 dt$$

$$= (2\pi)^n \sum_{\nu=1}^\infty A_\nu^2 \int_0^1 \alpha(t)\left|2^k\Gamma(k+1)\frac{J_k(t\sqrt{\nu})}{(t\sqrt{\nu})^k} - 1\right|^2 dt. \quad (3.5.20)$$

Assume that (3.5.16) holds. Since

$$J_k(t) = \sqrt{\frac{2}{\pi t}} \cos\left(kt - \frac{\pi}{2}k - \frac{\pi}{4}\right) + O\left(\frac{1}{t^{\frac{3}{2}}}\right),$$

as $t \to +\infty$, we know that there exists $M \geq 1$, such that

$$\left|2^k \Gamma(k+1)\frac{J_k(t\sqrt{\nu})}{(t\sqrt{\nu})^k} - 1\right|^2 > \frac{1}{2},$$

if $t\sqrt{\nu} \geq M$. Then we have

$$\sum_{\nu \geq M^2} A_\nu^2 \int_{\frac{M}{\sqrt{\nu}}}^1 \alpha(t)dt < +\infty.$$

It follows that

$$\omega(\nu) = \int_{\frac{1}{\sqrt{\nu}}}^1 \alpha(l)dt$$

$$= \left(\int_{\frac{1}{\sqrt{\nu}}}^{\frac{M}{\sqrt{\nu}}} + \int_{\frac{M}{\sqrt{\nu}}}^1\right)\alpha(t)dt$$

$$\leq \int_{\frac{M}{\sqrt{\nu}}}^1 \alpha(t)dt + \int_{\frac{1}{\sqrt{\nu}}}^{\frac{M}{\sqrt{\nu}}} (\sqrt{\nu}t)^4 \alpha(t)dt$$

$$\leq \int_{\frac{M}{\sqrt{\nu}}}^1 \alpha(t)dt + \nu^2 \int_0^{\frac{M}{\sqrt{\nu}}} t^4 \alpha(t)dt$$

$$\leq \int_{\frac{M}{\sqrt{\nu}}}^1 \alpha(t)dt + CM^4 \int_{\frac{M}{\sqrt{\nu}}}^1 \alpha(t)dt.$$

Consequently, (3.5.17) holds.

On the other hand, assume that (3.5.17) holds. By

$$\frac{2^k \Gamma(k+1) J_k(t)}{t^k} = 1 + O(t^2),$$

3.5 Almost everywhere convergence (the critical index)

as $t \to 0^+$, we can get that

$$\int_0^1 \alpha(t) \left| \frac{2^k \Gamma(k+1) J_k(t\sqrt{\nu})}{(t\sqrt{\nu})^k} - 1 \right|^2 dt$$
$$= \left(\int_0^{\frac{1}{\sqrt{\nu}}} + \int_{\frac{1}{\sqrt{\nu}}}^1 \right) \alpha(t) \left| \frac{2^k \Gamma(k+1) J_k(t\sqrt{\nu})}{(t\sqrt{\nu})^k} - 1 \right|^2 dt$$
$$\leq C \int_0^{\frac{1}{\sqrt{\nu}}} \alpha(t) t^4 \nu^2 dt + C \int_{\frac{1}{\sqrt{\nu}}}^1 \alpha(t) dt$$
$$= O(\omega(\nu)).$$

It hence follows from the above estimates and (3.5.20) that (3.5.16) holds. This completes the proof of Theorem 3.5.5. ∎

We can deduce two sufficient conditions of almost everywhere convergence from Theorem 3.5.5.

Corollary 3.5.1 Let $f \in L^2(Q^n)$ with $n > 1$. If

$$\int_0^1 \left(\omega(f;t)_2 \right)^2 \frac{\log \frac{1}{t}}{t} dt < \infty, \tag{3.5.21}$$

then we have

$$\lim_{R \to \infty} S_R^0(f;x) = f(x), \tag{3.5.22}$$

for a.e. $x \in Q^n$.

Proof. By

$$\|f_x(t) - f(x)\|_2^2 = \int_{Q^n} \left| \frac{1}{\omega_{n-1}} \int_{\mathbb{S}^{n-1}} (f(x+t\xi) - f(x)) d\sigma(\xi) \right|^2 dx$$
$$\leq C \big(\omega(f;t)_2 \big)^2,$$

and the inequality (3.5.21), we can easily obtain

$$\int_0^1 \|f_x(t) - f(x)\|_2^2 \frac{1}{t} \log \frac{1}{t} dt < \infty. \tag{3.5.23}$$

Denote
$$\omega(\nu) = \int_{\frac{1}{\sqrt{\nu}}}^{1} \frac{1}{t} \log \frac{1}{t} dt = \frac{1}{8} \log^2 \nu,$$

for $\nu \in \mathbb{N}$.

We obviously have that
$$\int_0^\delta t^3 \log \frac{1}{t} dt = O\left(\delta^4 \int_\delta^1 \frac{1}{t} \log \frac{1}{t} dt\right).$$

Let $\alpha(t) = \frac{1}{t} \log \frac{1}{t}$, for $t \in (0, 1]$. Clearly the function α satisfies the condition (3.5.15). And thus from Theorem 3.5.5, we can deduce that
$$\sum_{\nu=1}^\infty A_\nu^2 \log^2 \nu < \infty. \tag{3.5.24}$$

Let
$$\varphi_\nu(x) = \frac{1}{A_\nu} \sum_{|m|^2 = \nu} C_m(f) e^{im \cdot x}.$$

By (3.5.24) and the Menshoff-Rademacher theorem in theory of orthogonal series (see [KS1]), it implies that $\sum_{\nu=0}^\infty A_\nu \varphi_\nu(x)$ is convergent for a.e. $x \in Q^n$. That is to say, (3.5.22) is valid.

This finishes the proof of Corollary 3.5.1. ∎

Corollary 3.5.2 Let $f \in L^2(Q^n)$ $(n > 1)$. If there exists $\varepsilon > 0$, such that
$$\omega(f; \delta)_2 = O\left(\frac{1}{(\log 1/\delta)^{1+\varepsilon}}\right) \tag{3.5.25}$$

as $\delta \to 0^+$, then (3.5.22) holds.

Obviously, (3.5.25) implies (3.5.21). Thus Corollary 3.5.2 is a special case of Corollary 3.5.1.

Besides, Lu [Lu8] also obtained a result on the a.e.-convergence of lacunary sequence of the spherical Fourier partial sum for L^2.

Theorem 3.5.6 Let $f \in L^2(Q^n)$ with $n > 1$. If $R_1 > 0$, and $\frac{R_{j+1}}{R_j} \geq q > 1$, for all $j \in \mathbb{N}$, then
$$\lim_{j \to \infty} S_{R_j}^0(f; x) = f(x) \tag{3.5.26}$$

holds for a.e. $x \in Q^n$.

3.5 Almost everywhere convergence (the critical index)

Proof. Let
$$S_*(f)(x) = \sup_{j \in \mathbb{N}} \left| S_{R_j}^0(f; x) \right|.$$

Next it suffices to prove that S_* is of type $(2, 2)$. We have that
$$S_{R_j}^0(f; x) = S_{R_j}^n(f; x) + \sum_{|m| < R_j} \left[1 - \left(1 - \frac{|m|^2}{R_j^2} \right)^n \right] C_m(f) e^{im \cdot x},$$

and
$$[S_*(f)(x)]^2 \le 2 \left\{ [S_*^n(f)(x)]^2 + \sum_{j=1}^{\infty} \left| \sum_{|m| < R_j} \left[1 - \left(1 - \frac{|m|^2}{R_j^2} \right)^n \right] C_m(f) e^{im \cdot x} \right|^2 \right\}.$$

It follows that
$$\|S_*(f)\|_2^2 \le 2 \|S_*^n(f)\|_2^2 + 2 \sum_{j=1}^{\infty} (2\pi)^n \sum_{|m| < R_j} \left| 1 - \left(1 - \frac{|m|^2}{R_j^2} \right)^n \right|^2 |C_m(f)|^2$$
$$\le C \left\{ \|f\|_2^2 + \sum_{m \in \mathbb{Z}^n} \sum_{j=1}^{\infty} \frac{|m|^4}{R_j^4} |C_m(f)|^2 \chi_{[0, R_j)}(|m|) \right\}$$
$$\le C \left\{ \|f\|_2^2 + \sum_{m \ne 0} \left(\sum_{j: R_j > |m|} \frac{1}{R_j^4} \right) |m|^4 |C_m(f)|^2 \right\}.$$

By the property of $\{R_j\}$, we have
$$\sum_{j: R_j > |m|} \frac{1}{R_j^4} = O\left(\frac{1}{|m|^4} \right)$$

for $m \ne \mathbf{0}$. Therefore, we have
$$\|S_*(f)\|_2 \le C \|f\|_2.$$

This finishes the proof of Theorem 3.5.6. ∎

Among the results about the a.e. convergence, Stein's theorem, i.e., Theorem 3.5.2, is the most beautiful one, which was published in 1958. Since then, many scholars were searching for the possibility of finding the subclass of $L(Q^n)$, such that it is no smaller than $L(\log^+ L)^2$ which guarantees the a.e. convergence of the Bochner-Riesz means at the critical index. It did not get any great development until the proposal of concepts of entropy put forward by Fefferman [Fef1] in 1978 and that of Block spaces by Taibleson and Weiss [TW1] later on. Entropy space and the space generated by blocks both includes the Dini function class. To some extent, we can claim that, to the knowledge of all that we know at present, the block space is the biggest function space where the Bochner-Riesz means of the Fourier series at the critical index is a.e. convergent. In the next section, we will give a brief introduction about all those concepts mentioned above and some main results.

3.6 Spaces related to the a.e. convergence of the Fourier series

3.6.1 The concept of the entropy class

Let S is a measurable subset of Q^n. Define the entropy of S as

$$E(S) = \inf \left\{ \sum_k |I_k| \log \frac{|Q^n|e}{|I_k|} : \bigcup I_k \supset S \right\},$$

where I_k represents for the cube in Q^n, and $\{I_k\}$ is a sequence of the cubes.

By the definition above, the entropy of a set is not only related with the measure of the set, but also with its geometric shape. Fefferman [Fef1] put forward with such a kind of sequence of sets on one-dimensional Euclidean space, whose measure is zero, and yet entropy tends to 1.

Define the entropy of the integrable function f on Q^n as follows:

$$J(f) = \int_0^\infty E(\{x \in Q^n : |f(x)| < \lambda\}) d\lambda.$$

We have to point out that the entropies of two a.e. equal functions are not necessarily the same.

We define a function class denoted by J as follows.

$$f \in J \Longleftrightarrow \exists\, g,\ g = f, a.e.,\ \text{and}\ J(g) < +\infty.$$

If $n = 1$, it is easy to prove

$$J(f + g) \leq J(f) + J(g).$$

3.6 Spaces related to the a.e. convergence of the Fourier series

However if $n > 1$,

$$E(Q_1 \cap Q_2) + E(Q_1 \cup Q_2) \le E(Q_1) + E(Q_2)$$

is not necessarily valid, and then the Minkowski inequality no longer holds. But if we restrict the 'inf' in the definition of the entropy of a set is only taken over all covers of binary cubes I_k, then the entropy of a function is a norm and this definition is equivalent to the original one.

In order to illustrate the size of J, let us formulate the concept of the Dini class denoted by D.

Let $f \in L(Q^n)$. Define

$$\|f\|_D = \|f\|_{L(Q^n)} + \int\int_{x,y \in Q^n} \frac{|f(x) - f(y)|}{|x - y|^n} dx dy.$$

$$f \in D \iff \|f\|_D < +\infty.$$

Fefferman [Fef1] proved the following result for the case of $n = 1$:

Lemma 3.6.1 $J(f) \le C\|f\|_D$, and thus we have $D \subset J$.

Since the proof is tedious and troublesome, we do not quote the proof here.

Lemma 3.6.2 J is a subset of $L \log L$, that is, $J \subset L \log L$.

Proof. Let $f \in J$. Without loss of generality, we assume that f is nonnegative. Let

$$E_k = \{x \in Q^n : f(x) > 2^k\}$$

for $k \in \mathbb{Z}$.

We conclude that

$$J(f) = \int_0^\infty E(\{x \in Q^n : f(x) > \alpha\}) d\alpha$$

$$= \sum_{k \in \mathbb{Z}} \int_{2^{k-1}}^{2^k} E(\{x \in Q^n : f(x) > \alpha\}) d\alpha$$

$$\ge \sum_{k \in \mathbb{Z}} 2^{k-1} E(E_k)$$

$$\ge \frac{1}{2} \sum_{k \in \mathbb{Z}} 2^k |E_k| \log \frac{|Q^n|e}{|E_k|}.$$

Here we use the property of the entropy of a set

$$E(S) \geq |S| \log \frac{|Q^n|e}{|S|}.$$

We write

$$F_k = E_k - E_{k+1} = \left\{ x \in Q^n : 2^k < f(x) \leq 2^{k+1} \right\}.$$

Then we can get

$$J(f) \geq \frac{1}{4} \sum_{k=0}^{\infty} 2^{k+1} \log \frac{|Q^n|e}{|E_k|} |F_k|.$$

By $|E_k| \leq 2^{-k} \|f\|_1$, we have

$$J(f) \geq \frac{1}{4} \sum_{k=0}^{\infty} 2^{k+1} \left(\log 2^{k+1} + \log \frac{|Q^n|e}{2\|f\|_1} \right) |F_k|$$

$$\geq \frac{1}{4} \left\{ \int_{Q^n} f(x) \log^+ f(x) dx + \log \frac{|Q^n|e}{2\|f\|_1} \int_{E_0} f(x) dx \right\}.$$

This implies $f \in L \log^+ L$.

Consequently, we have the inclusion relationship,

$$D \subset J \subset L \log^+ L.$$

∎

Fefferman [Fef1] gave a conjecture as follows. "It seems that the entropy has close relation with the pointwise convergence of the Fourier series." Taibleson and Weiss [TW1] and Lu-Taibleson-Weiss [LTW1] verified the conjecture for the casees of unitary and multiple variables, respectively.

Theorem 3.6.1 *If $f \in J$, then*

$$S_R^{\frac{n-1}{2}}(f)(x) = f(x)$$

holds for a.e. $x \in Q^n$ with $n \in \mathbb{N}$.

Since Theorem 3.6.1 is included in Theorem 3.6.5 and Theorem 3.6.6 in the following, we do not give the proof here.

3.6 Spaces related to the a.e. convergence of the Fourier series

3.6.2 The concept of the block

The block becomes more and more important, since it can solve the problem of the a.e. convergence of $S_R^{\frac{n-1}{2}}$ well and the block space is bigger than the entropy space.

Definition 3.6.1 *Let $1 < q \leq \infty$. We say that measurable function b is a q-block, if there exists a block Q, where Q is a cube whose edge is parallel to the axis and $Q \subset Q^n$, such that*

(i) supp $b \subset Q$,

(ii) $\|b\|_{L^q} \leq |Q|^{\frac{1}{q}-1}$.

Definition 3.6.2 *If $f \in L(Q^n)$ can be expressed as*

$$f(x) = \sum_{k=1}^{\infty} c_k b_k(x), \qquad (3.6.1)$$

where b_k is a q-block, and the group of coefficients $\{c_k\} := c$ satisfies

$$N(c) := \sum_{k=1}^{\infty} |c_k| \left(1 + \log \frac{\sum_j |c_j|}{|c_k|}\right) < +\infty, \qquad (3.6.2)$$

then we say $f \in B_q$. We define the number as follows,

$$N_q(f) = \inf \left\{ N(c) : c = \{c_k\}_{k=1}^{\infty} \text{ such that } (3.6.1) \text{ holds} \right\}.$$

By (3.6.2) and the fact that b_k is a q-block, we say that the series of the right side of (3.6.1) must be absolutely convergent in the norm of L^1. In addition, by the condition of (ii) about the size of q-block, we have

$$\|f\|_1 \leq \sum_{k=1}^{\infty} |c_k| \|b\|_q |Q|^{1-\frac{1}{q}} \leq \sum_{k=1}^{\infty} |c_k| \leq N(c). \qquad (3.6.3)$$

Besides, it is obvious that B_q is a linear space.

Proposition 3.6.1 *If $f \in B_q$, and $g \in B_q$, then we have*

$$N_q(f+g) \leq (1 + \log 2)(N_q(f) + N_q(g)). \qquad (3.6.4)$$

Proof. Let
$$\varphi(x) = 1 - x\log x - (1-x)\log(1-x),$$
for $0 < x < 1$, and $\varphi(0) = \varphi(1) = 1$. We easily have
$$\varphi'(x) = \log\frac{1-x}{x}.$$
Obviously φ has the maximum at $x = \frac{1}{2}$ and
$$\varphi\left(\frac{1}{2}\right) = 1 + \log 2.$$
For every $\varepsilon > 0$, choose
$$f = \sum m_k b_k$$
and
$$g = \sum n_j c_j$$
such that
$$N(m) < N_q(f) + \varepsilon$$
and
$$N(n) < N_q(g) + \varepsilon.$$
Let $A = \sum |m_k|$, $B = \sum |n_j|$ and $x = \frac{A}{A+B}$. We thus have that
$$\begin{aligned}N_q(f+g) &\le \sum |m_k|\log\frac{A+B}{|m_k|} + \sum |n_k|\log\frac{A+B}{|n_k|} + (A+B)\\ &= (A+B) + \sum |m_k|\left(\log\frac{1}{x} + \log\frac{A}{|m_k|}\right)\\ &\quad + \sum |n_k|\left(\log\frac{1}{1-x} + \log\frac{B}{|n_k|}\right)\\ &= (A+B)\varphi(x) + \sum |m_k|\log\frac{A}{|m_k|} + \sum |n_k|\log\frac{B}{|n_k|}\\ &\le (1+\log 2)(A+B) + \sum |m_k|\log\frac{A}{|m_k|} + \sum |n_k|\log\frac{B}{|n_k|}\\ &\le (1+\log 2)\bigl(N(m) + N(n)\bigr)\\ &\le (1+\log 2)\bigl(N_q(f) + N_q(g) + 2\varepsilon\bigr).\end{aligned}$$

Consequently, we immediately get (3.6.4).
This completes the proof of Proposition 3.6.1. ∎

3.6 Spaces related to the a.e. convergence of the Fourier series

Besides, it is obvious that N_q has absolute homogeneity. That is to say,

$$N_q(af) = |a|N_q(f) \qquad (3.6.5)$$

for $a \in \mathbb{R}$ and $f \in B_q$.

According to (3.6.4), (3.6.5) and the obvious property,

$$N_q(f) = 0 \Longleftrightarrow f = 0,$$

we call N_q as the quasi-norm, inherited from Taibleson and Weiss [TW1]. Its difference from the norm is that the inequality (3.6.4) holds instead of Minkowski's inequality.

Proposition 3.6.2 B_q *forms a complete space if it is normed with the quasi-norm* N_q.

Proof. Assume that $\{f_k\}_{k=0}^\infty$ is a Cauchy sequence in B_q in the norm of quasi-norm N_q. Then we can take a subsequence $\{g_k\}_{k=1}^\infty$ from the sequence, such that

$$N_q(g_{k+1} - g_k) < 2^{-k}, \qquad (3.6.6)$$

for $k = 1, 2, \ldots$.

It follows from the inequality (3.6.3) that $\{f_k\}$ is also a Cauchy sequence in $L(Q^n)$. Then by the completeness of L^1, there exists $f \in L(Q^n)$, such that

$$f = g_1 + \sum_{k=1}^\infty (g_{k+1} - g_k) \qquad (3.6.7)$$

in the sense of L^1 norm.

Take some proper decomposition of $g_{k+1} - g_k$ in block

$$g_{k+1} - g_k = \sum c_j^{(k)} b_j^{(k)},$$

such that $N(c^{(k)}) < 2^{-k}$.

Assume that g_1 has the following decomposition

$$g_1 = \sum c_j b_j.$$

Then we have

$$f = \sum c_j b_j + \sum_{k=1}^\infty \sum_{j=1}^\infty c_j^{(k)} b_j^{(k)}. \qquad (3.6.8)$$

Therefore, we have $f \in B_q$. Both (3.6.7) and (3.6.8) implies that

$$N_q(f - g_k) = N_q\left\{\sum_{j=k+1}^{\infty}(g_{j+1} - g_j)\right\} \longrightarrow 0,$$

as $k \to \infty$, which leads to $g_k \to f$ and $f_k \to f$ in B_q, as $k \to \infty$.
This finishes the proof of Proposition 3.6.2. ∎

Proposition 3.6.3 *Let $f \in B_q$. If g is a measurable function defined on Q^n, and $|g| \leq |f|$, then we have $g \in B_q$ and $N_q(g) \leq N_q(f)$.*

Proof. Since $|g| \leq |f|$, there exists a measurable function $r(x)$, $|r| \leq 1$, such that $g = rf$. Suppose that f has a decomposition as

$$f = \sum m_j b_j.$$

By the fact that rb_j is also a block, then

$$g = \sum m_j(rb_j)$$

is a decomposition of g. And therefore $g \in B_q$ and $N_q(g) \leq N_q(f)$. ∎

From the above propositions, we can easily deduce that $f \in B_q$ implies $|f| \in B_q$ and $f, g \in B_q$ implies $\sup\{f, g\} \in B_q$ and $\inf\{f, g\} \in B_q$.

In Definition 3.6.2, under the condition of (3.6.2), the series (3.6.1) is not only convergent in the norm of L^1, but also a.e. convergent, which is contained in the following iteration principle. The principle of iteration is constructed by Stein and Weiss [SWe1], which is the foundation of the application of the whole block theory.

Lemma 3.6.3 *If $\{f_k\}$ is a sequence of measurable functions, such that*

$$|\{x : |f_k(x)| > \lambda\}| \leq \lambda^{-1} \tag{3.6.9}$$

for every $\lambda > 0$, and $m = \{m_k\}_{k=1}^{\infty}$ is a sequence of numbers, then

$$\left|\left\{x : \left|\sum_{k=1}^{\infty} m_k f_k(x)\right| > \lambda\right\}\right| \leq 4N(m)\lambda^{-1} \tag{3.6.10}$$

holds for every $\lambda > 0$.

3.6 Spaces related to the a.e. convergence of the Fourier series

Proof. Without loss of generality, we might as well think that $m_k > 0$ and $f_k(x) \geq 0$, for all k. Denote $M = \sum m_k$. Let $\lambda > 0$ and define

$$l_k(x) = \begin{cases} f_k(x) & \text{if } f_k(x) < \dfrac{\lambda}{2M}, \\ 0 & \text{if } f_k(x) \geq \dfrac{\lambda}{2M}, \end{cases}$$

$$u_k(x) = \begin{cases} f_k(x) & \text{if } f_k(x) > \dfrac{\lambda}{2m_k}, \\ 0 & \text{if } f_k(x) \leq \dfrac{\lambda}{2m_k}, \end{cases}$$

and

$$v_k(x) = f_k(x) - l_k(x) - u_k(x)$$
$$= \begin{cases} f_k(x) & \text{if } \dfrac{\lambda}{2M} \leq f_k(x) \leq \dfrac{\lambda}{2m_k}, \\ 0 & \text{if } f_k(x) \notin \left[\dfrac{\lambda}{2M}, \dfrac{\lambda}{2m_k}\right]. \end{cases}$$

In addition, we also denote

$$l(x) = \sum m_k l_k(x),$$
$$u(x) = \sum m_k u_k(x),$$

and

$$v(x) = \sum m_k v_k(x).$$

We have that

$$\{x : u(x) > 0\} = \bigcup_k \{x : u_k(x) > 0\} = \bigcup_k \left\{x : f_k(x) > \dfrac{\lambda}{2m_k}\right\},$$

and

$$|\{x : u(x) > 0\}| \leq \sum_k \left|\left\{x : f_k(x) > \dfrac{\lambda}{2m_k}\right\}\right| \leq \sum_k \dfrac{2m_k}{\lambda} = \dfrac{2M}{\lambda}.$$

Since

$$\left\{x : l(x) > \dfrac{\lambda}{2}\right\} \subset \bigcup_k \left\{x : l_k(x) > \dfrac{\lambda}{2M}\right\} = \emptyset,$$

we clearly have

$$\left|\left\{x : l(x) > \dfrac{\lambda}{2}\right\}\right| = 0.$$

Let $\mu_k(y) = |\{x : f_k(x) > y\}|$ for $y > 0$. We conclude that

$$\int v(x)dx = \sum m_k \int v_k(x)dx$$

$$= \sum m_k \int_{\frac{\lambda}{2M}}^{\frac{\lambda}{2m_k}} -y d\mu_k(y)$$

$$= \sum m_k \left(\int_{\frac{\lambda}{2M}}^{\frac{\lambda}{2m_k}} \mu_k(y)dy - y\mu_k(y) \Big|_{\frac{\lambda}{2M}}^{\frac{\lambda}{2m_k}} \right)$$

$$\leq \sum m_k \left(\log \frac{M}{m_k} + 1 \right)$$

$$= N(m).$$

Therefore it follows that

$$\left| \left\{ x : v(x) > \frac{\lambda}{2} \right\} \right| \leq \frac{2}{\lambda} \int_{\{x:v(x)>\frac{\lambda}{2}\}} v(x)dx \leq \frac{2}{\lambda} N(m).$$

Since

$$\left\{ x : \sum m_k f_k(x) > \lambda \right\} = \{x : l(x) + u(x) + v(x) > \lambda\}$$

$$\subset \{x : u(x) > 0\} \bigcup \left\{ x : v(x) > \frac{\lambda}{2} \right\} \bigcup \left\{ x : l(x) > \frac{\lambda}{2} \right\},$$

we have

$$\left| \left\{ x : \sum m_k f_k(x) > \lambda \right\} \right| \leq |\{x : u(x) > 0\}| + \left| \left\{ x : v(x) > \frac{\lambda}{2} \right\} \right|$$

$$\leq \frac{2M}{\lambda} + \frac{2}{\lambda} N(m)$$

$$\leq 4N(m)\lambda^{-1}.$$

This completes the proof. ∎

3.6.3 The structure of the block space

In this section, we will study the relation between B_q and L^q, the inclusion relations between B_q corresponding to different $q \in (1, \infty)$ and the properties of B_q as a complete linear distance space in the norm of N_q. The conjugate space of B_q is a collection of continuous linear functional defined on B_q, denoted by B_q'. The conclusion is as follows.

3.6 Spaces related to the a.e. convergence of the Fourier series

Theorem 3.6.2 *If $1 \leq p < q \leq \infty$, then we have*

(i) $L^q \subset B_q$,

(ii) *there exists $f \in B_p$, but $f \notin B_q$; that is, we have $B_p \supsetneq B_q$ and $B_1 = L^1$.*

Theorem 3.6.3 *Let $1 < q < \infty$. Then $B_q' = L^\infty(Q^n)$.*

By Theorem 3.6.3 and $(L^1)^* = L^\infty$, we can see that B_q is indeed very close to L^1.

Lemma 3.6.4 *Suppose that there exists two sequences of numbers: $m = \{m_k\}$ and $l = \{l_k\}$, satisfying $|m_k| \leq |l_k|$, for all k. Then $N(m) \leq N(l)$ holds.*

Proof. We might as well take it for granted that $0 < m_k \leq l_k$. In fact, it suffices to show that, for every $\varepsilon \geq 0$, for the sequence of numbers $m(\varepsilon) = \{m_1 + \varepsilon, m_2, m_3, \ldots\}$, we have $N(m(\varepsilon)) \geq N(m)$. Denote $\varphi(\varepsilon) = N(m(\varepsilon))$. We merely need to show that $\varphi(\varepsilon)$ is monotonically increasing.

Denote
$$A(\varepsilon) = \varepsilon + \sum_{j=1}^{\infty} m_j.$$

We have
$$\varphi(\varepsilon) = (m_1 + \varepsilon)\left(1 + \log \frac{A(\varepsilon)}{m_1 + \varepsilon}\right) + \sum_{k=2}^{\infty} m_k \left(1 + \log \frac{A(\varepsilon)}{m_k}\right).$$

Therefore, we conclude that
$$\varphi'(\varepsilon) = 1 + \log \frac{A(\varepsilon)}{m_1 + \varepsilon} + (m_1 + \varepsilon)\left(\frac{1}{A(\varepsilon)} - \frac{1}{m_1 + \varepsilon}\right) + \sum_{k=2}^{\infty} m_1 \frac{1}{A(\varepsilon)}$$
$$= 1 + \log \frac{A(\varepsilon)}{m_1 + \varepsilon} > 0,$$

for any $\varepsilon \geq 0$.

Consequently, the proof of the lemma is finished. ∎

Lemma 3.6.5 Suppose that $\nu \in \mathbb{N}$, $1 < q \le \infty$ and $\frac{1}{q} + \frac{1}{q'} = 1$. Let $N = \nu(1 + \log \nu)^{q'}$, $\delta = \frac{1}{\nu-1}\left(1 - \frac{1}{N}\right)$, and the intervals

$$I_i = \left[(i-1)\delta, (i-1)\delta + \frac{1}{N}\right]\lambda,$$

for $i = 1, 2, \ldots, \nu$ and $0 < \lambda < 1$. If the function $b_i(x) = \frac{1}{|I_i|}\chi_{I_i}(x)$ is defined on Q^1, and $f = \sum_{i=1}^{\nu} b_i$, then we have

$$N_q(f) = \nu(1 + \log \nu).$$

Proof. Every b_i is a q-block, and

$$\|b_i\|_q = \left(|I_i|^{-q+1}\right)^{\frac{1}{q}} = |I_i|^{-\frac{1}{q'}},$$

so we have $N_q(f) \le \nu(1 + \log \nu)$.

Now we assume that there exists another decomposition in blocks as $f = \sum m_k c_k$, where c_k is a q-block. It suffices to show that

$$\nu(1 + \log \nu) \le N(m).$$

Let J_k is a interval of Q^1 such that $\operatorname{supp} c_k \subset J_k$ and $\|c_k\|_q \le |J_k|^{-\frac{1}{q'}}$. If $\|c_k\|_q < |J_k|^{-\frac{1}{q}}$, of course $\|c_k\|_q > 0$, then we can expand c_k into new c'_k such that $\|c'_k\|_q = |J_k|^{-\frac{1}{q'}}$, at the same time, we shrink m_k into m'_k. Thus in the new decomposition of f, it always holds $\|c'_k\|_q = |J_k|^{-\frac{1}{q}}$. Yet by Lemma 3.6.4, $N(m') \le N(m)$, then our proof can switch to prove $\nu(1 + \log \nu) \le N(m')$. For this reason, we can assume $\|c_k\|_q = |J_k|^{-\frac{1}{q'}}$, for all $k \in \mathbb{N}$.

If every J_k can be contained in some I_i, then the proof is end at once. In this case, we can write

$$b_i = \sum_k m_k^i b_k^i, \quad f = \sum_{i=1}^{\nu} \sum_k m_k^i b_k^i, \qquad (3.6.11)$$

where each b_k^i is a q-block whose support is contained in $J_k^i \subset I_i$.

It follows

$$\int_{Q^1} b_i(x) dx = 1 \le \sum_k |m_k^i|,$$

for $i = 1, 2, \ldots, \nu$. Thus, we can choose $\{\widetilde{m_k^i}\}$, such that

$$0 \le \left|\widetilde{m_k^i}\right| \le \left|m_k^i\right|$$

3.6 Spaces related to the a.e. convergence of the Fourier series

and
$$\sum_k |m_k^i| = 1,$$
for $i = 1, 2, \ldots, \nu$.

Using Lemma 3.6.4, we have
$$N\left(\{\widetilde{m_k^i}\}\right) = \sum_{i,k} \left|\widetilde{m_k^i}\right| \left(1 + \log \frac{\nu}{\left|\widetilde{m_k^i}\right|}\right) \leq N\left(\{m_k^i\}\right) = N(m).$$

Since $\left|\widetilde{m_k^i}\right| \leq 1$ and
$$\log \frac{\nu}{\left|\widetilde{m_k^i}\right|} \geq \log \nu,$$
we have
$$N\left(\{\widetilde{m_k^i}\}\right) \geq \sum_i \sum_k \left|\widetilde{m_k^i}\right| (1 + \log \nu) = \nu(1 + \log \nu),$$
which leads to $N(m) \geq \nu(1 + \log \nu)$.

Hence, it remains to rewrite the decomposition $f = \sum_k m_k c_k$ as $f = \sum_k m'_k c'_k$, such that the interval J'_k corresponding to the support of c'_k is contained in some I_i while $N(m') \leq N(m)$ still holds. The rewriting step is carried out in the method of separating $m_k c_k$ into $m_k^1 c_k^1 + \cdots + m_k^\nu c_k^\nu$ one by one, where the support of c_k^i is contained in $J_k^i \subset I_i$ and $\|c_k^i\|_q = |J_k^i|^{-\frac{1}{q'}}$ still holds. And it only suffices to show that such a decomposition is applicable for $m_1 c_1$.

Suppose that the interval J_1 related to c_1 satisfies $|J_1 \cap I_i| > 0$ for $i = \mu, \ldots, \mu + l$, $1 \leq \mu < \mu + 1 \leq \mu + l \leq \nu$, and $|J_1 \cap I_i| = 0$ for $i < \mu$ or $i > \mu + l$.

Denote $L_j = J_1 \cap I_j$, $j = \mu, \ldots, \mu + l$. Set
$$\mu_j = |L_j|^{\frac{1}{q'}} \|b_1 \chi_{L_j}\|_q$$
and $b_1 \chi_{L_j} = b_1^j$ for $j = \mu, \ldots, \mu + l$, and $a_j = \mu_j^{-1} b_1^j$. Since $\|b_1^j\|_q > 0$, we have $\mu_j > 0$. Obviously a_j is a q-block.

By the definition of I_i, we have
$$|J_1| = \sum_{i=\mu}^{\mu+l} |L_i| + l\left(\delta - \frac{1}{N}\right)\lambda$$
$$= \sum_{i=\mu}^{\mu+l} |L_i| + l\frac{\nu}{\nu - 1}\left[(1 + \log \nu)^{q'} - 1\right]\frac{\lambda}{N}.$$

Since
$$\theta(t) = \frac{t}{t-1}\left((1+\log t)^{q'} - 1\right)$$
is monotonically increasing for $t > 1$ and $1 \leq q' < \infty$, we have
$$l\frac{\nu}{\nu-1}\left[(1+\log\nu)^{q'} - 1\right] \geq l\frac{l+1}{l}\left[(1+\log(l+1))^{q'} - 1\right],$$
and
$$|J_1| \geq \sum_{i=\mu}^{\mu+l} |L_i| + (l+1)\left[(1+\log(l+1))^{q'} - 1\right]\frac{\lambda}{N}$$
$$= (l+1)[1+\log(l+1)]^{q'}\frac{\lambda}{N} + |L_\mu| + |L_{\mu+l}| - 2\frac{\lambda}{N}.$$

It follows from the Hölder's inequality that
$$\sum_{j=\mu}^{\mu+l}\mu_j = \sum_{j=\mu}^{\mu+l}|L_j|^{\frac{1}{q'}}\|b_1^i\|_q$$
$$\leq \left(\sum_{j=\mu}^{\mu+l}|L_j|\right)^{\frac{1}{q'}}\left(\sum_{j=\mu}^{\mu+l}\|b_1^j\|_q^q\right)^{\frac{1}{q}}$$
$$= \left(\sum_{j=\mu}^{\mu+l}|L_j|\right)^{\frac{1}{q'}}\|b_1\|_q$$
$$= \left(\sum_{j=\mu}^{\mu+l}|L_j|\right)^{\frac{1}{q'}}|J_1|^{-\frac{1}{q'}}$$
$$\leq \left(\sum_{j=\mu}^{\mu+l}\frac{|L_j|}{(l+1)(1+\log(l+1))^{q'}\frac{\lambda}{N} + |L_\mu| + |L_{\mu+l}| - \frac{2\lambda}{N}}\right)^{\frac{1}{q'}}$$
$$= \left(\frac{1-\alpha}{[1+\log(l+1)]^{q'} - \alpha}\right)^{\frac{1}{q'}},$$

where
$$\alpha = \left(\frac{2\lambda}{N} - |L_\mu||L_{\mu+l}|\right)\frac{N}{(l+1)\lambda} \geq 0.$$
Since
$$\frac{1-\alpha}{[1+\log(l+1)]^{q'} - \alpha} \leq \frac{1}{[1+\log(l+1)]^{q'}},$$

3.6 Spaces related to the a.e. convergence of the Fourier series

we have
$$\sum_{j=\mu}^{\mu+l} \mu_j \leq \frac{1}{1+\log(l+1)}. \tag{3.6.12}$$

We know
$$f = \sum_{k=2}^{\infty} m_k c_k + \sum_{j=\mu}^{\mu+l} m_1 \cdot \mu_j a_j = m_1 \left(\sum_{k=2}^{\infty} m'_k c_k + \sum_{j=\mu}^{\mu+l} \mu_j a_j \right),$$

where $m'_k = \frac{m_k}{m_1}$.

Denote
$$A = \sum_{k=2}^{\infty} |m'_k|$$

and
$$B = \sum_{j=\mu}^{\mu+l} \mu_j.$$

It suffices to show that
$$\sum_{k=2}^{\infty} |m'_k| \left(1 + \log \frac{A+B}{|m'_k|}\right) + \sum_{j=\mu}^{\mu+l} \mu_j \left(1 + \log \frac{A+B}{\mu_j}\right)$$
$$\leq 1 + \log(1+A) + \sum_{k=2}^{\infty} |m'_k| \left(1 + \log \frac{1+A}{|m'_k|}\right)$$
$$= 1 + \log(1+A) + A[1 + \log(1+A)] - \sum_{k=2}^{\infty} |m'_k| \log |m'_k|.$$

That is,
$$(1+A)[1 + \log(1+A)] \geq (A+B)[\log(A+B) + 1] - \sum_{j=\mu}^{\mu+l} \mu_j \log \mu_j. \tag{3.6.13}$$

Since
$$\int_0^A (2 + \log(1+x)) \, dx = (1+A)(1 + \log(1+A)) - 1$$
$$\geq \int_0^A (2 + \log(B+x)) \, dx$$
$$= (B+A)(\log(B+A) + 1) - B(\log B + 1),$$

where we use $B \leq \frac{1}{1+\log(l+1)} < 1$ in (3.6.12). We merely need to prove

$$1 - B(\log B + 1) \geq -\sum_{j=\mu}^{\mu+l} \mu_j \log \mu_j. \tag{3.6.14}$$

Denote $\varphi_j = \frac{\mu_j}{B}$. Since $\sum_{j=\mu}^{\mu+l} \varphi_j = 1$, by the convexity of the exponential function, we can get

$$\exp\left(\sum_{j=\mu}^{\mu+l} \varphi_j \log \frac{1}{\varphi_j}\right) \leq \sum_{j=\mu}^{\mu+l} \varphi_j \exp\left(\log \frac{1}{\varphi_j}\right) = l + 1.$$

Therefore, we have

$$\sum_{j=\mu}^{\mu+l} \mu_j \log \frac{B}{\mu_j} \leq B \log(l+1)$$

and

$$-\sum_{j=\mu}^{\mu+l} \mu_j \log \mu_j \leq B \log(l+1) - B \log B.$$

We substitute the above results into (3.6.12) to obtain

$$B \leq \frac{1}{1 + \log(l+1)}.$$

This is equivalent to

$$B \log(l+1) \leq 1 - B.$$

We can immediately obtain (3.6.14).
This completes the proof of Lemma 3.6.5. ■

The function in Lemma 3.6.5 is

$$f = \sum_{i=1}^{\nu} \frac{1}{|I_i|} \chi_{I_i},$$

where

$$|I_i| = \frac{\lambda}{N} = \lambda \nu^{-1}(1 + \log \nu)^{-q}.$$

Denote $S = \bigcup_{i=1}^{\nu} I_i$. We obviously have $S \subset [0, \lambda]$. Set

$$\chi_S = \lambda \nu^{-1}(1 + \log \nu)^{-q'} f.$$

3.6 Spaces related to the a.e. convergence of the Fourier series

It follows
$$|S| = \lambda(1 + \log \nu)^{-q'}, \tag{3.6.15}$$
and
$$N_q(\chi_S) = \lambda(1 + \log \nu)^{1-q'}. \tag{3.6.16}$$

According to the property of translation invariance of the measure and the block, we can have from (3.6.15), (3.6.16) that Lemma 3.6.5 implies Lemma 3.6.6.

Lemma 3.6.6 *In any interval I with the length of λ for $0 < \lambda < 1$ in Q^1, there exists a subset S such that*
$$|S| = |I|(1 + \log \nu)^{-q'} \tag{3.6.17}$$
and
$$N_q(\chi_S) = |I|(1 + \log \nu)^{1-q'}, \tag{3.6.18}$$
where ν is a previously given arbitrary natural number with $\nu > 1$, and q' satisfies $1 \le q' < \infty$ with $\frac{1}{q} + \frac{1}{q'} = 1$.

Proof of Theorem 3.6.2. Firstly, the statement (i) is evident, because each function in $L^q(Q^n)$ is the multiple of a q-block.

For the statement (ii), we can only give a proof in the one-dimensional case with the help of Lemma 3.6.6. It is rather more troublesome for the case $n > 1$.

We consider the case of $q = \infty$ at first. In this case, $q' = 1$. Choose a series of intervals I_k in $[0, 1]$ which is not intersected with each other, such that $|I_k| = 2^{-k}$, for $k \in \mathbb{N}$. By Lemma 3.6.6, we can select the set S_k in I_k, such that
$$|S_k| = |I_k|(1 + \log n_k)^{-1},$$
$$n_k \ge e^{e^{2^{2k}}}$$
and
$$N_\infty(\chi_{S_k}) = |I_k| = 2^{-k}.$$
Let
$$f = \sum_{k=1}^{\infty} 2^{2k} \chi_{S_k}.$$
Proposition 3.6.3 implies that
$$N_\infty(f) \ge 2^{2k} N_\infty(\chi_{S_k}) = 2^k \to \infty.$$

This means $f \notin B_\infty$.

However, we have

$$\int_{Q^1} e^{f(x)} dx = \sum_{k=1}^\infty \int_{S_k} e^{2^{2k}} = \sum_{k=1}^\infty e^{2^{2k}} 2^{-k}(1+\log n_k)^{-1} \le \sum_{k=1}^\infty 2^{-k} = 1.$$

This implies $f \in L^p(Q^1)$ for $1 \le p < \infty$.

Now we suppose that $1 \le p < q < \infty$. Choose

$$\beta > \frac{q-1}{q-p}$$

and

$$\nu \in \left(\frac{p\beta-1}{q'}, \frac{\beta-1}{q'-1}\right)$$

and take

$$n_k = \left[e^{2^{\gamma k}}\right] + 1.$$

Let the intervals $I_k \subset [0,1]$ with $|I_k| = 2^{-k}$, and those I_k do not intersect with each other. By Lemma 3.6.6, we can take $S_k \subset I_k$, such that

$$|S_k| = |I_k|(1+\log n_k)^{-q'}$$

and

$$N_q(\chi_{S_k}) = |I_k|(1+\log n_k)^{1-q'}.$$

Let $f = \sum 2^{\beta k} \chi_{S_k}$. We have

$$\|f\|_p^p = \sum 2^{(p\beta-1)k}(1+\log n_k)^{-q'} \le \sum 2^{(p\beta-1)k-\gamma q'k} < +\infty,$$

and

$$\begin{aligned}
N_q(f) &\ge 2^{\beta k} N_q(\chi_{S_k}) \\
&\ge 2^{\beta k} 2^{-k}(2\log n_k)^{1-q'} \\
&\ge 2^{(\beta-1)k} 2^{(\gamma k+2)(1-q')} \\
&= 4^{1-q'} 2^{[(\beta-1)-(q'-1)\gamma]k} \to +\infty,
\end{aligned}$$

as $k \to \infty$.

This shows that $f \in L^p$, but $f \notin B_q$. In addition, since any q-block must be p-block, then we have $B_p \supset B_q$.

3.6 Spaces related to the a.e. convergence of the Fourier series

The statement (ii) in Theorem 3.6.2 is proven for $n = 1$. One can give the proof in the case of $n > 1$ for practise.

Before proving Theorem 3.6.3, we would like to clarify three facts as follows.

Fact I. It is easy to check directly that the continuity of the linear functional l on B_q about the topology generated by the quasi-norm N_q is equivalent to its boundedness. That is to say, the continuity of l is equivalent to the fact that there exists $K \geq 0$, such that

$$|l(f)| \leq K N_q(f),$$

for $f \in B_q$.

We denote the minimum of K as $\|l\|$. With this norm, B'_q becomes a Banach space.

Fact II. We have mentioned that

$$\|f\|_1 \leq N_q(f),$$

for $f \in B_q$.

Fact III. Let $f \in L^q$. We have

$$b := \left(\|f\|_q |Q^n|^{\frac{1}{q'}} \right)^{-1} f$$

is a q-block, and thus

$$N_q(f) \leq |Q^n|^{\frac{1}{q'}} \|f\|_q.$$

Proof of Theorem 3.6.3. Let $l \in B'_q$. Restricted on L^q, if $\{f_k\} \subset L^q$ and $f_k \to 0$ in L^q. Fact III implies that $f_k \to 0$ in B^q. Thus we have $l(f_k) \longrightarrow 0$. This shows that l restricted on L^q is a bounded linear functional on L^q. Since $1 < q < \infty$, and $(L^q)^* = L^{q'}$, there exists an unique $g \in L^{q'}$, such that

$$l(f) = \int gf, \qquad (3.6.19)$$

for all $f \in L^q$.

Let $N > \|l\|$ and $E = \{x \in Q^n : |g(x)| > N\}$. If $|E| > 0$, since almost every point of E is its Lebesgue point, then there exists a cube $Q \subset Q^n$, such that

$$\frac{\|l\|}{N} < \frac{|Q \cap E|}{|Q|} \leq 1.$$

Let $b = |Q|^{-1}\text{sgn}g\chi_{Q\cap E}$. b is a q-block. Therefore, we have

$$l(b) = \frac{1}{|Q|}\int_{Q\cap E}|g| \geq N\frac{|Q\cap E|}{|Q|} > \|l\|,$$

which contradicts with the following inequality

$$|l(b)| \leq \|l\|N_q(b) \leq \|l\|.$$

This implies $|E| = 0$. That is to say, $g \in L^\infty$ and

$$\|g\|_\infty \leq \|l\|. \tag{3.6.20}$$

Since $g \in L^\infty$, we can define

$$h(f) = \int gf$$

for $f \in B_q$. Obviously, we have $h \in B_q'$. However, by (3.6.19), we know that $h = l$ on L^q. Since L^q is dense in B_q, we have that h, as an element of B_q', equals to l. That is to say, (3.6.19) is actually valid on B_q.

It follows from (3.6.19) that

$$|l(f)| \leq \|g\|_\infty \|f\|_1,$$

for $f \in B_q$.

Fact II implies $\|f\|_1 \leq N_q(f)$. It follows

$$|l(f)| \leq \|g\|_\infty N_q(f),$$

for $f \in B_q$, which yields that

$$\|l\| \leq \|g\|_\infty. \tag{3.6.21}$$

Combining (3.6.20) with (3.6.21), we immediately have $\|l\| = \|g\|_\infty$. We can have that B_q' can be embedded into L^∞ with the same norm. This completes the proof of Theorem 3.6.3. ∎

Remark 3.6.1 *Theorem 3.6.9 is due to Meyer, Taibleson and Weiss [MTW1] or [LTW2].*

3.6.4 $B_\infty \supset J$

Theorem 3.6.4 *If $J(f) < \infty$, then we have $N_\infty(f) < \infty$.*

Proof. Let
$$S_k = \left\{x \in Q^n : 2^{k-1} < |f(x)| \leq 2^k\right\}, \quad \text{for } k \in \mathbb{Z}.$$
We have
$$J(f) = \int_0^\infty E(\{x \in Q^n : |f(x)| > \lambda\})d\lambda$$
$$= \sum_{k=-\infty}^\infty \int_{2^{k-1}}^{2^k} E(\{x \in Q^n : |f(x)| > \lambda\})d\lambda. \quad (3.6.22)$$

Since
$$\int_{2^{k-1}}^{2^k} E(\{|f| > \lambda\})d\lambda \geq (2^k - 2^{k-1})E(\{|f| > 2^k\})$$
$$\geq 2^{k-1}E(S_{k+1})$$
$$= \frac{1}{4}2^{k+1}E(S_{k+1}),$$

we have
$$J(f) \geq \frac{1}{4}\sum_{k=-\infty}^\infty 2^k E(S_k). \quad (3.6.23)$$

For any fixed $k > 0$, we choose a sequence of cubes $\{I_{k,l}\}_{l=1}^\infty$ such that each cube has the positive measure and does not intersect with each other, $S_k \subset \bigcup_l I_{k,l}$ and
$$E(S_k) \leq \sum_l |I_{k,l}| \log \frac{|Q^n|e}{|I_{k,l}|} \leq E(S_k) + \frac{1}{4^k}. \quad (3.6.24)$$

Let
$$b_{k,l}(x) = f(x)\chi_{S_k \cap I_{k,l}}(x)(2^k|I_{k,l}|)^{-1},$$
for $k \in \mathbb{N}$ and $l \in \mathbb{N}$. We thus have $b_{k,l}$ is a ∞-block whose support is contained in $I_{k,l}$.

Denote $m_{k,l} = 2^k|I_{k,l}|$ and
$$b_0(x) = \begin{cases} \dfrac{1}{|Q^n|}f(x), & \text{if } |f(x)| \leq 1, \\ 0, & \text{if } |f(x)| > 1. \end{cases}$$

We have that b_0 is a ∞-block whose support is contained in Q^n. We can get a decomposition of blocks as

$$f(x) = \sum_{k=1}^{\infty}\sum_{l=1}^{\infty} m_{k,l} b_{k,l}(x) + |Q^n| b_0(x). \tag{3.6.25}$$

By (3.6.24), we have that

$$A := \sum_{k=1}^{\infty}\sum_{l=1}^{\infty} m_{k,l}$$

$$\leq \sum_{k=1}^{\infty} 2^k \sum_{l=1}^{\infty} |I_{k,l}| \log \frac{|Q^n|e}{|I_{k,l}|}$$

$$\leq \sum_{k=1}^{\infty} 2^k E(S_k) + 1.$$

By (3.6.23), we have

$$A \leq 4J(f) + 1. \tag{3.6.26}$$

It follows that

$$N_\infty(f) \leq \sum_{k=1}^{\infty}\sum_{l=1}^{\infty} m_{k,l} \left(\log \frac{A+|Q^n|}{m_{k,l}} + 1\right) + |Q^n|\left(1 + \log \frac{A+|Q^n|}{|Q^n|}\right)$$

$$= A + |Q^n|\left(1 + \log \frac{A+|Q^n|}{|Q^n|}\right) + \sum_{k=1}^{\infty}\sum_{l=1}^{\infty} m_{k,l} \log \frac{A+|Q^n|}{m_{k,l}}. \tag{3.6.27}$$

Notice

$$\log \frac{A+|Q^n|}{m_{k,l}} = \log \frac{|Q^n|e}{|I_{k,l}|} + \log \frac{A+|Q^n|}{|Q^n|e} - k\log 2$$

$$\leq \log \frac{|Q^n|e}{|I_{k,l}|} + \log \frac{A+|Q^n|}{|Q^n|e},$$

for $k \in \mathbb{N}$. We can have that

$$\sum_{k=1}^{\infty}\sum_{l=1}^{\infty} m_{k,l} \log \frac{A+|Q^n|}{m_{k,l}} \leq \sum_{k=1}^{\infty} 2^k \sum_{l=1}^{\infty} |I_{k,l}| \log \frac{|Q^n|e}{|I_{k,l}|} + \left(\log \frac{A+|Q^n|}{|Q^n|e}\right) A$$

$$\leq A \log \frac{A+|Q^n|}{|Q^n|e} + 4J(f) + 1. \tag{3.6.28}$$

Substituting (3.6.28) into (3.6.27) yields that

$$N_\infty(f) \leq (A + |Q^n|)\log \frac{A+|Q^n|}{|Q^n|} + (1+|Q^n|) + 4J(f)$$

$$\leq [1 + |Q^n| + 4J(f)]\left(\log \frac{1+|Q^n|+4J(f)}{|Q^n|} + 1\right). \tag{3.6.29}$$

3.6 Spaces related to the a.e. convergence of the Fourier series

This finishes the proof of Theorem 3.6.4. ∎

3.6.5 The convergence of the Fourier series on B_q

For the case of one-dimension, Taibleson and Weiss [TW1] proved the following theorem.

Theorem 3.6.5 *Let $1 < q \leq \infty$. If $f \in B_q(Q^1)$, then*

$$S_k(f, x) \longrightarrow f(x)$$

holds for a.e. $x \in Q^1$.

Proof. For the case of one-dimension, we merely need to consider in the case of $R \in \mathbb{Z}_+$. At this time, we have

$$S_R(f, x) = \frac{1}{\pi} \int_{-\pi}^{\pi} f(x-t) D_R(t) dt,$$

for $R \in \mathbb{Z}_+$, where

$$D_R(t) = \frac{\sin\left(R + \frac{1}{2}\right)t}{2\sin(\frac{t}{2})}$$

is the Dirichlet kernel.

We first prove that there exists a constant $C = C_q > 0$, such that, for each q-block b,

$$|\{x \in Q^1 : S_*(b)(x) > \lambda\}| \leq \frac{C}{\lambda}, \qquad (3.6.30)$$

for $\lambda > 0$, where S_* is the maximal operator of the partial sum defined as

$$S_*(f)(x) = \sup_{R \geq 0} |S_R(f; x)|.$$

Suppose that $\operatorname{supp} b \subset I$ such that $\|b\|_q \leq |I|^{-\frac{1}{q'}}$, where I is an interval in Q^1. It suffices to consider under the situation $1 < q < \infty$.

It is well-known that S_* is of type (q, q), and thus, we have

$$|\{x \in Q^1 : S_*(b)(x) > \lambda\}| \leq C_q \left(\frac{\|b\|_q}{\lambda}\right)^q \leq C_q \frac{1}{(|I|\lambda)^{q-1}} \frac{1}{\lambda}.$$

Hence, when $\lambda \geq |I|^{-1}$, we have

$$|\{S_*(b) > \lambda\}| \leq \frac{C_q}{\lambda}.$$

We substitute $\{x \in Q^1 : S_*(b)(x) > \lambda\}$ by $\{S_*(b) > \lambda\}$ for simpleness, and we will adopt the brief form in the following.

Let $\lambda < |I|^{-1}$ and define $\rho(x, I) = \inf_{t \in I} |x - t|$. Since $|D_R(t)| \leq \frac{C}{|t|}$, we have that

$$|S_R(b, x)| = \frac{1}{\pi} \left| \int_I b(t) D_R(x-t) dt \right| \leq \|b\|_1 \frac{C}{\rho(x, I)} = \frac{C}{\rho(x, I)}.$$

And it follows that

$$S_*(b)(x) \leq \frac{C}{\rho(x, I)}.$$

Consequently, we conclude that

$$|\{S_*(b) > \lambda\}| \leq \left| \left\{ \frac{C}{\rho(x, I)} > \lambda \right\} \right| = \left| \left\{ \rho(x, I) < \frac{C}{\lambda} \right\} \right|.$$

Taking $|I| < \frac{1}{\lambda}$ into consideration, we can get that

$$\left| \left\{ \rho(x, I) < \frac{C}{\lambda} \right\} \right| < \frac{C}{\lambda}.$$

This means that (3.6.30) is valid.

Since $f \in B_q$, and by the decomposition in blocks as well as (3.6.30), we can use the iteration principle of Lemma 3.6.3 to deduce that

$$|\{x : |x| < \pi, S_*(f) > \lambda\}| \leq \frac{C N_q(f)}{\lambda},$$

for all $\lambda > 0$. ∎

Lu-Taibleson-Weiss [LTW1] proved the conclusions in the case of multi-dimension.

Theorem 3.6.6 *Let $n > 1$ and $1 < q \leq \infty$. If $f \in B_q \cap L \log^+ L(Q^n)$, then $S_R^{\frac{n-1}{2}}(f; x)$ converge to $f(x)$ for a.e. $x \in Q^n$.*

Proof. By Theorem 3.2.3 and the condition that $f \in L \log^+ L(Q^n)$, it suffices to prove that $B_R^{\frac{n-1}{2}}(\widetilde{f}; x) \to f(x)$.

It is well-known

$$B_R^{\frac{n-1}{2}}\left(\widetilde{f}; x\right) = \frac{1}{(2\pi)^n} \int_{Q^n} f(y) H_R^{\frac{n-1}{2}}(x-y) dy,$$

3.6 Spaces related to the a.e. convergence of the Fourier series

where
$$H_R^{\frac{n-1}{2}}(u) = (2\pi)^{\frac{n}{2}} 2^{\frac{n-1}{2}} \Gamma\left(\frac{n+1}{2}\right) \frac{J_{n-\frac{1}{2}}(R|u|)}{(R|u|)^{n-\frac{1}{2}}} R^n.$$

Similar to the method of the proof in Theorem 3.6.5, we merely need to construct the weak-type inequality

$$|\{B_*(f)(x) > \lambda\}| < \frac{C}{\lambda}, \tag{3.6.31}$$

for $\lambda > 0$, where the maximal operator B_* is defined by

$$B_*(f)(x) = \sup_{R>0} \left| B_R^{\frac{n-1}{2}}(f;x) \right|.$$

Since $f \in B_q$, so does $\widetilde{f}$. By the decomposition in blocks and the iteration principle, it suffices to establish (3.6.31) for the q-block b.

Here, we directly quote the conclusion that the maximal operator B_* is of type (q,q) with $q \in (1,\infty)$. Thus we merely need notice that the kernel $H_R^{\frac{n-1}{2}}(u)$ satisfies the following inequality

$$\left| H_R^{\frac{n-1}{2}}(u) \right| \leq \frac{C}{|u|^n}.$$

Using the method of Theorem 3.6.5, we can complete the proof. This finishes the proof of Theorem 3.6.6. ∎

In summary, we can have that the function classes (or spaces) mentioned above have the inclusion relationships as follows.

$$D \subset J \subset L \log^+ L$$

and

$$J \subset B_\infty \bigcap L \log^+ L \subset B_q \bigcap L \log^+ L \subset L,$$

where $1 < q < \infty$.

Theorem 3.6.6 established the almost everywhere convergence property of the Bochner-Riesz means of multiple Fourier series at the critical index on the function class of $(\cup_{q>1} B_q) \cap L \log^+ L$. It is a beautiful result, which includes Theorem 3.6.1. By the way, we have to mention that Theorem 3.6.1 was independently proven by Sato [Sa1]. In Sato's paper, he also constructed such a function $f \in L(Q^1)$, $J(f) < \infty$, but $f \notin (L \log^+ L \log^+ \log^+ L \bigcup D)$. The example illustrates that D is a proper subset of J, and on the other hand, Sjölin's theorem about the a.e.-convergence in [Sj1] cannot contain Theorem 3.6.1.

Finally, we want to point out that the block space is not only applicable for the research of the a.e.-convergence problem about the triangle Fourier series, but for the Walsh-Fourier series as well. Wang [Wa4] have a notation about this.

In the following, we introduce certain building blocks called smooth blocks, and define spaces generated by smooth blocks. A smooth block is obtained by imposing certain smoothness on a block. The reason for studying spaces generated by smooth blocks is to investigate the relation between the smoothness imposed on the blocks and the rate of convergence of Bochner-Riesz means of multiple Fourier integral at the critical index.

Let us now turn to the definition of smooth blocks. A (q, λ)-block, $1 < q \leq \infty$, is a function b that is supported on a cube Q satisfying

$$\|b\|_{\mathcal{L}_\lambda^q} \leq |Q|^{\frac{1}{q}-1},$$

where $\mathcal{L}_\lambda^q$ denotes the Bessel potential space. Let $B_q^\lambda(\mathbb{R}^n)$ be the function space that consists of all functions f of the form, $f = \sum_k m_k b_k$, where each b_k is a (q, λ)-block, and $N(\{m_k\}) < \infty$.

Note that $B_q^0(\mathbb{R}^n) = B_q(\mathbb{R}^n)$. A natural conjecture on the relation between the smoothness and the rate of convergence is formulated as follows. For $0 \leq \lambda < 2$, $f \in B_q^\lambda(\mathbb{R}^n)$ implies

$$\left(B_R^{\frac{n-1}{2}} f\right)(x) - f(x) = o\left(\frac{1}{R^\lambda}\right)$$

for a.e. $x \in \mathbb{R}^n$, as $R \to \infty$.

Lu and Wang in [LW1] only obtain an affirmative answer for $\lambda = 1$. Their results are stated as follows.

Theorem 3.6.7 *If $f \in B_q^1(\mathbb{R}^n)$ with $1 < q \leq \infty$, then*

$$\left(B_R^{\frac{n-1}{2}} f\right)(x) - f(x) = o\left(\frac{1}{R}\right)$$

holds for a.e. $x \in \mathbb{R}^n$, as $R \to \infty$.

The proof of Theorem 3.6.7 is based on L^p-estimates of a maximal operator. Let $M_\lambda^\alpha f$ be the maximal function defined by

$$(M_\lambda^\alpha f)(x) = \sup_{R>0} \left| R^\lambda \{(B_R^\alpha f)(x) - f(x)\} \right|.$$

3.6 Spaces related to the a.e. convergence of the Fourier series

Theorem 3.6.8 *Let* $0 \leq \lambda \leq 2$, $1 < p < \infty$, $\alpha = \sigma + i\tau$, *and*

$$\sigma > \frac{n-1}{2}\left|\frac{2}{p} - 1\right|.$$

If $f \in \mathcal{L}^p_\lambda(\mathbb{R}^n)$, *then we have*

$$\|M^\alpha_\lambda f\|_p \leq C\|f\|_{\mathcal{L}^p_\lambda}.$$

To prove Theorem 3.6.8, we need some lemmas.

Lemma 3.6.7 *Let* $1 < p < \infty$, $\alpha = \sigma + i\tau$, *and* $\sigma > \frac{n-2}{2}$. *If* $f \in L^p_2(\mathbb{R}^n)$, *then we have*

$$\|M^\alpha_2 f\|_p \leq C_{n,\sigma,p} e^{\pi|\tau|^2}\|f\|_{L^p_2}.$$

Proof. We write

$$R^2[(B^\alpha_R f)(x) - f(x)]$$
$$= C_{n,\alpha} \int_{\mathbb{R}^n} \big(f(x+y) + f(x-y) - 2f(x)\big) R^{n+2} \frac{J_{(n/2)+\alpha}(R|y|)}{(R|y|)^{(n/2)+\alpha}} dy,$$

where

$$|C_{n,\alpha}| = \left|2^{\alpha-(n-2)}\pi^{-\frac{n}{2}}\Gamma(\alpha+1)\right| \leq C_{n,\sigma}.$$

Let

$$g(x,t) = \int_{\mathbb{S}^{n-1}} \big(f(x+ty) + f(x-ty) - 2f(x)\big) ds(y),$$

where ds is surface measure on $\mathbb{S}^{n-1}$. Thus we have

$$R^2[(B^\alpha_R f)(x) - f(x)] = C_{n,\alpha} \int_0^\infty R^2 g\left(x, \frac{t}{R}\right) t^{n-1} \frac{J_{(n/2)+\alpha}(t)}{t^{(n/2)+\alpha}} dt. \quad (3.6.32)$$

Denote

$$A(t) = \int_t^\infty r^{n-1} \frac{J_{(n/2)+\alpha}(t)}{t^{(n/2)+\alpha}} dr$$

and

$$B(t) = \int_t^\infty A(r) dr.$$

Clearly we have that

$$g\left(x, \frac{t}{R}\right)\bigg|_{t=0} = 0$$

and
$$\left(\frac{d}{dt}\right)g\left(x,\frac{t}{R}\right)\bigg|_{t=0}=0.$$
Thus, it follows from (3.6.32) and integration by parts that
$$R^2[(B_R^\alpha f)(x)-f(x)]=C_{n,\alpha}\int_0^\infty R^2\frac{d^2}{dt^2}g\left(x,\frac{t}{R}\right)B(t)dt.$$
Let
$$g_{ij}(x,t)=\int_{\mathbb{S}^{n-1}}|D_{ij}f(x+ty)|ds(y),$$
where
$$D_{ij}f(x)=\left(\frac{\partial}{\partial x_j}\right)\left(\frac{\partial}{\partial x_i}\right)f(x).$$
We hence have
$$(M_2^\alpha f)(x)\leq C_{n,\sigma}\sum_{i,j=1}^n \sup_{R>0}\int_0^\infty g_{ij}\left(x,\frac{t}{R}\right)|B(t)|dt, \tag{3.6.33}$$
$$B(t)=-\left(\frac{2}{\pi}\right)^{\frac{1}{2}}\frac{\cos(t-\theta)}{t^{\alpha+1-(n-1)/2}}+O\left(\frac{1}{t^{\sigma+2-(n-1)/2}}\right), \tag{3.6.34}$$
for $t\geq 1$, and
$$|B(t)|\leq C_{n,\sigma}e^{|\tau|^2}, \tag{3.6.35}$$
for $0<t<1$, where
$$\theta=\frac{\pi(n+2\alpha+1)}{4}$$
and
$$\left|O\left(\frac{1}{t^{\sigma+2-(n-1)/2}}\right)\right|\leq C_{n,\sigma}e^{\pi|\tau|}\frac{1}{t^{\sigma+2-(n-1)/2}}.$$
It follows from (3.6.35) that
$$\int_0^1 g_{ij}\left(x,\frac{t}{R}\right)|B(t)|dt\leq C_{n,\sigma}e^{|\tau|^2}\int_0^1 g_{ij}\left(x,\frac{t}{R}\right)dt$$
$$\leq C_{n,\sigma}e^{|\tau|^2}R\int_{|y|<1/R}|y|^{1-n}|D_{ij}f(x+y)|dy$$
$$\leq C_{n,\sigma}e^{|\tau|^2}M(D_{ij}f)(x),$$
where Mf is the Hardy-Littlewood maximal function of f. Thus, we have
$$\left\|\sup_{R>0}\int_0^1 g_{ij}\left(x,\frac{t}{R}\right)|B(t)|dt\right\|_p\leq C_{n,\sigma,p}e^{|\tau|^2}\|f\|_{L_2^p}. \tag{3.6.36}$$

3.6 Spaces related to the a.e. convergence of the Fourier series

Meanwhile, by (3.6.34), we have

$$\int_1^\infty g_{ij}\left(x, \frac{t}{R}\right) |B(t)| dt \leq C_{n,\sigma} e^{\pi|\tau|} \int_1^\infty g_{ij}\left(x, \frac{t}{R}\right) t^{\frac{n-1}{2}-\sigma-1} dt$$

$$= C_{n,\sigma} e^{\pi|\tau|} R^{\frac{n-1}{2-\sigma}} \int_{\frac{1}{R}}^\infty g_{ij}(x,t) t^{\frac{n-1}{2}-\sigma-1} dt.$$

Using integration by parts and the following inequality

$$\int_0^t \tau^{n-1} g_{ij}(x,\tau) d\tau \leq C_n t^n M(D_{ij}f)(x),$$

we have

$$\int_1^\infty g_{ij}\left(x, \frac{t}{R}\right) |B(t)| dt \leq C_{n,\sigma} e^{\pi|\tau|} M(D_{ij}f)(x).$$

Thus we obtain

$$\left\| \sup_{R>0} \int_1^\infty g_{ij}\left(x, \frac{t}{R}\right) |B(t)| dt \right\|_p \leq C_{n,\sigma,p} e^{\pi|\tau|} \|f\|_{L_2^p}. \quad (3.6.37)$$

Consequently, the conclusion of Lemma 3.6.7 is immediately derived from (3.6.33), (3.6.36) and (3.6.37). ∎

Lemma 3.6.8 *Let $0 \leq \lambda \leq 2$, $1 < p < \infty$, $\alpha = \sigma + i\tau$, and $\sigma > (n-1)/2$. If $f \in \mathcal{L}_\lambda^p(\mathbb{R}^n)$, then*

$$\|M_\lambda^\alpha f\|_p \leq C_{n,\sigma,\lambda,p} e^{\pi|\tau|^2} \|f\|_{\mathcal{L}_\lambda^p}.$$

Proof. Let $\{r_k\}$ be a sequence consisting of all positive rational numbers and $\Lambda_k = \{r_1, ..., r_k\}$. Define

$$\left(F_\lambda^{\alpha,k} f\right)(x) = \sup_{r_j \in \Lambda_k} r_j^\lambda \left|(B_{r_j}^\alpha f)(x) - f(x)\right|.$$

Thus we have

$$\left(F_\lambda^{\alpha,k} f\right)(x) \leq \left(F_\lambda^{\alpha,k+1} f\right)(x)$$

and

$$(M_\lambda^\alpha f)(x) = \lim_{k \to \infty} \left(F_\lambda^{\alpha,k} f\right)(x).$$

Fix $f \in \mathcal{L}_\lambda^p(\mathbb{R}^n)$ and k. Let S_j with $1 \leq j \leq k$ be a set such that, for $x \in S_j$,

$$\left(F_\lambda^{\alpha,k} f\right)(x) = r_j^\lambda \left|(B_{r_j}^\alpha f)(x) - f(x)\right|,$$

and
$$\left(F_\lambda^{\alpha,k} f\right)(x) > r_i^\lambda |(B_{r_i}^\alpha f)(x) - f(x)|,$$
for $i < j$.

It is easy to note that the sets S_j do not intersect each other. Let
$$\Omega = \{z \in \mathbb{C} : 0 \leq \operatorname{Re} z \leq 1\}.$$
Define
$$\psi_j(x) = \operatorname{sign}((B_{r_j}^\alpha f)(x) - f(x)),$$
and
$$(T_z g)(x) = \sum_{r_j \in \Lambda_k} r_j^{2z} \chi_{S_j}(x) \left(B_{r_j}^\alpha (J_{2z}g)(x) - (J_{2z}g)(x)\right) \psi_j(x).$$

Then $\{T_z\}$ is a family of linear operators. It is easy to verify that $\{T_z\}$ is an admissible family of operators (see [SW1]).

Now we can write $f = G_\lambda * g$, where
$$\hat{G}_\lambda(x) = \left(1 + 4\pi^2 |x|^2\right)^{-\lambda/2}.$$

Since $f \in \mathcal{L}_\lambda^p$, a multiplier theorem (see [St4]) implies that
$$\|J_{i\eta} g\|_p \leq C_{n,p} P(\eta) \|g\|_p,$$
where P is a polynomial of degree $k > n/2$.

Note $\alpha > (n-2)/2$. We have
$$\|T_{i\eta} g\|_p \leq \left\|F_0^{\alpha,k}(J_{2i\eta}g)\right\|_p$$
$$\leq \|M_0^\alpha(J_{2i\eta}g)\|_p$$
$$\leq C_{n,\sigma,p} e^{2\pi|\tau|} \|J_{2i\eta}g\|_p$$
$$\leq C_{n,\sigma,p} e^{2\pi|\tau|} |P(2\eta)| \|g\|_p.$$

On the other hand, it follows from Lemma 3.6.7 that
$$\|T_{1+i\eta} g\|_p \leq \left\|F_2^{\alpha,k}(J_{2+2i\eta}g)\right\|_p$$
$$\leq \|M_2^\alpha(J_{2+2i\eta}g)\|_p$$
$$\leq C_{n,\sigma,p} e^{|\tau|^2} \|J_{2+2i\eta}g\|_p$$
$$\leq C_{n,\sigma,p} e^{|\tau|^2} \|J_{2i\eta}g\|_p$$
$$\leq C_{n,\sigma,p} e^{|\tau|^2} |P(2\eta)| \|g\|_p.$$

3.6 Spaces related to the a.e. convergence of the Fourier series

Using the Stein's interpolation theorem of analytic operators, we have

$$\begin{aligned}
\|F_\lambda^{\alpha,k} f\|_p &= \|T_{\lambda/2} g\|_p \\
&\leq C_{n,\sigma,p} e^{|\tau|^2} \|g\|_p \\
&\leq C_{n,\sigma,\lambda,p} e^{|\tau|^2} \|f\|_{\mathcal{L}_\lambda^p}.
\end{aligned}$$

Finally, by Lebesgue's monotonic convergence theorem, we obtain the conclusion of Lemma 3.6.8. ∎

Let

$$(N_\lambda^\alpha f)(x) = \sup_{R>0} R^\lambda \left| (B_R^{\alpha+1} f)(x) - (B_R^\alpha f)(x) \right|.$$

Lemma 3.6.9 *Let $\alpha = \sigma + i\tau$, $\sigma > 0$, and $0 \leq \lambda \leq 2$. If $f \in \mathcal{L}_\lambda^2(\mathbb{R}^n)$, then we have*

$$\|N_\lambda^\alpha f\|_2 \leq \begin{cases} C_n \exp\{C_n(\sigma + |\tau|)\} 1/\sqrt{\sigma}(1 + 1/\sqrt{2-\lambda} + 1/\sqrt{\sigma}) \|f\|_{\mathcal{L}_\lambda^2}, & 0 \leq \lambda < 2, \\ C_n \exp\{C_n(\sigma + |\tau|)\} 1/\sigma \|f\|_{L_2^2}, & \lambda = 2. \end{cases}$$

Proof. Let $\beta \in \mathbb{C}$, $\mathrm{Re}(\beta) > \frac{1}{2}$, $\delta > -\frac{1}{2}$, and $0 \leq \lambda < 2$, Using the formula at the page 278 in [SW1], taking the the Fourier transform, and doing some algebra, we get the identity:

$$\begin{aligned}
&\left(B_R^{\beta+\delta+1} f\right)(x) - \left(B_R^{\beta+\delta} f\right)(x) \\
&= \frac{2}{B(\beta,\delta+1)} \frac{1}{R^{2(\beta+\delta+1)}} \int_0^R (R^2 - u^2)^{\beta-1} u^{2\delta+3} \left[\left(B_u^{\delta+1} f\right)(x) - \left(B_u^\delta f\right)(x)\right] du,
\end{aligned}$$

where

$$B(\beta,\delta) = \frac{\Gamma(\beta)\Gamma(\delta)}{\Gamma(\beta+\delta)}.$$

Thus we conclude that

$$R^\lambda \left| \left(B_R^{\beta+\delta+1} f\right)(x) - \left(B_R^{\beta+\delta} f\right)(x) \right|$$

$$\leq \frac{2}{|B(\beta,\delta+1)|} \frac{1}{R^{2(\operatorname{Re}\beta+\delta+1)-\lambda}}$$

$$\times \left(\int_0^R (R^2 - u^2)^{2\operatorname{Re}\beta - 2} u^{4\delta+7-2\lambda} du \right)^{\frac{1}{2}}$$

$$\times \left(\int_0^R u^{2\lambda - 1} \left| \left(B_u^{\delta+1} f\right)(x) - \left(B_u^\delta f\right)(x) \right|^2 du \right)^{\frac{1}{2}}$$

$$\leq \frac{2}{|B(\beta,\delta+1)|} \left\{ \int_0^1 (1-t^2)^{2\operatorname{Re}\beta-2} t^{4\delta+7-2\lambda} dt \right\}^{\frac{1}{2}} \left(G_\lambda^\delta f\right)(x)$$

$$= \frac{\{2B(2\operatorname{Re}\beta - 1, 2\delta - \lambda + 4)\}^{\frac{1}{2}}}{B(\beta,\delta+1)} \left(G_\lambda^\delta f\right)(x),$$

where

$$\left(G_\lambda^\delta f\right)(x) = \left(\int_0^\infty u^{2\lambda - 1} \left| \left(B_u^{\delta+1} f\right)(x) - (B_u^\delta f)(x) \right|^2 du \right)^{\frac{1}{2}}.$$

Let $\delta = \frac{\sigma-1}{2} > -\frac{1}{2}$, $\beta = \alpha - \delta = \frac{\sigma+1}{2} + i\tau$. We conclude that

$$\|N_\lambda^\alpha f\|_2 \leq \frac{[2B(\sigma, \sigma - \lambda + 3)]^{\frac{1}{2}}}{|B(\frac{\sigma+1}{2} + i\tau, \frac{\sigma+1}{2})|} \left\| G_\lambda^{\frac{\sigma-1}{2}} f \right\|_2.$$

It follows from the Plancherel theorem that

$$\left\| G_\lambda^{\frac{\sigma-1}{2}} f \right\|_2^2$$

$$= \int_{\mathbb{R}^n} \int_0^\infty u^{2\lambda - 1} \left| \left(B_u^{\frac{\sigma+1}{2}} f\right)(x) - \left(B_u^{\frac{\sigma-1}{2}} f\right)(x) \right|^2 du\, dx$$

$$= \int_0^\infty u^{2\lambda - 1} \int_{|y|<u} \left| \left(1 - \frac{|y|^2}{u^2}\right)^{\frac{\sigma+1}{2}} - \left(1 - \frac{|y|^2}{u^2}\right)^{\frac{\sigma-1}{2}} \right|^2 |\hat{f}(y)|^2 dy\, du$$

$$= \int_0^\infty u^{2\lambda - 1} \int_{|y|<u} \frac{|y|^4}{u^4} \left(1 - \frac{|y|^2}{u^2}\right)^{\sigma - 1} |\hat{f}(y)|^2 dy\, du$$

$$= \int_{\mathbb{R}^n} |y|^4 |\hat{f}(y)|^2 \left(\int_{|y|}^\infty u^{2\lambda - 5} \left(1 - \frac{|y|^2}{u^2}\right)^{\sigma - 1} du \right) dy$$

3.6 Spaces related to the a.e. convergence of the Fourier series

$$= \frac{1}{2} \left(\int_{\mathbb{R}^n} |y|^{2\lambda} |\hat{f}(y)|^2 dy \right) \left(\int_0^1 (1-v)^{\sigma-1} v^{1-\lambda} dv \right)$$

$$\leq C \left(1 + \frac{1}{\sigma} + \frac{1}{2-\lambda} \right) \|f\|_{\mathcal{L}^2_\lambda}^2.$$

Note that

$$\left| \left\{ B\left(\frac{\sigma+1}{2} + i\tau, \frac{\sigma+1}{2} \right) \right\}^{-1} \right| \leq C_n \exp\{C_n(\sigma + |\tau|)\}$$

and

$$[2B(\sigma, \sigma - \lambda + 3)]^{\frac{1}{2}} \leq \frac{C_n}{\sigma^{\frac{1}{2}}}.$$

Hence we have

$$\|N_\lambda^\alpha f\|_2 \leq C_n \exp\{C_n(\sigma + |\tau|)\} \frac{1}{\sqrt{\sigma}} \left(1 + \frac{1}{\sqrt{\sigma}} + \frac{1}{\sqrt{2-\lambda}} \right) \|f\|_{\mathcal{L}^2_\lambda},$$

for $0 \leq \lambda < 2$.

When $\lambda = 2$, we have

$$R^2((B_R^{\alpha+1} f)(x) - (B_R^\alpha f)(x)) = -B_R^\alpha(\Delta f)(x).$$

Thus, by the method similar to the proof of Lemma 5.10 in [SW1], we have

$$\|N_2^\alpha f\|_2 = \left\| \sup_{R>0} B_R^\alpha(\Delta f) \right\|_2$$
$$\leq C_n e^{C_n(\sigma+|\tau|)} (1/\sigma) \|\Delta f\|_2$$
$$\leq C_n e^{C_n(\sigma+|\tau|)} (1/\sigma) \|f\|_{L^2_2}.$$

∎

Next let us turn to the proof of Theorem 3.6.8.

Set $f = G_\lambda * g \in L^p(\mathbb{R}^n)$. For $\sigma > 0$, we choose a $k \in \mathbb{N}$, such that $\sigma + k > (n-1)/2$. Thus, by Lemma 3.6.7, 3.6.8, and the inequality

$$(M_\lambda^\alpha f)(x) \leq \sum_{j=0}^{k-1} (N_\lambda^{\alpha+1} f)(x) + (M_\lambda^{\alpha+k} f)(x),$$

it follows that

$$\|M_\lambda^\alpha f\|_2 \leq C_{n,\sigma,\lambda} e^{|\tau|^2} \|f\|_{L^2_\lambda} = C_{n,\sigma,\lambda} e^{|\tau|^2} \|g\|_2. \qquad (3.6.38)$$

Now, let $p_1 > 1$ and $\sigma > (n-1)/2$. It follows from Lemma 3.6.7 that

$$\|M_\lambda^\alpha f\|_{p_1} \leq C_{n,\sigma,\lambda,p_1} e^{|\tau|^2} \|f\|_{\mathcal{L}_\lambda^{p_1}} = C_{n,\sigma,\lambda,p_1} e^{|\tau|^2} \|g\|_{p_1}. \qquad (3.6.39)$$

Consider $p_1 \leq p \leq 2$, and write $1/p = (1-t)/2 + t/p_1$. Let $\mu_0 > 0, \mu_1 > (n-1)/2$, and $\delta(z) = \mu_0(1-z) + \mu_1 z$ for $0 \leq \operatorname{Re} z \leq 1$. If $\mu_0 \to 0$, $\mu_1 \to (n-1)/2$, and $p_1 \to 1$, then we have

$$\delta(t) \to \frac{n-1}{2\left(\frac{2}{p}-1\right)}.$$

Therefore, if

$$\sigma > \frac{n-1}{2\left(\frac{2}{p}-1\right)},$$

then there exist μ_0, μ_1 and p_1 satisfying the above condition such that $\delta(t) = \sigma$, where

$$t = \frac{\frac{1}{p} - \frac{1}{2}}{\frac{1}{p_1} - \frac{1}{2}}.$$

Let such μ_0, μ_1, and p_1 be fixed later. Let $\{G_j\}$ be a sequence consisting of all positive rational numbers. Denote $A_k = \{R_1, ..., R_k\}$, and

$$\left(F_\lambda^{\alpha,k} f\right)(x) = \sup_{R \in A_k} \left\{ R^\lambda |(B_R^\alpha f)(x) - f(x)| \right\}.$$

We have

$$\left(F_\lambda^{\alpha,k} f\right)(x) \leq \left(F_\lambda^{\alpha,k+1} f\right)(x),$$

and

$$(M_\lambda^\alpha f)(x) = \lim_{k \to \infty} \left(F_\lambda^{\alpha,k} f\right)(x).$$

For $1 \leq j \leq k$, let

$$E_j = \left\{ x \in \mathbb{R}^n : \sup_{R \in A_k} \left\{ R^\lambda |(B_\lambda^\alpha f)(x) - f(x)| \right\} = R_j^\lambda |(B_{R_j}^\alpha f)(x) - f(x)| \right\},$$

and $F_1 = E_1, F_j = E_j \setminus \bigcup_{i=1}^{j-1} E_i$ for $j = 2, ..., k$. Define

$$(T_z h)(x) = \sum_{j=1}^k R_j^\lambda \chi_{F_j}(x) \left(B_{R_j}^{\delta(z)}(G_\lambda * h)(x) - (G_\lambda * h)(x) \right) \psi_j(x),$$

3.6 Spaces related to the a.e. convergence of the Fourier series

where
$$\psi_j(x) = \text{sign}\left(B^\sigma_{R_j}(G_\lambda * g)(x) - G_\lambda * g(x)\right).$$

It is easy to verify that $\{T_z\}$ is an admissible family of linear operator (see [SW1].) Using (3.6.38) and (3.6.39) implies that

$$\|T_{i\tau}g\|_2 \le \left\|F_\lambda^{\delta(i\tau),k}f\right\|_2 \le \left\|M_\lambda^{\delta(i\tau)}f\right\|_2 \le C_{n,\mu_0,\lambda}e^{(\mu_1-\mu_0)^2\tau^2}\|g\|_2,$$

and
$$\|T_{1+i\tau}g\|_{p_1} \le \left\|F_\lambda^{\delta(1+i\tau),k}f\right\|_{p_1} \le C_{n,\mu_1,\lambda,p_1}e^{(\mu_1-\mu_0)^2\tau^2}\|g\|_{p_1}.$$

Thus, by the Stein's interpolation theorem of analytic operators, we have that
$$\left\|F_\lambda^{\sigma,k}f\right\|_p = \left\|F_\lambda^{\delta(t),k}f\right\|_p = \|T_tg\|_p \le C\|g\|_p = C\|f\|_{\mathcal{L}^p_\lambda}.$$

It follows from the monotonic convergence theorem that

$$\|M_\lambda^\sigma f\|_p \le C\|f\|_{\mathcal{L}^p_\lambda},$$

for $1 < p \le 2$. We can obtain the similar estimate for M_λ^α, $\alpha = \sigma + i\tau$, as in [SW1]. Finally, it should be pointed out that the proof in the case of $2 < p < \infty$ is similar to the above.

To prove Theorem 3.6.7, we first need to establish a weak type estimate of the maximal operator $M_1^{(n-1)/2}$ on any block.

Lemma 3.6.10 *If b is a $(q,1)$-block, then we have*

$$\left|\left\{x : \left(M_1^{(n-1)/2}b\right)(x) > \lambda\right\}\right| \le C\lambda^{-1},$$

where C is independent of λ and b.

Proof. We write

$$R\left\{\left(B_R^{(n-1)/2}b\right)(x) - b(x)\right\}$$

$$= CR^{n+1}\int_{\mathbb{R}^n}[b(x+y)-b(x)]\frac{J_{n-\frac{1}{2}}(R|y|)}{(R|y|)^{n-\frac{1}{2}}}dy$$

$$= CR^{n+1}\int_0^\infty\left\{\int_{\mathbb{S}^{n-1}}[b(x+t\xi)-b(x)]d\sigma(\xi)\right\}\frac{J_{n-\frac{1}{2}}(Rt)}{(Rt)^{n-\frac{1}{2}}}t^{n-1}dt$$

$$= CR\int_0^\infty\left\{\int_{\mathbb{S}^{n-1}}\left[b\left(x+\frac{\tau}{R}\xi\right)-b(x)\right]d\sigma(\xi)\right\}\frac{J_{n-\frac{1}{2}}(\tau)}{(\tau)^{n-\frac{1}{2}}}\tau^{n-1}d\tau.$$

Since $t^{-1/2}J_{n-\frac{1}{2}}(t) \notin L^1(\mathbb{R}^n)$, these integrals should be interpreted as
$\lim_{T\to+\infty}\int_{|y|\leq T}$ and $\lim_{T\to+\infty}\int_0^T$, respectively.

Denote
$$g(x,\tau) = \int_{\mathbb{S}^{n-1}}[b(x+\tau\xi) - b(x)]d\sigma(\xi).$$

We have
$$R\left\{\left(B_R^{(n-1)/2}b\right)(x) - b(x)\right\} = CR\int_0^\infty g(x,\tau/R)\frac{J_{n-\frac{1}{2}}(\tau)}{(\tau)^{n-\frac{1}{2}}}\tau^{n-1}d\tau.$$

Using integration by parts, we obtain
$$R\left\{\left(B_R^{(n-1)/2}b\right)(x) - b(x)\right\} = CR\int_0^\infty \frac{d}{d\tau}g(x,\tau/R)A(\tau)d\tau,$$

where
$$A(\tau) = \int_\tau^\infty x^{n-1}\frac{J_{n-\frac{1}{2}}(x)}{(x)^{n-\frac{1}{2}}}dx.$$

By the properties of Bessel function, it follows that $|A(\tau)| \leq \frac{C}{\tau}$ for $\tau \geq 1$, and $|A(\tau)| \leq C$ for $0 < \tau < 1$. Thus, we have

$$\left|R\left\{\left(B_R^{(n-1)/2}b\right)(x) - b(x)\right\}\right|$$
$$\leq C\int_0^\infty \left\{\int_{\mathbb{S}^{n-1}}\left|\nabla_x b\left(x+\frac{\tau}{R}\xi\right)\right|d\sigma(\xi)\right\}|A(\tau)|d\tau$$
$$= C\int_0^\infty \left\{\int_{\mathbb{S}^{n-1}}|\nabla_x b(x+t\xi)|d\sigma(\xi)\right\}|A(Rt)|Rdt$$
$$\leq C\int_0^\infty \left\{\int_{\mathbb{S}^{n-1}}|\nabla_x b(x+t\xi)|d\sigma(\xi)\right\}t^{-1}dt$$
$$= C\int_{\mathbb{R}^n}|\nabla_x b(x+y)|\frac{dy}{|y|^n}$$
$$= C\int_{\mathbb{R}^n}\frac{|\nabla b(u)|}{|u-x|^n}du.$$

It follows that
$$\left(M_1^{(n-1)/2}b\right)(x) \leq C\int_Q\frac{|\nabla b(u)|}{|u-x|^n}du, \tag{3.6.40}$$

where supp $b \subset Q$.

3.6 Spaces related to the a.e. convergence of the Fourier series

Let $\tilde{Q} = 2Q$. Then it follows from (3.6.40) that

$$\left(M_1^{(n-1)/2} b\right)(x) \leq \frac{C}{|x - x_Q|^n},$$

provided that $x \notin \tilde{Q}$, where x_Q is the center of Q. Thus, we have

$$\left|\left\{x \notin \tilde{Q} : \left(M_1^{(n-1)/2} b\right)(x) > \lambda, \lambda \leq 1/|Q|\right\}\right| \leq C\lambda^{-1}. \tag{3.6.41}$$

It follows from Theorem 3.6.8 that

$$\|M_1^{(n-1)/2} f\|_q \leq C\|f\|_{L_1^q}, \tag{3.6.42}$$

for $1 < q < \infty$. Thus, we have

$$\left|\left\{x \in \mathbb{R}^n : (M_1^{(n-1)/2} b)(x) > \lambda, \lambda > 1/|Q|\right\}\right| \leq C\left(\frac{\|b\|_{L_1^q}}{\lambda}\right)^q \leq C\lambda^{-1}. \tag{3.6.43}$$

Combining (3.6.41) with (3.6.43) yields the conclusion of Lemma 3.6.10. ∎

Proof of Theorem 3.6.7. Let $f \in B_q^1(\mathbb{R}^n)$. We have

$$f(x) = \sum_k m_k b_k(x) = \sum_{k=1}^N m_k b_k(x) + \sum_{k=N+1}^\infty m_k b_k(x) := g(x) + h(x),$$

where each b_k is a $(q,1)$-block and $N(\{m_k\}) < \infty$.

To complete the proof of Theorem 3.6.7, we must prove

$$\left|\left\{x : \limsup_{R \to \infty} \left|R\left\{(B_R^{(n-1)/2} f)(x) - f(x)\right\}\right| > \lambda\right\}\right| = 0.$$

Since (3.6.42) implies

$$\lim_{R \to \infty} \left\{\left(B_R^{(n-1)/2} g\right)(x) - g(x)\right\} = o\left(\frac{1}{R}\right)$$

for a.e. $x \in \mathbb{R}^n$ and $g \in L_1^q(\mathbb{R}^n)$, we have

$$\left|\left\{x : \limsup_{R \to \infty} \left|R\left\{\left(B_R^{(n-1)/2} g\right)(x) - g(x)\right\}\right| > \lambda/2\right\}\right| = 0.$$

Thus, we obtain that

$$\left|\left\{x : \limsup_{R \to \infty} \left|R\left\{\left(B_R^{(n-1)/2}f\right)(x) - f(x)\right\}\right| > \lambda\right\}\right|$$
$$\leq \left|\left\{x : \left(M_R^{(n-1)/2}h\right)(x) > \frac{\lambda}{2}\right\}\right|$$
$$\leq C\lambda^{-1} \sum_{k=N+1}^{\infty} |m_k|\left(1 + \log \frac{\sum_{s=1}^{\infty} |m_s|}{|m_k|}\right).$$

This completes the proof of Theorem 3.6.7. ∎

3.7 The uniform convergence and approximation

Concerning the problems of the convergence and approximation in the scale of uniform, here, using $C(\mathbb{R}^n)$ instead of $L^\infty(\mathbb{R}^n)$, we have mentioned in Section 3.4. However, the order is $\alpha > \alpha_\infty = \frac{n-1}{2}$. In this section, we mainly discuss the situation of the critical index.

Parallel to Theorem 3.5.4 about the pointwise convergence, Lu [Lu2] obtained the following result.

Theorem 3.7.1 *Let $f \in C(Q^n)$ with $n \geq 2$. The following two assertions hold.*

(a) If the condition (3.5.10) uniformly holds for $x \in Q^n$ and $r \in [h, r_0]$, then we have

$$\left\|S_R^{\frac{n-1}{2}}(f) - f\right\|_C \to 0$$

as $R \to \infty$.

(b) If

$$\widetilde{\omega}(f; \delta) = o\left(\frac{1}{\log \frac{1}{\delta}}\right)$$

as $\delta \to 0^+$, then the condition of (a) is satisfied, where

$$\widetilde{\omega}(f; \delta) = \sup\{|f_x(t+h) - f_x(t)| : x \in Q^n,\ 0 \leq h \leq \delta,\ t > 0\}$$

is the modulus of continuity introduced by Golubov [Go1].

3.7 The uniform convergence and approximation

Proof. For any fixed point $x_0 \in Q^n$, we denote

$$Q_{x_0} = \left\{ x : x - x_0 \in \frac{1}{2} Q^n \right\}.$$

Obviously, the following propositions are equivalent:

(I) $\left\| S_R^{\frac{n-1}{2}}(f) - f \right\|_C \to 0$, as $R \to \infty$.

(II) $\left\| S_R^{\frac{n-1}{2}}(f) - f \right\|_{C(Q_0)} \to 0$, as $R \to \infty$, for $x_0 \in Q^n$.

(III) For $x_0 \in Q^n$, define

$$g(x) = g^{(x_0)}(x) = \begin{cases} f(x), & \text{if } x - x_0 \in Q^n, \\ 0, & \text{if } x - x_0 \notin Q^n. \end{cases}$$

$$\sup_{x \in Q_0} \left| B_R^{\frac{n-1}{2}}(g; x) - g(x) \right| \to 0, \text{ as } R \to \infty.$$

(IV) For $x_0 \in Q^n$, we have

$$\lim_{R \to \infty} \sup_{s \in Q_0} \left| \int_{\frac{\pi}{R}}^{2\sqrt{n\pi}} [g_x(t) - g(x)] \frac{\cos\left(tR - \frac{n\pi}{2}\right)}{t} dt \right| = 0.$$

Here $\|g\|_{C(Q_0)} = \sup\{|g(x)| : x \in Q_0\}$.

We have to point out that it is quite obvious that (I) $\iff$ (II) and (II) $\iff$ (III) follows from Stein's theorem (see Theorem 3.2.3). (III) $\iff$ (IV) follows from the following facts, the integral expression of $B_R^{\frac{n-1}{2}}(g)$, the uniform continuity of g on $\overline{Q}_0$, the asymptotic formula of Bessel functions and the conclusion that when $t > 2\sqrt{n\pi}$, $g_x(t) = 0$, for $x \in Q_0$.

Hence, to prove the conclusion of (I), it suffices to prove (IV). We may take it for granted that $x_0 = 0$, and in the condition (3.5.10), $r_0 \in (0, \frac{1}{2})$.

Firstly, we will show that

$$\lim_{R \to \infty} \sup_{x \in Q_0 - \frac{1}{2} Q^n} \left| \int_{r_0}^{2\sqrt{n\pi}} [g_x(t) - g(x)] \frac{\cos\left(tR - \frac{n\pi}{2}\right)}{t} dt \right| = 0. \quad (3.7.1)$$

Now let $x_0 = 0$ and

$$g(x) = f(x) \chi_{Q^n}(x).$$

We have
$$|g(x)| \leq \max_{x \in Q^n} |f(x)| = M.$$
Therefore, we have
$$\left| \int_{r_0}^{\sqrt{n}\pi} g(x) \frac{\cos\left(tR - \frac{n\pi}{2}\right)}{t} dt \right| = \left| g(x) \int_{Rr_0}^{R2\sqrt{n}\pi} \frac{\cos\left(u - \frac{n\pi}{2}\right)}{u} du \right|$$
$$\leq M \left| \int_{Rr_0}^{R2\sqrt{n}\pi} \frac{\cos\left(u - \frac{n\pi}{2}\right)}{u} du \right| \to 0. \quad (3.7.2)$$

For any $\varepsilon > 0$, there exists $\varphi \in C^\infty(\mathbb{R}^n)$ with $|\varphi| \leq 1$, such that $\varphi(x) = 1$, if $x \in Q^n$, and $\varphi(x) = 0$, if $x \notin (1+\varepsilon) \cdot Q^n$. Let $h = f\varphi$. Then we have

$$\left| \int_{r_0}^{2\sqrt{n}\pi} [g_x(t) - h_x(t)] \frac{\cos\left(tR - \frac{n\pi}{2}\right)}{t} dt \right|$$
$$\leq \int_{r_0}^{2\sqrt{n}\pi} \int_{\mathbb{S}^{n-1}} |g(x+t\xi) - h(x+t\xi)| d\sigma(\xi) \frac{dt}{t}$$
$$\leq \frac{1}{r_0^n} \int_{r_0 < |y-x| < 2\sqrt{n}\pi} |g(y) - h(y)| dy$$
$$\leq \frac{M}{r_0^n} |(1+\varepsilon)Q^n - Q^n| = \frac{M}{r_0^n} ((1+\varepsilon)^n - 1) |Q^n|$$
$$< C\varepsilon, \quad (3.7.3)$$

where we may assume that $0 < \varepsilon < 1$.

Since h is uniformly continuous, if R is big enough, we have
$$\left| \int_{r_0}^{\sqrt{n}} h_x(t) \frac{\cos\left(tR - \frac{n\pi}{2}\right)}{t} dt \right| < \varepsilon, \quad (3.7.4)$$

for $x \in \mathbb{R}^n$. By (3.7.3) and (3.7.4), we can easily get (3.7.2).

Now it remains to prove the following equality.
$$\lim_{R \to \infty} \sup_{x \in \frac{1}{2}Q^n} \left| \int_{\frac{\pi}{R}}^{r_0} [g_x(t) - g(x)] \frac{\cos\left(tR - \frac{n\pi}{2}\right)}{t} dt \right| = 0. \quad (3.7.5)$$

Denote the spherical integral of g centered at x with radius r by
$$G(x, r) = \int_{B(x,r)} [g(y) - g(x)] dy = \omega_{n-1} \int_0^r [g_x(t) - g(x)] t^{n-1} dt.$$

3.7 The uniform convergence and approximation

Set

$$\varphi_x(t) = [g_x(t) - g(x)]t^{n-1} = \frac{1}{\omega_{n-1}} \int_{|\xi|=t} [g(x+\xi) - g(x)]d\sigma(\xi).$$

Then

$$G(x,r) = \omega_{n-1} \int_0^r \varphi_x(t)dt.$$

Since g and f coincide in Q^n, for $x \in \frac{1}{2}Q^n$ and $r \in [h, r_0]$, $0 < h < r_0 \leq \frac{1}{2}$, the following equality

$$\frac{1}{r^{n-1}} \int_0^h \left(\varphi_x(r+h+t) - \varphi_x(r+h-t)\right) dt = o\left(\frac{h}{\log \frac{1}{h}}\right) \quad (3.7.6)$$

holds uniformly, as $h \to 0^+$.

The remaining arguments are the same as that of Theorem 3.5.4. However, the estimate we can get at this time is uniformly valid for $x \in \frac{1}{2}Q^n$. Thus we have proven (3.7.5), from which we can induce the conclusion of (I) directly.

In order to prove (II), it suffices to show that when

$$\widetilde{\omega}(f; \delta) = o\left(\frac{1}{\log \frac{1}{\delta}}\right)$$

as $\delta \to 0^+$,

$$\frac{1}{r^{n-1}} \int_0^h \Big\{ (r+h+t)^{n-1}[f_x(r+h+t) - f(x)]$$
$$-(r+h-t)^{n-1}[f_x(r+h-t) - f(x)] \Big\} dt$$
$$= o\left(\frac{h}{\log \frac{1}{h}}\right) \quad (3.7.7)$$

holds uniformly for $x \in Q^n$ and $r \in [h, r_0]$, as $h \to 0^+$.

The integral on the left side of (3.7.7) can be divided into the sum of the following three terms:

$$J_1 = \int_0^h [f_x(r+h+t) - f_x(r+h-t)]dt,$$

$$J_2 = \int_0^h \left[\left(1 + \frac{h+t}{r}\right)^{n-1} - 1\right][f_x(r+h+t) - f(x)]dt,$$

and

$$J_3 = -\int_0^h \left[\left(1 - \frac{h+t}{r}\right)^{n-1} - 1\right][f_x(r+h-t) - f(x)]dt.$$

Obviously, we can have that

$$J_1 = \int_0^h \mathrm{o}(\tilde{\omega}(f, 2t))dt = \mathrm{o}\left(\frac{h}{\log \frac{1}{h}}\right),$$

$$J_2 = \int_0^h \mathrm{O}\left(\frac{t}{r}\right)\mathrm{O}(\tilde{\omega}(f; r+h+t))dt$$

$$= \int_0^h \mathrm{O}\left[\frac{h}{r}\left(1 + \frac{r}{h}\right)\right] \cdot \tilde{\omega}(f; h)dt$$

$$= \mathrm{o}\left(\frac{h}{\log \frac{1}{h}}\right)$$

and

$$J_3 = \mathrm{o}\left(\frac{h}{\log \frac{1}{h}}\right)$$

hold uniformly for $x \in Q^n$ and $r \in [h, r_0]$.

This completes the proof of Theorem 3.7.1. ∎

The conclusion of (II) shows that Theorem 3.7.1 is stronger than the corresponding result obtained by Golubov. Of course, the condition (3.5.10) in Theorem 3.7.1 (and Theorem 3.5.4) can be weakened into unilateral condition. The unilateral condition means that: if $|A(t)| = \mathrm{o}(\alpha(t))$ as $t \to t_0$, then $A(t) \le \varepsilon(t)\alpha(t)$ or $A(t) \ge -\varepsilon(t)\alpha(t)$, where $\varepsilon(t) \ge 0$ and $\varepsilon(t) = \mathrm{o}(1)$ as $t \to t_0$. Jiang has obtained such results and extended these conclusions to the situation of the conjugate series (see Jiang [J1]).

Making use of the Lebesgue constant, one can deduce the following results immediately (see Wang [Wa6]).

Theorem 3.7.2 Let $f \in C(Q^n)$ with $n > 1$. Then we have

$$\left\|S_R^{\frac{n-1}{2}}(f) - f\right\|_C = O(\log R)\omega_2\left(f; \frac{1}{R}\right), \qquad (3.7.8)$$

where ω_2 is the 2-ordered continuous norm of f.

3.7 The uniform convergence and approximation

Proof. Denote g by the best R-ordered triangle polynomial approximation of f. We have that

$$S_R^{\frac{n-1}{2}}(f) - f = S_R^{\frac{n-1}{2}}(f-g) - (f-g) + S_R^{\frac{n-1}{2}}(g) - g.$$

It follows that

$$\left\|S_R^{\frac{n-1}{2}}(f) - f\right\|_C \le L_R\|f-g\|_C + \left\|S_R^{\frac{n-1}{2}}(g) - g\right\|_C,$$

where L_R is the Lebesgue constant of $S_R^{\frac{n-1}{2}}$, and $L_R = O(\log R)$.

At the same time, just as we do in Section 3.4,

$$\|f-g\|_C = E_R(f)_c \le C\omega_2\left(f; \frac{1}{R}\right),$$

we conclude that

$$\left\|S_R^{\frac{n-1}{2}}(g) - g\right\|_C \le \left\|S_R^{\frac{n-1}{2}}(g) - S_R^{\frac{n+1}{2}}(g)\right\|_C + \left\|S_R^{\frac{n+1}{2}}(g) - g\right\|_C$$

$$= \frac{1}{R^2}\left\|S_R^{\frac{n-1}{2}}(\Delta g)\right\|_C + C\omega_2\left(g; \frac{1}{R}\right)$$

$$\le \frac{1}{R^2}L_R\|\Delta g\|_C + C\omega_2\left(f; \frac{1}{R}\right)$$

$$\le C\log R \cdot \omega_2\left(f; \frac{1}{R}\right),$$

for $R > 2$.

In the above estimate, we have utilized the relation between the modulus of continuity of the derivative of the best triangle polynomial approximation and that of the function which is to be approximated. This relation can be easily extended from the corresponding result in the unitary case.

Combining all the above results, we can have (3.7.8) which concludes the proof. ∎

Remark 3.7.1 *From the proof, we can see that if n is an odd number with $n \ge 3$, there is a more accurate estimate*

$$\left\|f - S_R^{\frac{n-1}{2}}(f)\right\|_C = O(\log R)E_R(f) + O\left(\omega_2\left(f; \frac{1}{R}\right)\right). \tag{3.7.9}$$

For functions with higher derivatives, the approximation order can be increased.

Definition 3.7.1 *Let $f \in C(Q^n)$. If f is absolutely continuous with respect to all x_j and its partial derivative $\frac{\partial f}{\partial x_j}$ are essentially bounded on Q^n, for $j = 1, 2, \ldots, n$, then we call $f \in W^1 L^\infty$, where L^∞ does not refer to C, but is essentially bounded). If $\frac{\partial f}{\partial x_j} \in W^1 L^\infty$, $j = 1, 2, \ldots, n$, then it implies that $f \in W^2 L^\infty$.*

Besides, Lipα refers to the function class of $\omega(f; \delta) = O(\delta^\alpha)$ as $\delta \to 0^+$ where $\delta \in (0, 1]$. Here it should be pointed out that we limit to consider the periodical functions only. It is obvious that $W^1 L^\infty = \text{Lip}1$.

Obviously, if $f \in W^2 L^\infty$, then $\frac{\partial^2 f}{\partial x_i \partial x_j}$ is essentially bounded, for $i, j = 1, 2, \ldots, n$.

We also denote the subclass of the functions in $C(Q^n)$ whose j-th partial derivatives are continuous in $C^j(Q^n)$.

Theorem 3.7.3 *If $f \in W^2 L^\infty$, then we have*

$$\left\| S_R^{\frac{n-1}{2}}(f) - f \right\|_C = O\left(\frac{\log R}{R^2}\right) \tag{3.7.10}$$

and

$$S_R^{\frac{n-1}{2}}(f; x) - f(x) = O\left(\frac{1}{R^2}\right) \tag{3.7.11}$$

for a.e. $x \in Q^n$.

Proof. Since $f \in W^2 L^\infty$, we have

$$M = Mf := \|\triangle f\|_\infty = \left\| \frac{\partial^2 f}{\partial x_1^2} + \cdots + \frac{\partial^2 f}{\partial x_n^2} \right\|_\infty < \infty,$$

and

$$\omega_2\left(f; \frac{1}{R}\right) = O\left(\frac{1}{R^2}\right).$$

Then,

$$S_R^{\frac{n-1}{2}}(f; x) - f(x) = S_R^{\frac{n-1}{2}}(f; x) - S_R^{\frac{n+1}{2}}(f; x) + S_R^{\frac{n+1}{2}}(f; x) - f(x).$$

For

$$S_R^{\frac{n-1}{2}}(f; x) - S_R^{\frac{n+1}{2}}(f; x) = \frac{-1}{R^2} S_R^{\frac{n-1}{2}}(\triangle f; x),$$

we have

$$\left\|S_R^{\frac{n-1}{2}}(f) - S_R^{\frac{n+1}{2}}(f)\right\|_C = \frac{1}{R^2}O(\log R) \cdot \|\triangle f\|_\infty = O\left(\frac{\log R}{R^2}\right), \quad (3.7.12)$$

and

$$\left|S_R^{\frac{n-1}{2}}(f;x) - S_R^{\frac{n+1}{2}}(f;x)\right| \le \frac{1}{R^2}S_*^{\frac{n-1}{2}}(\triangle f;x), \quad (3.7.13)$$

where $S_*^{\frac{n-1}{2}}$ is the maximal operator, and since it is type (p,p) with $1 < p < \infty$, we can get that $S_*^{\frac{n-1}{2}}(\triangle f;x)$ is finite almost everywhere.

Besides of the above, by the result in Section 3.4,

$$\left\|S_R^{\frac{n+1}{2}}(f) - f\right\|_C = O\left(\omega_2\left(f;\frac{1}{R}\right)\right) = O\left(\frac{1}{R^2}\right). \quad (3.7.14)$$

Therefore, combining (3.7.12) with (3.7.14), we can get (3.7.10), (3.7.13) and (3.7.14) together giving (3.7.11).

This completes the proof of Theorem 3.7.3. ∎

3.8 $(C,1)$ means

Since the role of $S_R^{\frac{n-1}{2}}$ is the similar as that of Fourier partial sum in unitary variable in the spaces of $L^1(Q^n)$ and $C(Q^n)$, it is natural to ask whether

$$\sigma_R^{\frac{n-1}{2}} := \frac{1}{R}\int_0^R S_r^{\frac{n-1}{2}}\,dr$$

is considerably equivalent to the Fejér means or not. And also, we can consider

$$V_R = \frac{1}{R}\int_R^{2R} S_r^{\frac{n-1}{2}}\,dr.$$

As an analogy of the Vallée Poussin means, we call $\sigma_R^{\frac{n-1}{2}}$ as $(C,1)$ means.

Jiang discussed the approximation problem of continuous functions by the $(C,1)$ means (see Jiang [J1] or [J2]).

Lemma 3.8.1 Let $\mathrm{Re}\alpha > \frac{n-3}{2}$ and $f \in L(Q^n)$. Then we have

$$\sigma_R^\alpha(f;x) := \frac{1}{R}\int_0^R S_r^\alpha(f;x)dr$$
$$= \frac{2^{\alpha+1-\frac{n}{2}}\Gamma(\alpha+1)}{\Gamma\left(\frac{n}{2}\right)} \int_0^\infty f_x(t)\frac{1}{R}\int_0^R \frac{J_{\frac{n}{2}+\alpha}(tr)}{(tr)^{\frac{n}{2}+\alpha}}r^n dr\, t^{n-1}dt.$$
(3.8.1)

Proof. Denote
$$C_n(\alpha) = \frac{2^{\alpha+1-\frac{n}{2}}\Gamma(\alpha+1)}{\Gamma\left(\frac{n}{2}\right)}.$$

Clearly C_n is analytic function with respect to α in the domain $\mathrm{Re}\alpha > -1$. By the Bochner formula (see (3.1.2)), if $\mathrm{Re}\alpha > \frac{n-1}{2}$, we have

$$S_r^\alpha(f;x) = C_n(\alpha)r^n \int_0^\infty f_x(t)\frac{J_{\frac{n}{2}+\alpha}(rt)}{(rt)^{\frac{n}{2}+\alpha}}t^{n-1}dt,$$

for $r > 0$. Thus we immediately deduce that (3.8.1) holds, provided that $\mathrm{Re}\alpha > \frac{n-1}{2}$.

Denote
$$K_R^\alpha(t) = \frac{1}{R}\int_0^R \frac{J_{\frac{n}{2}+\alpha}(tr)}{(tr)^{\frac{n}{2}+\alpha}}r^n dr\, t^{n-1}.$$

This is equivalent to
$$K_R^\alpha(t) = \frac{1}{t^2 R}\int_0^{tR} J_{\frac{n}{2}+\alpha}(s)s^{\frac{n}{2}-\alpha}ds. \qquad (3.8.2)$$

By the formula
$$\frac{d}{dt}\left(t^\nu J_\nu(t)\right) = t^\nu J_{\nu-1}(t),$$

and integration by parts, we can have

$$K_R^\alpha(t) = \frac{1}{t^2 R}\left\{(tR)^{\frac{n}{2}-\alpha}J_{\frac{n}{2}+\alpha+1}(tR) + (1+2\alpha)\int_0^{tR} s^{-1-\alpha+\frac{n}{2}}J_{\frac{n}{2}+\alpha+1}(s)ds\right\}$$
$$= \frac{1}{t^2 R}\left\{(tR)^{\frac{n}{2}-\alpha}J_{\frac{n}{2}+\alpha+1}(tR) + (1+2\alpha)(tR)^{\frac{n}{2}-1-\alpha}J_{\frac{n}{2}+\alpha+2}(tR)\right.$$
$$\left. + (1+2\alpha)(3+2\alpha)\int_0^{tR} s^{-2-\alpha+\frac{n}{2}}J_{\frac{n}{2}+\alpha+2}(s)ds\right\}. \qquad (3.8.3)$$

3.8 $(C,1)$ means

Denote $\alpha = \frac{n-3}{2} + \delta + i\tau$, for $\delta > 0$ and $\tau \in \mathbb{R}$. When $t \geq R^{-1}$, we have

$$|K_R^\alpha(t)| \leq Me^{2\pi|\tau|}\frac{1}{t}\left\{\frac{1}{(tR)^\delta} + \frac{1}{(tR)^{1+\delta}} + \frac{1}{tR}\int_0^{tR}(1+s)^{-1-\delta}ds\right\}$$
$$\leq Me^{2\pi|\tau|}\frac{1}{t}\left\{\frac{1}{(tR)^\delta} + \frac{\log(1+tR)}{tR}\right\}. \quad (3.8.4)$$

When $0 < t < R^{-1}$, we have

$$|K_R^\alpha(t)| \leq Me^{2\pi|\tau|}t^{n-1}R^n. \quad (3.8.5)$$

According to (3.8.4) and (3.8.5), if we fix x and R, then the following integral

$$F(\alpha) := C_n(\alpha)\int_0^\infty f_x(t)K_R^\alpha(t)dt$$

is uniformly convergent about α on the compact subset of $\text{Re}\alpha > \frac{n-3}{2}$. It is obvious that $K_R^\alpha(t)$ is an analytic function with respect to α. Consequently, $F(\alpha)$ is analytic on $\text{Re}\alpha > \frac{n-3}{2}$.

Meanwhile, $\sigma_R^\alpha(f;x)$ is obviously analytic about α.

Since $\text{Re}\alpha > \frac{n-1}{2}$, $F(\alpha) = \sigma_R^\alpha(f;x)$, as the analytic function on $\text{Re}\alpha > \frac{n-3}{2}$, the equation is valid on this whole scale.

This finishes the proof of Lemma 3.8.1. ∎

Theorem 3.8.1 Let $f \in C(Q^n)$ and $\text{Re}\alpha > \frac{n-3}{2}$. Then

$$\sigma_R^\alpha(f;x) - f(x)$$
$$= C_n(\alpha)\lambda_n(\alpha)\int_1^\infty t^{-2}\left[f_x\left(\frac{t}{R}\right) - f(x)\right]lt + O\left(\tilde{\omega}\left(f;\frac{1}{R}\right)\right) \quad (3.8.6)$$

holds uniformly about x, where the definition of $\tilde{\omega}$ can be checked in II of Theorem 3.7.1, and

$$\lambda_n(\alpha) = \frac{2^{\frac{n}{2}-\alpha-1}\Gamma\left(\frac{n+1}{2}\right)}{\Gamma(\alpha+3/2)}(1+2\alpha),$$

hence,

$$C_n(\alpha)\lambda_n(\alpha) = \frac{\Gamma\left(\frac{n+1}{2}\right)\Gamma(\alpha+1)}{\Gamma\left(\frac{n}{2}\right)\Gamma(\alpha+3/2)}(1+2\alpha). \quad (3.8.7)$$

Proof. By (3.8.1), we have

$$\sigma_R^\alpha(f;x) - f(x) = C_n(\alpha) \int_0^\infty [f_x(t) - f(x)] K_R^\alpha(t) dt$$

$$= C_n(\alpha) \int_0^\infty \left[f_x\left(\frac{t}{R}\right) - f(x) \right] \frac{1}{t} K_t^\alpha(1) dt. \quad (3.8.8)$$

It follows from (3.8.3) that

$$tK_t^\alpha(1) = t^{\frac{n}{2}-\alpha} J_{\frac{n}{2}+\alpha+1}(t) + (1+2\alpha) t^{\frac{n}{2}-\alpha-1} J_{\frac{n}{2}+\alpha+2}(t)$$

$$+ (1+2\alpha)(2+2\alpha) \int_0^t s^{\frac{n}{2}-\alpha-2} J_{\frac{n}{2}+\alpha+2}(s) ds. \quad (3.8.9)$$

Thus we have

$$\left| \int_0^1 \left[f_x\left(\frac{t}{R}\right) - f(x) \right] \frac{1}{t} K_t^\alpha(1) dt \right| \le \int_0^1 \widetilde{\omega}\left(f; \frac{1}{R}\right) A_{n,\alpha} t^{n-1} dt$$

$$\le A_{n,\alpha} \widetilde{\omega}\left(f; \frac{1}{R}\right). \quad (3.8.10)$$

When $t \ge 1$, we still write $\alpha = \frac{n-3}{2} + \delta + i\tau$, $\delta > 0$. We have

$$t^{\frac{n}{2}-\alpha} J_{\frac{n}{2}+\alpha+1}(t) = \sqrt{\frac{2}{\pi}} \frac{\cos(t-\theta)}{t^{\delta-1+i\tau}} + O\left(\frac{1}{t^\delta}\right)$$

and

$$t^{\frac{n}{2}-\alpha-1} J_{\frac{n}{2}+\alpha+2}(t) = O\left(\frac{1}{t^\delta}\right),$$

where $\theta = \frac{\pi}{2}\left(\frac{n}{2}+\alpha+1\right) + \frac{\pi}{4}$.

Consequently, we have

$$\int_0^t s^{\frac{n}{2}-\alpha-2} J_{\frac{n}{2}+\alpha+2}(s) ds$$

$$= \int_0^\infty s^{\frac{n}{2}-\alpha-2} J_{\frac{n}{2}+\alpha+2}(s) ds - \int_t^\infty s^{\frac{n}{2}-\alpha-2} J_{\frac{n}{2}+\alpha+2}(s) ds.$$

Substituting the two equalities

$$\int_0^\infty s^{\frac{n}{2}-\alpha-2} J_{\frac{n}{2}+\alpha+2}(s) ds = \frac{2^{\frac{n}{2}-\alpha-2} \Gamma\left(\frac{n+1}{2}\right)}{\Gamma\left(\alpha+\frac{5}{2}\right)}$$

and

$$\int_t^\infty s^{\frac{n}{2}-\alpha-2} J_{\frac{n}{2}+\alpha+2}(s) ds = O\left(\frac{1}{t^\delta}\right)$$

3.8 (C,1) means

into (3.8.9) yields that

$$tK_t^\alpha(1) = \sqrt{\frac{2}{\pi}} \frac{\cos(t-\theta)}{t^{\delta-1+i\tau}} + \frac{2^{\frac{n}{2}-\alpha-2}\Gamma\left(\frac{n+1}{2}\right)}{\Gamma\left(\alpha+\frac{5}{2}\right)} + O\left(\frac{1}{t^\delta}\right).$$

Therefore, we conclude that

$$C_n(\alpha)\int_1^\infty \left[f_x\left(\frac{t}{R}\right) - f(x)\right]\frac{1}{t}K_t^\alpha(1)dt$$

$$= C_n(\alpha)(1+2\alpha)(3+2\alpha)2^{\frac{n}{2}-\alpha-2}\frac{\Gamma\left(\frac{n+1}{2}\right)}{\Gamma(\alpha+5/2)}$$

$$\times \int_1^\infty \left[f_x\left(\frac{t}{R}\right) - f(x)\right]t^{-2}dt + C_n(\alpha)\sqrt{\frac{2}{\pi}}\int_1^\infty \left[f_x\left(\frac{t}{R}\right) - f(x)\right]$$

$$\times \frac{\cos(t-\theta)}{t^{\delta+1+i\tau}}dt + O\left(\tilde{\omega}\left(f;\frac{1}{R}\right)\right).$$

It is obvious that

$$\int_1^\infty \left[f_x\left(\frac{t}{R}\right) - f(x)\right]\frac{\cos(t-\theta)}{t^{\delta+1+i\tau}}dt = O\left(\tilde{\omega}\left(f;\frac{1}{R}\right)\right).$$

We substitute the above into (3.8.8), and immediately obtain (3.8.6). This completes the proof of Theorem 3.8.1. ∎

Remark 3.8.1 *Theorem 3.8.1 is also valid for $n = 1$. When $n = 1$ and $\alpha = 0$, we can get the well-known Fejér approximation estimate. It seems interesting $\lambda_1\left(-\frac{1}{2}\right) = 0$, when $n = 1$ and $\alpha = \frac{n-2}{2} = -\frac{1}{2}$. Then we can get the uniform estimate from (3.8.6)*

$$\sigma_R^{-\frac{1}{2}}(f;x) - f(x) = O\left(\omega\left(f;\frac{1}{R}\right)\right).$$

Generally, the result is better than the one in the case $\alpha > 0$.

From Theorem 3.8.1, we can directly obtain the following corollary.

Corollary 3.8.1 *Let $f \in C(Q^n)$ and $\text{Re}\alpha > \frac{n-3}{2}$. Then we have*

$$\|\sigma_R^\alpha(f) - f\|_C \leq \left|(1+2\alpha)\frac{\Gamma(\alpha+1)}{\Gamma(\alpha+\frac{3}{2})}\right|\frac{\Gamma\left(\frac{n+1}{2}\right)}{\Gamma(\frac{n}{2})}$$

$$\times \frac{1}{R}\int_0^R \tilde{\omega}\left(f;\frac{1}{r}\right)dr + O\left(\tilde{\omega}\left(f;\frac{1}{R}\right)\right).$$

(3.8.11)

Suppose that $\omega(\delta)$ is a modulus of continuity. Define a function class

$$H^\omega = \{f \in C(Q^n) : \widetilde{\omega}(f;\delta) \leq \omega(\delta)\}.$$

As far as the approximation on H^ω is concerned, there is the following theorem (see [J1]).

Theorem 3.8.2 *There exists constants $C_1 > C_2 > 0$, only related to the variable number n, such that*

$$\frac{C_2}{R}\int_0^R \omega\left(\frac{1}{r}\right)dr \leq \sup_{f\in H^\omega}\left\|\sigma_R^{\frac{n-1}{2}}(f)-f\right\|_C \leq \frac{C_1}{R}\int_0^R \omega\left(\frac{1}{r}\right)dr \quad (3.8.12)$$

for $R > 0$.

Proof. It only suffices to prove the left half of the above inequality. Let us take it for granted that ω is upper convex. Take

$$f_0(x) = \begin{cases} \omega(|x|) & 0 \leq |x| \leq \pi, \\ \omega(\pi) & x \in Q^n \setminus B(0,\pi). \end{cases}$$

Obviously, $f_0 \in H^\omega$.

According to (3.8.6), we have

$$\sigma_R^{\frac{n-1}{2}}(f;x) - f(x) = 2\left(\frac{\Gamma\left(\frac{n+1}{2}\right)}{\Gamma\left(\frac{n}{2}\right)}\right)^2 \frac{1}{R}\int_0^R\left[f_x\left(\frac{1}{r}\right)-f(x)\right]dr$$

$$+ O\left(\widetilde{\omega}\left(f;\frac{1}{R}\right)\right).$$

Thus we have

$$\left|\sigma_R^{\frac{n-1}{2}}(f;x)-f(x)\right| \geq 2\left|\frac{1}{R}\int_0^R\left(f_x\left(\frac{1}{r}\right)-f(x)\right)dr\right| - A\widetilde{\omega}\left(f;\frac{1}{R}\right), \quad (3.8.13)$$

where $A > 0$.

Substituting into the inequality (3.8.13) with f_0 and $x = 0$, it follows that

$$\left|\sigma_R^{\frac{n-1}{2}}(f_0;0)-f_0(0)\right| \geq 2\frac{1}{R}\int_0^R \omega\left(\frac{1}{r}\right)dr - A\omega\left(\frac{1}{R}\right). \quad (3.8.14)$$

If

$$\omega\left(\frac{1}{R}\right) < \frac{1}{A}\frac{1}{R}\int_0^R \omega\left(\frac{1}{r}\right)dr$$

3.8 $(C,1)$ means

holds for fixed $R > 0$, then the above inequality shows that

$$\left|\sigma_R^{\frac{n-1}{2}}(f_0;0) - f_0(0)\right| > \frac{1}{R}\int_0^R \omega\left(\frac{1}{r}\right)dr.$$

If

$$\omega\left(\frac{1}{R}\right) \geq \frac{1}{A}\frac{1}{R}\int_0^R \omega\left(\frac{1}{r}\right)dr$$

holds for fixed $R \geq 2$, then, setting

$$f_R(x) = \frac{1}{3A}\frac{1}{R}\int_0^R \omega\left(\frac{1}{r}\right)dr\, e^{i[\frac{R}{2}]x_1},$$

it follows that

$$S_r^{\frac{n-1}{2}}(f_R;x) = \begin{cases} 0, & 0 \leq r \leq \left[\frac{R}{2}\right], \\ \dfrac{1}{3A}\dfrac{1}{R}\displaystyle\int_0^R \omega\left(\frac{1}{r}\right)dr\left(1 - \dfrac{[\frac{R}{2}]^2}{r^2}\right)^{\frac{n-1}{2}} e^{i[\frac{R}{2}]x_1}, & r > \left[\dfrac{R}{2}\right]. \end{cases}$$

Therefore, we conclude that

$$\sigma_R^{\frac{n-1}{2}}(f_R;0) - f_R(0)$$

$$= \frac{1}{3AR}\int_0^R \omega\left(\frac{1}{r}\right)dr \cdot \left[\frac{1}{R}\int_{[\frac{R}{2}]}^R \left(1 - \frac{[\frac{R}{2}]^2}{r^2}\right)^{\frac{n-1}{2}} dr - 1\right]$$

$$\leq \frac{1}{3AR}\int_0^R \omega\left(\frac{1}{r}\right)dr \left[\frac{1}{R}\left(R - \left[\frac{R}{2}\right]\right) - 1\right]$$

$$< -\frac{1}{12A}\frac{1}{R}\int_0^R \omega\left(\frac{1}{r}\right)dr.$$

When $0 < \delta \leq \frac{1}{[\frac{R}{2}]}$, we have

$$\omega(f_R;\delta) \leq \frac{1}{3A}\frac{1}{R}\int_0^R \omega\left(\frac{1}{r}\right)dr \left[\frac{R}{2}\right] \cdot \delta$$

$$\leq \frac{1}{3}\omega\left(\frac{1}{R}\right)\left[\frac{R}{2}\right]\delta$$

$$\leq \frac{1}{3}\omega(\delta)\left(\frac{1}{\delta R} + 1\right)\left[\frac{R}{2}\right]\delta$$

$$\leq \omega(\delta).$$

When $\delta > \frac{1}{[\frac{R}{2}]}$, we obtain

$$\omega(f_R;\delta) \le \frac{2}{3}\frac{1}{A}\int_0^R \omega\left(\frac{1}{r}\right)dr \le \frac{2}{3}\omega\left(\frac{1}{R}\right) \le \omega(\delta),$$

which implies $f_R \in H^\omega$.

The above results imply that when $R \ge 2$,

$$\sup_{f\in H^\omega}\left\|\sigma_R^{\frac{n-1}{2}}(f)-f\right\|_C \ge \frac{1}{12A}\frac{1}{R}\int_0^R \omega\left(\frac{1}{r}\right)dr,$$

where the constant A only depends on n, and we take it for granted that $A \ge 1$.

When $0 < R < 2$, if

$$\omega\left(\frac{1}{R}\right) \ge \frac{1}{A}\frac{1}{R}\int_0^R \omega\left(\frac{1}{r}\right)dr,$$

then we take $f(x) = \frac{\omega(1)}{4}e^{i2x_1}$. Thus we have

$$\omega(f;\delta) \le \begin{cases} \dfrac{\omega(1)}{4}\cdot 2\delta < \omega(\delta) & 0 < \delta \le 1, \\ \dfrac{\omega(1)}{4}\cdot 2 < \omega(\delta) & \delta > 1. \end{cases}$$

Therefore, $f \in H^\omega$. Meanwhile, when $0 < R < 2$, we have that

$$\left|\sigma_R^{\frac{n-1}{2}}(f;0)-f(0)\right| = f(0)$$

$$= \frac{1}{4}\omega(1)$$

$$\ge \frac{1}{16\sqrt{n\pi}}\omega(2\sqrt{n\pi})$$

$$\ge \frac{1}{16\sqrt{n\pi}}\frac{1}{R}\int_0^R \omega\left(\frac{1}{r}\right)dr.$$

This completes the proof of Theorem 3.8.2. ∎

From Lemma 3.8.1 and the estimate of the kernel $K_R^\alpha(t)$, we can directly get the estimate of the maximal operator

$$\sigma_*^\alpha(f)(x) := \sup_{R>0}|\sigma_R^\alpha(f;x)|$$

for $\operatorname{Re}\alpha > \frac{n-3}{2}$.

Theorem 3.8.3 *If* $\operatorname{Re}\alpha > \frac{n-3}{2}$ *and* $f \in L(Q^n)$, *then*

$$\sigma_*^\alpha(f)(x) \leq C_{n,\alpha} M f(x)$$

holds, where M is the Hardy-Littlewood maximal operator.

3.9 The saturation problem of the uniform approximation

Let $f \in C(Q^n)$. Using the non-zero-ordered Bochner-Riesz means to approximate to f, we have that the order of the best approximation of f is R^{-2} as $R \to \infty$, except for the trivial case when f is a constant. That is to say, the saturation of the uniform approximation is R^{-2}. It is well-known that if

$$\|S_R^\alpha(f) - f\|_C = o\left(R^{-2}\right)$$

holds for $\alpha \neq 0$, then we have

$$\lim_{R \to \infty} R^2 \frac{1}{(2\pi)^n} \int_{Q^n} [S_R^\alpha(f;x) - f(x)] e^{-im \cdot x} dx = 0,$$

for every $m \in \mathbb{Z}^n$. That is,

$$\lim_{R \to \infty} R^2 \left[\left(1 - \frac{|m|^2}{R^2}\right)^\alpha - 1\right] a_m(f) = 0$$

holds.

Since $\alpha \neq 0$, from the following equality

$$\lim_{R \to \infty} R^2 \left[\left(1 - \frac{|m|^2}{R^2}\right)^\alpha - 1\right] = -|m|^2 \alpha,$$

we can easily conclude that $a_m(f) = 0$, if $m \neq 0$. consequently, f is a constant.

The first question which we will discuss is how to identify the saturation class of $\{S_R^\alpha\}$. In [J1], Jiang gave the characterization of the saturation class if $\alpha > \frac{n-1}{2}$.

The second question is to investigate the saturation problem of the operator $\{\sigma_R^\alpha\}$ given by the previous section. In the case of $\alpha = \frac{n-1}{2}$ and the dimension $n > 1$, Jiang [J1] also obtained the characterization of the saturation class. He carried out general discussion about the saturation problem, referring to L^p, $1 \le p < \infty$ and C. Here, we only discuss in the context of $C(Q^n)$. The same steps can be applied to the situation of L^p.

Definition 3.9.1 *Suppose that $\{T_R\}$ with $R > 0$ and $R \to \infty$ is a family of bounded linear operators mapping from $C(Q^n)$ to itself, and $\varphi(\cdot)$ is a positive function monotonically decreasing to zero. We say $\{T_R\}$ is saturated on $C(Q^n)$, with the saturation of $\varphi(R)$, if the following two conditions hold.*

(1) $\|f - T_R(f)\|_C = o(\varphi(R)) \iff f = \text{constant}$,

(2) *there exists $f_0 \in C(Q^n)$, f_0 is not a constant, and*

$$\|f_0 - T_R(f_0)\|_C = O(\varphi(R)).$$

We denote the collection of all f_0 which satisfies the condition (2) by $F(T, C)$, and call it the saturation class of $\{T_R\}$.

Now suppose that, for $f \in C(Q^n)$ and the fixed $\lambda_R(m)$ with $m \in \mathbb{Z}^n$ and $R > 0$, the series $\sum \lambda_R(m) C_m(f) e^{im \cdot x}$ is uniformly convergent with respect to x, where $C_m(f)$ is the coefficient of f. Thus it is the Fourier series of the function g in $C(Q^n)$, We denote g by $T_R(f)$. In this way, we have defined a linear operator T_R, the multiplier operator, which is identified by $\{\lambda_R(m)\}$ defined by

$$T_R(f)(x) = \sum \lambda_R(m) C_m(f) e^{im \cdot x} \qquad (3.9.1)$$

for $R > 0$.

Definition 3.9.2 *Suppose that $\{\psi(m)\}_{m \in \mathbb{Z}^n}$ is a sequence of numbers.*

$$V(C, \{\psi(m)\}) = \Big\{ f \in C(Q^n) : f \text{ is not a constant, and there exists an}$$

essentially bounded function g, such that $\psi(m) C_m(f) = C_m(g)$, $\forall\, m \in \mathbb{Z}^n \Big\}.$

Theorem 3.9.1 *Suppose that the linear operator T_R defined by (3.9.1) is bounded, $\lambda_R(0) = 1$, and there exists a positive function $\varphi(R)$ which is monotonically decreasing to zero, such that*

$$\lim_{R \to \infty} \frac{1 - \lambda_R(m)}{\varphi(R)} = \psi(m) \ne 0, \qquad (3.9.2)$$

3.9 The saturation problem of the uniform approximation

for $m \in \mathbb{Z}^k$ with $m \neq \mathbf{0}$. Then the following two conclusions I and II are satisfied.

(I) $\{T_R\}$ is saturated on $C(Q^n)$, with the saturation of $\varphi(R)$.

(II) Let $f \in C(Q^n)$. We have the implication relationship for the following three propositions.

(a) f is not a constant and there uniformly holds about R

$$\left\|\sum \lambda_R(m)\psi(m)C_m(f)e^{im\cdot x}\right\|_\infty = O(1),$$

(b) $f \in V(C, \{\psi(m)\})$,
(c) $f \in F(T, C)$.

Then, we have (a) $\Longrightarrow$ (b). If we also know that the norm of the operator satisfies

$$\|T_R\| \leq M < +\infty, \tag{3.9.3}$$

for $R > 0$, then we have (c) $\Longrightarrow$ (a).

Proof. Choose $f \in C(Q^n)$ such that $\|f - T_R(f)\|_C = o(\varphi(R))$. Then for $m \in \mathbb{Z}^n$ with $m \neq \mathbf{0}$, we have

$$C_m(f)(1 - \lambda_R(m)) = (2\pi)^{-n} \int_{Q^n} [f(x) - T_R(f)(x)] e^{-imx} dx$$
$$= o(\varphi(R)),$$

as $R \to \infty$.

By the condition (3.9.2), we have $C_m(f) = 0$ for $m \neq \mathbf{0}$, and thus $f = $ constant.

On the other hand, we choose $f_0(x) = e^{ix_1}$ and denote $e_1 = (1, 0, \ldots, 0)$. Then it follows that

$$\|f_0 - T_R(f_0)\|_C = |1 - \lambda_R(e_1)| = O(\varphi(R)),$$

as $R \to \infty$, which implies that $\{T_R\}$ is saturated, with the saturation of $\varphi(R)$.

Next we show the conclusions II. Assume (a) is valid. Denote

$$F_R(x) = \sum \lambda_R(m)\psi(m)C_m(f)e^{im\cdot x}.$$

Since L^∞ is weak $*$ sequentially compact, as the conjugate space of L^1, in the bounded family $\{F_R\}$ of L^∞, there exists a subsequence $\{F_{R_j}\}_{j=1}^\infty$, with $R_j \to +\infty$ as $j \to \infty$, such that

$$F_{R_j} \to g \in L^\infty(Q^n)$$

in the sense of weak $*$ topology. Thus we have

$$\lim_{j\to\infty} (2\pi)^{-n} \int_{Q^n} F_{R_j}(x) e^{-im\cdot x} dx = (2\pi)^{-n} \int_{Q^n} g(x) e^{-im\cdot x} dx = C_m(g),$$

for all $m \in \mathbb{Z}^n$. That is,

$$\lim_{j\to\infty} \lambda_{R_j}(m) \psi(m) C_m(f) = C_m(g).$$

It follows from (3.9.2) that $\lambda_R(m) \to 1$ with $m \neq 0$ as $R \to \infty$ and $\lambda_R(0) = 1$. Hence, we have

$$\psi(m) C_m(f) = C_m(g),$$

for all $m \in \mathbb{Z}^n$, which shows (b) is valid.

Assume (3.9.3) holds. Let $f \in F(T, C)$, that is to say, (c) is valid. Since

$$\|f - T_R(f)\|_C = O(\varphi(R)),$$

we have that

$$\|T_R(f - T_R(f))\|_C = O(\varphi(R)).$$

uniformly holds for $R > R' > 0$. That is to say,

$$\left\|\sum \lambda_{R'}(m) \left(\frac{1 - \lambda_R(m)}{\varphi(R)}\right) C_m(f) e^{im\cdot x}\right\|_C \leq M < +\infty.$$

For the fixed R', the function

$$h_R(x) := \sum \lambda_{R'}(m) \frac{1 - \lambda_R(m)}{\varphi(R)} C_m(f) e^{im\cdot x}$$

is also a bounded family in the norm of $C(Q^n)$, for $R > R'$. By the property of being $*$ weak sequentially compact, there exists a monotone increasing sequence $R_j \to \infty$, such that

$$h_{R_j} \to h_{R'}^* \in L^\infty(Q^n)$$

in the sense of $*$ weak topology, keeping $\|h_{R'}^*\|_\infty \leq M$ for $R' > 0$.

Since

$$\lim_{j\to\infty} \frac{1 - \lambda_{R_j}(m)}{\varphi(R_j)} = \psi(m),$$

we have

$$\lambda_{R'}(m) \psi(m) C_m(f) = C_m(h_{R'}^*).$$

3.9 The saturation problem of the uniform approximation

That is to say,
$$\sum \lambda_R(m)\psi(m)C_m(f)e^{im\cdot x}$$
is the Fourier series of h_R^* in L^∞, while $\|h_R^*\|_\infty \leq M$, for $R > 0$, which is exactly the conclusion (a).

This completes the proof of Theorem 3.9.1. ∎

Now we investigate two concrete operators.

Firstly, we investigate the Abel-Poisson means. Let $f \in L(Q^n)$. Define the Abel-Poisson means of f as

$$P_\varepsilon(f;x) = \sum_{m\in\mathbb{Z}^n} e^{-\varepsilon|m|} C_m(f)e^{im\cdot x}, \quad (3.9.4)$$

for $\varepsilon > 0$.

Obviously we have that

$$P_\varepsilon(f;x) = \frac{1}{(2\pi)^n} \int_{Q^n} f(x-y) P_\varepsilon(y) dy, \quad (3.9.5)$$

where

$$P_\varepsilon(y) = \sum_{m\in\mathbb{Z}^n} e^{-\varepsilon|m|} e^{im\cdot y}.$$

We all know that the the Fourier transform of the function $\psi(x) = e^{-|x|}$, $x \in \mathbb{R}^n$, is

$$\widehat{\psi}(y) = \frac{\Gamma\left(\frac{n+1}{2}\right)}{[\pi(1+|y|^2)]^{\frac{n+1}{2}}} := P(y) \in L(\mathbb{R}^n). \quad (3.9.6)$$

Thus we conclude that

$$\psi(x) = \int_{\mathbb{R}^n} P(y) e^{ix\cdot y} dy$$
$$= \frac{1}{(2\pi)^n} \int_{\mathbb{R}^n} [(2\pi)^n P(y)] e^{-ix\cdot y} dy$$
$$= \mathscr{F}((2\pi)^n P)(x).$$

It follows that

$$\psi(\varepsilon x) = \mathscr{F}\big((2\pi)^n \varepsilon^{-n} P(\varepsilon^{-1} y)\big)(x).$$

Since the function $(2\pi)^n \varepsilon^{-n} P(\varepsilon^{-1} y)$ satisfies the condition of the Poisson summation formula, the kernel P_ε here has the following expression,

$$P_\varepsilon(y) = \sum_{m \in \mathbb{Z}^n} (2\pi)^n \varepsilon^{-n} P\left(\frac{y + 2\pi m}{\varepsilon}\right)$$

$$= 2^n \pi^{\frac{n-1}{2}} \Gamma\left(\frac{n+1}{2}\right) \sum \varepsilon \left(\varepsilon^2 + |y + 2\pi m|^2\right)^{-\frac{n+1}{2}}. \quad (3.9.7)$$

We notice that the kernel is positive and

$$\int_{Q^n} P_\varepsilon(y) = \sum_m \int_{Q^n} (2\pi)^n \varepsilon^{-n} P\left(\frac{y + 2\pi m}{\varepsilon}\right) dy$$

$$= (2\pi)^n \varepsilon^{-n} \int_{\mathbb{R}^n} P\left(\frac{y}{\varepsilon}\right) dy$$

$$= (2\pi)^n \int_{\mathbb{R}^n} P(y) dy$$

$$= (2\pi)^n.$$

Therefore, the operator P_ε is uniformly bounded on L^∞. For $f \in L^\infty(Q^n)$, we have

$$|P_\varepsilon(f; x)| \leq \|f\|_\infty \frac{1}{(2\pi)^n} \int_{Q^n} P_\varepsilon(y) dy = \|f\|_\infty. \quad (3.9.8)$$

From (3.9.5), (3.9.6) and (3.9.7), we can obtain that

$$P_\varepsilon(f; x) = \frac{2\Gamma\left(\frac{n+1}{2}\right)}{\sqrt{n}\,\Gamma\left(\frac{n}{2}\right)} \int_0^\infty \frac{t^{n-1}}{(1+t^2)^{\frac{n+1}{2}}} f_x(t\varepsilon) dt$$

and

$$P_\varepsilon(f; x) - f(x) = \frac{2\Gamma\left(\frac{n+1}{2}\right)}{\sqrt{n}\,\Gamma\left(\frac{n}{2}\right)} \int_0^\infty \frac{t^{n-1}}{(1+t^2)^{\frac{n+1}{2}}} (f_x(t\varepsilon) - f(x)) dt. \quad (3.9.9)$$

for $f \in L(Q^n)$.

Consequently, we can immediately have that

$$\|P_\varepsilon(f) - f\|_C \leq \frac{2}{B\left(\frac{n}{2}, \frac{1}{2}\right)} \int_0^\infty \frac{t^{n-1}}{(1+t^2)^{\frac{n+1}{2}}} \widetilde{\omega}(f; t\varepsilon) dt. \quad (3.9.10)$$

It follows that

$$\|P_\varepsilon(f) - f\|_C \leq A_n \omega\left(f; \varepsilon \log \frac{1}{\varepsilon}\right), \quad (3.9.11)$$

for $\varepsilon < e^{-1}$.

3.9 The saturation problem of the uniform approximation

Theorem 3.9.2 *The Abel-Poisson means $\{P_\varepsilon\}$ is saturated in $C(Q^n)$, with a saturation of ε, $\varepsilon \to 0^+$, and the following three properties are equivalent.*

(a) $\left\| \sum e^{-\varepsilon|m|} |m| C_m(f) e^{im\cdot x} \right\|_C = O(1).$

(b) *there exists $g \in L^\infty(Q^n)$, such that*
$$|m| C_m(f) = C_m(g),$$

for $m \in \mathbb{Z}^n$.

(c) $f \in F(P, C)$ *or f is a constant.*

Proof. Since
$$\lim_{\varepsilon \to 0^+} \frac{1 - e^{-\varepsilon|m|}}{\varepsilon} = |m|,$$

for $m \in \mathbb{Z}^n$, by Theorem 3.9.1, we can have that the saturation of $\{P_\varepsilon\}$ is ε. In addition, by (3.9.8) we can see that (3.9.3) is also satisfied. And thus by Theorem 3.9.1, it follows that $(c) \Longrightarrow (a)$ and $(a) \Longrightarrow (b)$.

Assume that (b) is valid. By (3.9.8) and $|P_\varepsilon(g; x)| \le \|g\|_\infty$, it follows that
$$\frac{1}{(2\pi)^n} \int_{Q^n} \left(\int_0^\varepsilon P_\eta(g; x) d\eta \right) e^{im\cdot x} dx = \int_0^\varepsilon \left\{ \frac{1}{(2\pi)^n} \int_{Q^n} P_\eta(g; x) e^{im\cdot x} dx \right\} d\eta$$
$$= \int_0^\varepsilon e^{-\eta|m|} C_m(g) d\eta, \qquad (3.9.12)$$

for $m \in \mathbb{Z}^n$.

On the other hand, we have
$$\frac{1}{(2\pi)^n} \int_{Q^n} [f(x) - P_\varepsilon(f; x)] e^{-imx} dx = \left(1 - e^{-\varepsilon|m|} \right) C_m(f)$$
$$= C_m(f)|m| \int_0^\varepsilon e^{-\eta|m|} d\eta, \qquad (3.9.13)$$

for $m \in \mathbb{Z}^n$.

By the condition that $|m| C_m(f) = C_m(g)$, (3.9.12) and (3.9.13), we can deduce that
$$f(x) - P_\varepsilon(f; x) = \int_0^\varepsilon P_\eta(g; x) d\eta,$$

for $\varepsilon > 0$.

It follows that

$$\|f - P_\varepsilon(f)\|_C \le \|g\|_\infty \varepsilon,$$

which follows (c) is valid.

This finishes the proof of Theorem 3.9.2. ∎

Another example is the Gauss-Weierstrass means.
Define

$$W_R(f;x) = \sum e^{-\frac{|m|^2}{R^2}} C_m(f) e^{im\cdot x}, \qquad (3.9.14)$$

for $f \in L(Q^n)$. By the method similar as the Abel-Poisson means, we can obtain the expression as follows,

$$\begin{aligned} W_R(f;x) &= \frac{1}{2^{n-1}\Gamma\left(\frac{n}{2}\right)} \int_0^\infty e^{-\frac{1}{4}t^2} t^{n-1} f_x\left(\frac{t}{R}\right) dt \\ &= \frac{1}{\Gamma\left(\frac{n}{2}\right)} \int_0^\infty e^{-t} t^{\frac{n}{2}-1} f_x\left(\frac{2\sqrt{t}}{R}\right) dt. \end{aligned} \qquad (3.9.15)$$

Theorem 3.9.3 *The Gauss-Weierstrass means $\{W_R\}$ is saturated in $C(Q^n)$, with the saturation of R^{-2}, and the following three are equivalent.*

(a) $\left\|\sum e^{-\frac{|M|^2}{|R|^2}} |m|^2 C_m(f) e^{im\cdot x}\right\|_C = O(1)$, and f is not a constant,

(b) $f \in V\left(C, \{|m|^2\}\right)$,

(c) $f \in F(W, C)$.

For the application later on, we can define the operator V_ε^l for $\varepsilon > 0$ and $l = 1, 2$ as follows

$$V_\varepsilon^l(f;x) = \sum e^{-(\varepsilon|m|)^l} \left(1 - (\varepsilon|m|)^l\right)^{n+1} C_m(f) e^{im\cdot x}.$$

Lemma 3.9.1 V_ε^l *is an uniformly bounded linear operator mapping from $C(Q^n)$ to itself. That is to say, for any $\varepsilon > 0$, the operator norm satisfies*

$$\|V_\varepsilon^l\| \le M < +\infty. \qquad (3.9.16)$$

3.9 The saturation problem of the uniform approximation

Proof. By the Poisson summation formula, it suffices to show that

$$\mathscr{F}\left(e^{-|x|^l}(1-|x|^l)^{n+1}\right)(y) \in L(\mathbb{R}^n).$$

However, by the results about the the Fourier transform of the radial function, we merely need to prove the unitary-variable function

$$s^{\frac{n}{2}} \int_0^\infty e^{-t^l}\left(1-t^l\right)^{n+1} t^{\frac{n}{2}} J_{\frac{n}{2}-1}(ts)dt \in L_{(\mathbb{R}^+)}.$$

Denote

$$\varphi_l(s) = \int_0^\infty e^{-t^l}(1-t^l)^{n+1} t^{\frac{n}{2}} J_{\frac{n}{2}-1}(ts)dt,$$

for $s \geq 0$. It suffices to show that there exists a positive number η, such that

$$\varphi(s) = O\left(-\frac{1}{s^{n/2+1+\eta}}\right), \qquad (3.9.17)$$

as $s \to \infty$.

We first consider the case $l = 1$.

Denote

$$I_j(s) = \int_0^\infty e^{-t} t^{\frac{n}{2}+j} J_{\frac{n}{2}-1}(ts)dt,$$

for $j = 0, 1, \ldots, n$. Directly using a formula in [Ba2], we can have that

$$I_0(s) = C_n s^{\frac{n}{2}-1}\left(1+s^2\right)^{-\frac{n+1}{2}} = O\left(s^{-(\frac{n}{2}+2)}\right),$$

as $s \to +\infty$. We can get that

$$I_j(s) = \frac{\Gamma(n+j)}{(1+s^2)^{\frac{1}{2}(\frac{n}{2}+1+j)}} P_{\nu+1+j}^{-\nu}\left(\frac{1}{\sqrt{1+y^2}}\right),$$

for $j = 1, 2, \ldots, n$, where $\nu = \frac{n}{2} - 1$ and $P_{\nu+1+j}^{-\nu}$ is the Legendre function. Thus we have

$$I_j(s) = O\left(s^{-(\frac{n}{2}+1+j)}\right),$$

for $j = 1, 2, \ldots, n$.

Since

$$\varphi_1(s) = \sum_{j=0}^{n+1} C_{1+n}^j (-1)^j I_j(s),$$

we have

$$\varphi_1(s) = O\left(s^{-\frac{n}{2}-2}\right).$$

as $s \to +\infty$.

Now we consider the case of $l = 2$. Denote

$$h_0(t) = e^{-t^2}\left(1 - t^2\right)^{n+1}$$

and

$$h_{j+1}(t) = \frac{h'_j(t)}{t},$$

for $j = 0, 1, 2, \ldots$.

Obviously, we have

$$h_j(t) \in C^\infty[0, \infty)$$

and

$$|h_j(t)| \le M_{n,j} e^{-t^2}\left(1 + t^{2n}\right),$$

for $j = 0, 1, 2, \ldots$.

Using the formula

$$\frac{d}{dt}[t^{\nu+1} J_{\nu+1}(ts)] = t^{\nu+1} J_\nu(ts) s,$$

we can obtain that

$$\varphi_2(s) = (-s)^{-j} \int_0^\infty h_j(t) t^{\frac{n}{2}+j} J_{\frac{n}{2}+j-1}(ts) dt,$$

for $j = 0, 1, 2, \ldots$.

Consequently, it follows that $\varphi_2(s) = O(s^{-j})$ holds for every natural number j. Of course, we have

$$\varphi_2(s) = O\left(s^{-\frac{n}{2}-2}\right)$$

as $\to +\infty$, which gets (3.9.17) and thus completes the proof of Lemma 3.9.1. ∎

Lemma 3.9.2 *Define a linear operator H_ε^l with $\varepsilon > 0$ and $l = 1, 2$ on $L^\infty(Q^n)$ as*

$$H_\varepsilon^l(f; x) = \sum_{m \ne 0} e^{-(\varepsilon|m|)^l} \left\{1 - \left[1 - (\varepsilon|m|)^l\right]^{n+1}\right\} \frac{1}{(\varepsilon|m|)^l} C_m(f) e^{im \cdot x}$$
$$+ (n+1) C_0(f). \qquad (3.9.18)$$

3.9 The saturation problem of the uniform approximation

Then, the norm of H^l_ε is uniformly bounded from $L^\infty(Q^n)$ to $C(Q^n)$, that is,

$$\left\|H^l_\varepsilon(f)\right\|_C \leq M\|f\|_\infty, \tag{3.9.19}$$

for $\varepsilon > 0$, where the positive number M is independent of f, l and ε.

Proof. By the Poisson summation formula, it suffices to show that

$$\mathscr{F}\left(e^{-|x|^l}\left\{1-\left[1-|x|^l\right]^{n+1}\right\}\frac{1}{|x|^l}\right)(\cdot) \in L(\mathbb{R}^n).$$

Hence, we merely need to prove that

$$\psi_l(s) := \int_0^\infty e^{-t^l}\frac{1-(1-t^l)^{n+1}}{t^l}t^{\frac{n}{2}}J_{\frac{n}{2}-1}(ts)dt s^{\frac{n}{2}} \in L(\mathbb{R}^+).$$

It follows from Lemma 3.9.1 that

$$\psi_1(s) = s^{\frac{n}{2}}\sum_{j=0}^n C_{n+1}^{j+1}(-1)^j I_j(s).$$

Therefore, we have

$$\psi_1(s) = O\left(s^{-2}\right),$$

as $s \to +\infty$. This follows $\psi_1 \in L(\mathbb{R}^+)$.

Similar to the estimate of φ_2 in Lemma 3.9.1, we can get that $\psi_2(s) = O(s^{-j})$ for all natural number j as $s \to +\infty$. And therefore, $\psi_2 \in L(\mathbb{R}^+)$.

This completes the proof of Lemma 3.9.2. ∎

Theorem 3.9.4 *A family of operators $\{V^l_\varepsilon\}$ with $l = 1, 2$ and $\varepsilon \to 0$ is saturated on C, with a saturation of ε^l. The saturation class is*

$$F(V^l, C) = V\left(C, \left\{|m|^l\right\}\right).$$

Proof. It is obvious that

$$\lim_{\varepsilon \to 0}\frac{1 - e^{-(\varepsilon|m|)^l}\left[1 - (\varepsilon|m|)^l\right]^{n+1}}{\varepsilon^l} = (n+2)|m|^l.$$

Therefore, according to Theorem 3.9.1, we have the saturation of $\{V^l_\varepsilon\}$ is ε^l. By Lemma 3.9.1, we know that the condition (3.9.3) is valid (see (3.9.16)). Then by Theorem 3.9.1, we have

$$F(V^l, C) \subset V\left(C, \left\{|m|^l\right\}\right).$$

Now suppose that $f \in V(C, \{|m|\})$ in the case of $l = 1$. By Theorem 3.9.2, we have
$$\|f - P_\varepsilon(f)\|_C = O(\varepsilon)$$
as $\varepsilon \to 0$.

We conclude that
$$\begin{aligned} P_\varepsilon(f;x) - V_\varepsilon^l(f;x) &= \sum e^{-\varepsilon|m|} \left\{1 - (1-\varepsilon|m|)^{n+1}\right\} C_m(f) e^{im \cdot x} \\ &= \varepsilon \sum_{m \neq 0} e^{-\varepsilon|m|} \left\{1 - (1-\varepsilon|m|)^{n+1}\right\} \frac{1}{\varepsilon|m|} C_m(g) e^{im \cdot x}, \end{aligned}$$
where $g \in L^\infty(Q^n)$.

By the definition of H_ε^l, we can write
$$P_\varepsilon(f;x) - V_\varepsilon^1(f;x) = \varepsilon \left(H_\varepsilon^1(g;x) - (n+1)C_0(g)\right).$$

By Lemma 3.9.2, we can get
$$\|P_\varepsilon(f) - V_\varepsilon^1(f)\|_C \leq \varepsilon M \|g\|_\infty, \qquad (3.9.20)$$
which follows
$$\|f - V_\varepsilon^1(f)\|_C \leq \|f - P_\varepsilon(f)\|_C + \|P_\varepsilon(f) - V_\varepsilon^1(f)\| = O(\varepsilon).$$

Thus we have that $f \in F(V^1, C)$, And this implies
$$F(V^1, C) = V(C, \{|m|\}).$$

In the case of $l = 2$, we need the help of Theorem 3.9.3, and the other steps are the same as that of $l = 1$.

This completes the proof of Theorem 3.9.4. ∎

Lemma 3.9.3 *Let $0 < \varepsilon \leq 1$ and $l = 1, 2$. Define an operator $\widetilde{V}_\varepsilon^l$:*
$$\widetilde{V}_\varepsilon^l(f;x) = \sum_{|m| < 1/\varepsilon} e^{-|\varepsilon m|^l} \left(1 - |\varepsilon m|^l\right)^{n+1} \frac{1}{|\varepsilon m|^l} C_m(f) e^{im \cdot x}.$$

Then $\widetilde{V}_\varepsilon^l$ is an uniformly bounded linear operator mapping from $L^\infty(Q^n)$ to $C(Q^n)$. That is,
$$\left\|\widetilde{V}_\varepsilon^l(f)\right\|_C \leq M \|f\|_\infty, \qquad (3.9.21)$$
for every $\varepsilon \in (0, 1]$.

3.9 The saturation problem of the uniform approximation

Proof. It suffices to show that

$$\widetilde{\varphi^l}(s) := s^{\frac{n}{2}} \int_1^\infty e^{-t^l} \left(1 - t^l\right)^{n+1} \frac{1}{t^l} t^{\frac{n}{2}} J_{\frac{n}{2}-1}(ts) dt \in L(\mathbb{R}^+).$$

Denote

$$h_0(t) = e^{-t^l} \left(1 - t^l\right)^{n+1} \frac{1}{t^l}$$

and

$$h_{j+1}(t) = \frac{h'_j(t)}{t},$$

for $j = 0, 1 \ldots, n$.

Obviously we have $h_j \in C^\infty[1, \infty)$, $h_j(1) = 0$ and

$$|h_j(t)| \le M_{n,j} e^{-t^l}(1 + t^{2n}),$$

for $j = 0, 1, \ldots, n+1$.

After executing the partial integration for $n+1$ times, we can have that

$$\widetilde{\varphi^l}(s) = s^{\frac{n}{2}-(n+1)} \int_1^\infty (-1)^{n+1} h_{n+1}(t) t^{\frac{n}{2}+n+1} J_{\frac{n}{2}+n}(ts) dt.$$

This implies that $\widetilde{\varphi^l}(s) \in L(\mathbb{R}^+)$.
This is the end of the proof. ∎

Now we begin to consider the $(C,1)$ means $\sigma_R^{\frac{n-1}{2}}$. According to the definition,

$$\sigma_R^{\frac{n-1}{2}}(f; x) = \frac{1}{R} \int_0^R \sum_{|m|<r} \left(1 - \frac{|m|^2}{r^2}\right)^{\frac{n-1}{2}} C_m(f) e^{im \cdot x} dr$$

$$= \sum_{|m|<R} C_m(f) e^{im \cdot x} \frac{1}{R} \int_0^R \left(1 - \frac{|m|^2}{r^2}\right)^{\frac{n-1}{2}} \chi(|m|, R)(r) dr$$

$$= \sum_{|m|<R} \frac{1}{R} \int_{|m|}^R \left(1 - \frac{|m|^2}{r^2}\right)^{\frac{n-1}{2}} dr C_m(f) e^{im \cdot x},$$

we have

$$\lambda_R(m) = \begin{cases} \frac{1}{R} \int_{|m|}^R \left(1 - \frac{|m|^2}{r^2}\right)^{\frac{n-1}{2}} dr, & |m| < R, \\ 0, & |m| \ge R. \end{cases} \quad (3.9.22)$$

If $n = 1$, then we have
$$\lambda_R(m) = 1 - \frac{|m|}{R},$$
for $|m| < R$, and
$$\lim_{R \to \infty} R(1 - \lambda_R(m)) = |m|.$$

If $n > 1$, then we can compute it directly by the method of integration by parts. When $|m| < R$, we have

$$\lambda_R(m) = \left(1 - \frac{|m|^2}{R^2}\right)^{\frac{n-1}{2}} - (n-1)\frac{|m|}{R} \int_{\frac{|m|}{R}}^{1} (1-t^2)^{\frac{n-3}{2}} dt$$

$$= \left(1 - \frac{|m|^2}{R^2}\right)^{\frac{n-1}{2}} - \frac{n-1}{2} \frac{|m|}{R} \int_{\frac{|m|^2}{R^2}}^{1} (1-t)^{\frac{n-3}{2}} t^{-\frac{1}{2}} dt.$$

Thus it follows that

$$\lim_{R \to \infty} R(1 - \lambda_R(m)) = \frac{n-1}{2}|m| \int_0^1 (1-t)^{\frac{n-3}{2}} \cdot t^{-\frac{1}{2}} dt$$

$$= \frac{|m|\Gamma\left(\frac{n+1}{2}\right)\Gamma\left(\frac{1}{2}\right)}{\Gamma\left(\frac{n}{2}\right)},$$

for every $n \in \mathbb{N}$.

From the above results and Theorem 3.9.1, we can think that $\left\{\sigma_R^{\frac{n-1}{2}}\right\}$ is saturated in $C(Q^n)$ with a saturation of R^{-1} as $R \to \infty$. We denote its saturation class as $F(\sigma, C)$.

Lemma 3.9.4 *Let*

$$\xi(t) = t^{-1}\left(1 - t\int_t^1 (1-r^2)^{\frac{n-1}{2}} r^{-2} dr\right)$$

for $0 < t \leq 1$,
$$\eta(t) = e^{-t}(1-t)^{n+1}$$

for $0 \leq t \leq 1$, and

$$\varphi(t) = \begin{cases} \xi(t)\eta(t), & 0 \leq t < 1, \\ 0, & t \geq 1. \end{cases}$$

3.9 The saturation problem of the uniform approximation

Define an operator $\widetilde{H^l_\varepsilon}$ as,

$$\widetilde{H^l_\varepsilon}(f;x) = \sum \varphi(\varepsilon|m|)C_m(f)e^{im\cdot x},$$

for $\varepsilon > 0$. Then, $\widetilde{H^l_\varepsilon}$ is an uniformly bounded linear operator mapping from $L^\infty(Q^n)$ to $C(Q^n)$. That is,

$$\left\|\widetilde{H^l_\varepsilon}(f)\right\|_C \leq M\|f\|_\infty, \qquad (3.9.23)$$

for $\varepsilon > 0$.

Proof. It suffices to show that

$$h(s) := s^{\frac{n}{2}} \int_0^\infty \varphi(t) t^{\frac{n}{2}} J_{\frac{n}{2}-1}(ts) dt \in L(\mathbb{R}^+).$$

Consider

$$h(s) = s^{\frac{n}{2}} \int_0^1 \xi(t)\, \eta(t) t^{\frac{n}{2}} J_{\frac{n}{2}-1}(ts) dt.$$

Since

$$\xi'(t) = t^{-2}\left((1-t^2)^{\frac{n-1}{2}} - 1\right) = \sum_{j=0}^\infty a_j t^{2j},$$

for $0 \leq t < 1$, we can see that $\xi \in C^\infty[0,1)$.

On the other hand, since

$$\varphi^{(k)}(t) = \sum_{j=0}^k C_k^j\, \xi^{(j)}(t)\eta^{(k-j)}(t), \qquad (3.9.24)$$

for $k \in \mathbb{Z}_+$, and $\eta \in C^\infty[0,1)$, we have $\varphi \in C^\infty[0,1)$.

It is obvious that when $k = 0, 1, \ldots, n$, we have

$$\varphi^{(k)}(1) = \lim_{t \to 1^-} \varphi^{(k)}(t) = 0$$

and

$$\varphi^{(n+1)}(1) = \lim_{t \to 1^-} \varphi^{(n+1)}(t).$$

Hence, we have $\varphi \in C^{n+1}[0,1]$ and $\varphi^{(k)}(1) = 0$ for $k = 0, 1, \ldots, n$.

Define $h_0(t) = \varphi(t)$ and $h_{j+1}(t) = h_j(t)t^{-1}$ for $j = 0, 1, \ldots, n$. Obviously, when $j = 1, 2, \ldots, n$, we have

$$h_j(t) = \frac{\varphi^{(j)}(t)}{t^j} + b_{j,1}\frac{\varphi^{(j-1)}(t)}{t^{j+1}} + \cdots + b_{j,j-1}\frac{\varphi'(t)}{t^{2j-1}}.$$

Hence, it follows that
$$h_j(t) = \mathrm{O}\left(t^{-2j}\right),$$
as $t \to 0^+$, and $h_j(1) = 0$, for $j = 0, 1, \ldots, n$.

Since
$$t^{\frac{n}{2}+j} J_{\frac{n}{2}+j}(ts) = \mathrm{O}\left(t^{n+2j}\right),$$
as $s > 0$, we can have
$$\left. h_j(t)\, t^{\frac{n}{2}+j} J_{\frac{n}{2}+j}(ts) \right|_0^1 = 0,$$
for $j = 0, 1, \ldots, n$. Thus by the method of integration by parts for $n+1$ times, we can deduce that
$$h(s) = \frac{(-1)^{n+1}}{s^{\frac{n}{2}+1}} \int_0^1 h_{n+1}(t)\, t^{\frac{n}{2}+n+1} J_{\frac{n}{2}+n}(ts)\, dt$$
$$= \mathrm{O}\left(\frac{1}{s^{\frac{n}{2}+1}}\right),$$
as $s \to +\infty$.

Therefore, we have $h \in L(\mathbb{R}^+)$, and finish the proof of the lemma. ∎

Lemma 3.9.5 *Define an operator $\triangle_R^1$ as*
$$\triangle_R^1(f;x) = V_{\frac{1}{R}}^1(f;x) - \sigma_R^{\frac{n-1}{2}}\left(V_{\frac{1}{R}}^1(f);\, x\right). \tag{3.9.25}$$

If $f \in V(C, \{|m|\})$, then we have
$$\|\triangle_R^1(f)\|_C = \mathrm{O}\left(\frac{1}{R}\right).$$

Proof. By the definition, we have
$$\triangle_R^1(f;x) = \sum e^{-\frac{|m|}{R}}\left(1 - \frac{|m|}{R}\right)^{n+1} C_m(f) e^{im\cdot x}$$
$$- \sum_{|m|<R} \lambda_R(m) e^{-\frac{|m|}{R}}\left(1 - \frac{|m|}{R}\right)^{n+1} C_m(f) e^{im\cdot x},$$
where $\lambda_R(m)$ is given by (3.9.22). It follows from direct computation that
$$\lambda_R(m) = \frac{|m|}{R} \int_{\frac{|m|}{R}}^1 (1-r^2)\, r^{-2}\, dr,$$

3.9 The saturation problem of the uniform approximation

for $0 < |m| < R$, and $\lambda_R(0) = 1$.

Since $f \in V(C, \{|m|\})$, then there exists $g \in L^\infty(Q^n)$, such that

$$|m|C_m(f) = C_m(g),$$

for $m \in \mathbb{Z}^n$.

We conclude that

$$\Delta_R^1(f;x) = \frac{1}{R} \sum_{|m|>R} e^{-\frac{|m|}{R}} \left(1 - \frac{|m|}{R}\right)^{n+1} \frac{R}{|m|} C_m(g) e^{im\cdot x}$$

$$+ \frac{1}{R} \sum_{0<|m|<R} (1 - \lambda_R(m)) e^{-\frac{|m|}{R}} \left(1 - \frac{|m|}{R}\right)^{n+1} \frac{R}{|m|} C_m(g) e^{im\cdot x}$$

$$= \frac{1}{R} \left\{ \widetilde{V}^1_{\frac{1}{R}}(g;x) - \widetilde{H}^1_{\frac{1}{R}}(g;x) + \frac{\Gamma\left(\frac{n+1}{2}\right)\Gamma\left(\frac{1}{2}\right)}{\Gamma\left(\frac{n}{2}\right)} C_0(g) \right\},$$

for $R \geq 1$.

By Lemma 3.9.3 and 3.9.4, we can obtain that

$$\|\Delta_R^1(f)\|_C = O\left(\frac{1}{R}\right).$$

This completes the proof of Lemma 3.9.5. ∎

Theorem 3.9.5 *The saturation of the family of operators $\left\{\sigma_R^{\frac{n-1}{2}}\right\}$ is R^{-1}, and the saturation class is $F(\sigma, C) = V(C, \{|m|\})$.*

Proof. We have calculated the saturation in the previous. Concerned about the saturation class, by Theorem 3.9.1, we can notice that

$$F(\sigma, C) \subset V(C, \{|m|\}_{m \in \mathbb{Z}^n}).$$

Next we will prove the reverse inclusion relationship. Assume $f \in V(C, \{|m|\})$. We conclude that

$$f - \sigma_R^{\frac{n-1}{2}}(f) = f - V^1_{\frac{1}{R}}(f) + V^1_{\frac{1}{R}}(f) - \sigma_R^{\frac{n-1}{2}}\left(V^1_{\frac{1}{R}}(f)\right)$$

$$+ \sigma_R^{\frac{n-1}{2}}\left(V^1_{\frac{1}{R}}(f)\right) - \sigma_R^{\frac{n-1}{2}}(f)$$

$$= f - V^1_{\frac{1}{R}}(f) + \Delta_R^1(f) + \sigma_R^{\frac{n-1}{2}}\left(V^1_{\frac{1}{R}}(f) - f\right).$$

It follows from Theorem 3.9.4 that
$$\left\|f - V^1_{\frac{1}{R}}(f)\right\|_C = O\left(\frac{1}{R}\right).$$

By the uniform boundedness of $\sigma_R^{\frac{n-1}{2}}$, we can have
$$\left\|\sigma_R^{\frac{n-1}{2}}\left(V^1_{\frac{1}{R}}(f) - f\right)\right\|_C = O\left(\left\|V^1_{\frac{1}{R}}(f) - f\right\|_C\right) = O\left(\frac{1}{R}\right).$$

By Lemma 3.9.5, we immediately have
$$\|\triangle^1_R(f)\|_C = O\left(\frac{1}{R}\right).$$

In another words, $f \in F(\sigma, C)$, which finishes the proof of the theorem. ∎

Now let us consider the saturation class of the Bochner-Riesz means $\{S_R^\alpha\}$ with $\alpha > \frac{n-1}{2}$, that is the case over the critical index.

Lemma 3.9.6 Let $\alpha > \frac{n-1}{2}$,
$$\xi(t) = t^{-2}\left[1 - (1 - t^2)^\alpha\right]$$
for $0 < t \leq 1$, and
$$\xi(0) = \xi(0^+) = a.$$
Set
$$\eta(t) = e^{-t^2}(1 - t^2)^{n+1}$$
for $0 \leq t \leq 1$, and
$$\varphi(t) = \begin{cases} \xi(t)\eta(t), & 0 \leq t < 1, \\ 0, & t \geq 1. \end{cases}$$

Define the operator $\widetilde{H^2_\varepsilon}$ as
$$\widetilde{H^2_\varepsilon}(f; x) = \sum \varphi(\varepsilon|m|)C_m(f)e^{im\cdot x},$$
for $\varepsilon > 0$. Then $\widetilde{H^2_\varepsilon}$ is a uniformly bounded linear operator mapping from $L^\infty(Q^n)$ to $C(Q^n)$. That is,
$$\left\|\widetilde{H^2_\varepsilon}(f)\right\|_C \leq M\|f\|_\infty, \tag{3.9.26}$$
for $\varepsilon > 0$.

3.9 The saturation problem of the uniform approximation

The proof of lemma 3.9.6 is the same as that of Lemma 3.9.4 and thus we omit the proof here.

Lemma 3.9.7 *Define an operator* $\triangle_R^2$ *as*

$$\triangle_R^2(f;\, x) = V_{\frac{R}{2}}^2(f;\, x) - S_R^\alpha\left(V_{\frac{R}{2}}^2(f);x\right).$$

If $f \in V(C, \{|m|^2\})$, *then we have*

$$\left\|\triangle_R^2(f)\right\|_C = O\left(\frac{1}{R^2}\right).$$

The steps of the proof are the same as that of Lemma 3.9.5 with the help of Lemmas 3.9.3 and 3.9.6, so we skip without the proof here.

Theorem 3.9.6 *Let* $\alpha > \frac{n-1}{2}$. *The saturation of the family of the operators* $\{S_R^\alpha\}$ *is* R^{-2}, *and the saturation class is* $V(C, \{|m|^2\})$.

We have to point out that when $\alpha > \frac{n-1}{2}$, Theorem 3.9.6 gave a characterization of the saturation class of $\{S_R^\alpha\}$. According to Theorem 3.9.1, this conclusion can be written as, $f \in F(S^\alpha, C)$ if and only if f is not a constant and

$$\left\|\sum_{|m|<R}\left(1 - \frac{|m|^2}{R^2}\right)^\alpha |m|^2 C_m(f)e^{im\cdot x}\right\|_C = O(1), \qquad (3.9.27)$$

for $R > 0$.

This means the condition (3.9.27) can be considered as a characterization of the saturation class of the Riesz means with a positive order $\alpha > 0$.

The above discussion is also suitable for the generalized Riesz means $S_R^{l,\alpha}(f;\, x)$ for $l \in \mathbb{N}$ (see [J1]). The definition of the generalized Riesz means is as follows,

$$S_R^{l,\alpha}(f;\, x) = \sum_{|m|<R}\left(1 - \frac{|m|^l}{R^l}\right)^\alpha C_m(f)e^{im\cdot x}.$$

Cheng and Chen [ChC1] discussed about the uniform approximation problem of $S_R^{l,\alpha}$ with $l \in \mathbb{N}$. Wang [Wa5] went further to carry out more general discussions about the operator.

Finally, we will characterize the constructive property of the functions in $F(S^\alpha, C)$ with $\alpha > \frac{n-1}{2}$.

Theorem 3.9.7 Let $f \in C(Q^n)$, f is not a constant and $\alpha > \frac{n-1}{2}$. Then $f \in F(S^\alpha, C)$ if and only if either of the following two conditions holds.

(a) $\left\|\Delta S_R^\alpha(f)\right\|_C = O(1)$, for $R \geq 0$;

(b) $\left\|\frac{d}{dt}f_x(t)\right\|_C = \sup_{x \in Q^n}\left|\frac{d}{dt}f_x(t)\right| = O(t)$, for $t > 0$.

Proof. In fact, the condition (a) is exactly (3.9.27). Obviously it is the necessary and sufficient condition for $f \in F(S^\alpha, C)$ with $\alpha > \frac{n-1}{2}$.

Now let $f \in F(S^\alpha, C)$ firstly. It follows from the property of uniform convergence that

$$f(x) = S_1^\alpha(f; x) + \sum_{j=0}^{\infty}\left(S_{2^{j+1}}^\alpha(f; x) - S_{2^j}^\alpha(f; x)\right) \qquad (3.9.28)$$

uniformly holds.

By the well-known Bernstein inequality, we have

$$\left\|\frac{\partial}{\partial x_k}\left(S_{2^{j+1}}^\alpha(f; x) - S_{2^j}^\alpha(f; x)\right)\right\|_C \leq 2^{j+1}\left\|S_{2^{j+1}}^\alpha(f) - S_{2^j}^\alpha(f)\right\|_C$$
$$\leq 2^{j+1}\left(\left\|S_{2^{j+1}}^\alpha(f) - f\right\|_C + \left\|f - S_{2^j}^\alpha(f)\right\|_C\right)$$
$$= M 2^{-j},$$

for $j \in \mathbb{Z}_+$ and $k = 1, 2, \ldots, n$.

Then the series of the right side of (3.9.28) are uniformly convergent. Thus we have

$$\frac{\partial f(x)}{\partial x_k} = \frac{\partial}{\partial x_k}S_1^\alpha(f; x) + \sum_{j=0}^{\infty}\frac{\partial}{\partial x_k}\left[S_{2^{j+1}}^\alpha(f; x) - S_{2^j}^\alpha(f; x)\right] \qquad (3.9.29)$$

uniformly with respect to x. This implies $f \in C^1(Q^n)$.

Thus we have

$$f_x'(t) = \frac{\Gamma(n/2)}{2\pi^{n/2}}\frac{d}{dt}\int_{|\xi|=1}f(x + t\xi)d\sigma(\xi)$$
$$= \frac{\Gamma(n/2)}{2\pi^{n/2}}\int_{|\xi|=1}\sum_{j=1}^{n}\left.\frac{\partial f(u)}{\partial u_j}\right|_{u=x+t\xi}\xi_j d\sigma(\xi).$$

Set

$$I_R(t) = \int_{|\xi|=1}\sum_{j=1}^{n}\left(\frac{\partial f(u)}{\partial u_j} - \frac{\partial S_R^\alpha(f; u)}{\partial u_j}\right)\Big|_{u=x+t\xi}$$

3.9 The saturation problem of the uniform approximation

and
$$J_R(t) = \int_{|\xi|=1} \sum_{j=1}^{n} \frac{\partial S_R^\alpha(f;u)}{\partial u_j}\bigg|_{u=x+t\xi} \xi_j \, d\sigma(\xi).$$

Making use of the Green formula, we have
$$J_R(t) = \int_{|\xi|<1} \triangle S_R^\alpha(f; x+t\xi) t d\xi.$$

It follows from (a) that
$$|J_R(t)| \leq Mt,$$
for $t \geq 0$.

By (3.9.29), we can obtain that
$$|f'_x(t)| \leq \limsup_{j\to\infty} \left| \frac{\Gamma(n/2)}{2\pi^{n/2}} \left(I_{2^j}(t) + J_{2^j}(t)\right) \right|$$
$$\leq C_n \limsup_{j\to\infty} |J_{2^j}(t)|$$
$$\leq Mt,$$
for any x. this implies that (b) holds.

On the contrary, we assume that (b) holds. It follows from the Bochner formula that
$$S_R^\alpha(f;x) - f(x) = C \int_0^\infty \left[f_x\left(\frac{t}{R}\right) - f(x) \right] J_{\frac{n}{2}+\alpha}(t) t^{\frac{n}{2}-\alpha-1} dt,$$

and the mean value formula
$$f_x\left(\frac{t}{R}\right) - f(x) = f'_x\left(\frac{\theta}{R}\right) \cdot \frac{t}{R} = O\left(\frac{t^2}{R^2}\right),$$

and setting $\alpha = \frac{n}{2} + 4$, we have that
$$|S_R^\alpha(f;x) - f(x)| \leq M \int_0^\infty |J_{n+4}(t)| t^{-3} dt \frac{1}{R^2}$$
$$\leq M \frac{1}{R^2}.$$

Hence, we have $f \in F(S^{\frac{n}{2}+4}, C)$. By Theorem 3.9.6, for every $\alpha > \frac{n-1}{2}$, so is $F(S^\alpha, C)$.

This completes the proof of Theorem 3.9.7. ∎

3.10 Strong summation

The strong summation problem is to find certain conditions for the limit

$$\lim_{R\to\infty} \frac{1}{R} \int_0^R \left|S_r^{\frac{n-1}{2}}(f;x) - f(x)\right|^q dr = 0 \qquad (3.10.1)$$

to hold, where $q > 0$.

When $n > 1$, Bochner and Chandrasekharan [BoC1] initiated to investigate the problem. They obtained that the validity of (3.10.1) is a local property of the functions, that is to say, if f vanishes at the neighborhood of x_0, then (3.10.1) holds for $q = 2$ at the point of x_0. Besides of the above, they also characterized a sufficient condition for (3.10.1) to hold in the case $q = 2$, that is to say, if $f \in L^2(B(x_0, \eta))$ for some $\eta > 0$ and

$$\int_0^t |f_{x_0}(\tau) - f(x_0)|^2 d\tau = o(t), \qquad (3.10.2)$$

as $t \to 0$, then the equality (3.10.1) holds for $q = 2$ at x_0.

In 1958, using the method of interpolation of operators, Stein proved that when

$$\delta > \frac{n-1}{2}\left(\frac{2}{p}-1\right) - \left(1-\frac{1}{p}\right)$$

and $1 < p \le 2$, the maximal operator

$$\Lambda_\delta(f;x) := \sup_{R>0} \left(\frac{1}{R}\int_0^R \left|S_u^\delta(f;x)\right|^2 du\right)^{\frac{1}{2}}$$

is of type (p,p). And thus we can deduce that

$$\lim_{R\to\infty} \frac{1}{R}\int_0^R \left|S_u^\delta(f;x) - f(x)\right|^2 du = 0$$

holds almost everywhere $x \in Q^n$ for $f \in L^p(Q^n)$.

The case of L^1 is especially important. However, the related conclusion has not been proven yet.

In 1985, Lu [Lu7] achieved breakthrough result about the strong summation problem.

Theorem 3.10.1 *Let $f \in L(Q^n)$ with $n \ge 2$. If there exists $\delta > 0$, such that f vanishes in $B(x_0, \delta)$, then we have*

$$\lim_{R\to\infty} \frac{1}{R}\int_0^R \left|S_r^{\frac{n-1}{2}}(f;x_0)\right|^q dr = 0,$$

for every $q > 0$.

3.10 Strong summation

Proof. Choose $\beta \in (0, \frac{1}{2}]$. We have

$$\frac{1}{R}\int_1^R \left|S_r^{\frac{n-1}{2}}(f;x_0)\right|^q dx = \lim_{\beta \to 0^+} \frac{1}{R}\int_1^R \left|S_r^{\frac{n-1}{2}+\beta}(f;x_0)\right|^q dr,$$

for $R > 1$.

Consequently, it suffices to show that

$$\sup_{\beta \in (0, \frac{1}{2}]} \int_1^R \left|S_r^{\frac{n-1}{2}+\beta}(f;x_0)\right|^q dx = o(R), \qquad (3.10.3)$$

for $R \to \infty$. By the Bochner formula

$$S_R^{\frac{n-1}{2}+\beta}(f;x_0) = \frac{\left(\frac{1}{2}+\beta\right)\Gamma\left(\frac{n+1}{2}+\beta\right)}{\Gamma(\frac{n}{2})} \int_{R\delta}^{\infty} f_{x_0}\left(\frac{t}{R}\right) \frac{J_{n-\frac{1}{2}+\beta}(t)}{t^{\frac{1}{2}+\beta}} dt.$$

We might as well take it for granted that $x_0 = 0$.

Denote

$$C_\beta = \frac{\left(\frac{1}{2}+\beta\right)\Gamma\left(\frac{n+1}{2}+\beta\right)}{\Gamma(\frac{n}{2})}.$$

Obviously, $0 < C_\beta \leq \frac{n}{2}$, for $\beta \in (0, \frac{1}{2}]$.

By the asymptotic formula

$$J_\nu(t) = \sqrt{\frac{2}{\pi t}} \cos\left(t - \frac{\pi}{2}\nu - \frac{\pi}{4}\right) + O\left(\frac{1}{t^{\frac{3}{2}}}\right),$$

where the remainder $O\left(\frac{1}{t^{\frac{3}{2}}}\right)$ satisfies

$$\left|O\left(\frac{1}{t^{\frac{3}{2}}}\right)\right| \leq M_n \cdot \frac{1}{t^{\frac{3}{2}}}$$

and $\nu = n - \frac{1}{2} + \beta$ with $\beta \in \left(0, \frac{1}{2}\right)$, we can get that

$$S_r^{\frac{n-1}{2}+\beta}(f;0) = \sqrt{\frac{2}{\pi}} C_\beta \int_{r\delta}^{\infty} f_0\left(\frac{t}{r}\right) \frac{\cos\left(t - \frac{\pi}{2}(n+\beta)\right)}{t^{1+\beta}} dt + I_r^\beta,$$

where

$$\left|I_r^\beta\right| \leq M_n \int_{r\delta}^{\infty} \left|f_0\left(\frac{t}{r}\right)\right| \frac{1}{t^2} dt \leq M_n \frac{1}{r}. \qquad (3.10.4)$$

Thus, it suffices to prove

$$\sup_{\beta \in (0, \frac{1}{2}]} \int_1^R \left|\int_{r\delta}^{\infty} f_0\left(\frac{t}{r}\right) \frac{\cos\left(t - \frac{\pi}{2}(n+\beta)\right)}{t^{1+\beta}} dt\right|^q dr = o(R).$$

Since
$$\cos\left(t - \frac{\pi}{2}(n+\beta)\right) = a_\beta \cos t - b_\beta \sin t,$$
with $|a_\beta| + |b_\beta| \leq 2$, we merely need to prove the following two equations,
$$\sup_{\beta \in (0,\frac{1}{2})} \int_1^R \left| \int_\delta^\infty f_0(t) \frac{\cos rt}{t^{1+\beta} r^\beta} dt \right|^q dr = o(R) \tag{3.10.5}$$
and
$$\sup_{\beta \in (0,\frac{1}{2})} \int_1^R \left| \int_\delta^\infty f_0(t) \frac{\sin rt}{t^{1+\beta} r^\beta} dt \right|^q dr = o(R). \tag{3.10.6}$$

Next we only prove (3.10.5), for the proof of either equation above is quite the same.

In addition, when $0 < q < q'$, it follows from the Hölder's inequality that
$$\int_1^R |h(t)|^q dt \leq \left(\int_1^R |h(t)|^{q'} dt \right)^{\frac{q}{q'}} (R-1)^{1-\frac{q}{q'}}.$$

To prove (3.10.5) is valid for every $q > 0$, we merely need to prove its validity for every index in the form of $q = 2^N$ for $N = 1, 2, \cdots$.

To this end, we assume $q = 2^N$. Without loss of generality, we may assume $x_0 = 0$. Since the integral on the left side of (3.10.5) is
$$\int_1^R \left| \int_\delta^\infty f_0(t) \frac{\cos rt}{t^{1+\beta} r^\beta} dt \right|^q dr$$
$$= \int_\delta^\infty \cdots \int_\delta^\infty \frac{f_0(t_1) \cdots f_0(t_q)}{(t_1 \cdots t_q)^{1+\beta}} \left\{ \int_1^R \frac{1}{r^{q\beta}} \left(\cos rt_1 \cdots \cos rt_q \right) dr \right\} dt_1 \cdots dt_q,$$
the second mean value theorem implies that
$$\int_1^R \frac{1}{r^{q\beta}} (\cos rt_1 \cdots \cos rt_q) dr = \int_1^\xi (\cos rt_1 \cdots \cos rt_q) dr$$
$$:= a(\xi; t_1, \ldots, t_q),$$
with $\xi \in (1, R)$.

We line up all the q-dimensional vectors whose coordinate is either 1 or -1 to denote by $\{e_k\}_{k=1}^{2^q}$, where $e_k = (e_{k,1}, \ldots, e_{k,q})$, and $e_{k,j} = 1$ or -1 for $j = 1, 2, \ldots, q$.

Denote by $t = (t_1, \ldots, t_q)$ and $(e_k \cdot t) = e_{k_1} t_1 + \cdots + e_{k_q} t_q$. Then, we have
$$\cos rt_1 \cdots \cos rt_q = 2^{-q} \sum_{k=1}^{2^q} \cos r(e_k \cdot t).$$

3.10 Strong summation

Thus it follows that

$$a(\xi; t_1, \ldots, t_q) = 2^{-q} \sum_{k=1}^{2^q} \frac{1}{(e_k \cdot t)} \big[\sin \xi(e_k \cdot t) - \sin(e_k \cdot t) \big].$$

We have that

$$\sup_{\beta \in (0, \frac{1}{2})} \left| \int_1^R \frac{1}{r^{q\beta}} \big(\cos r t_1 \cdots \cos r t_q \big) dr \right| \leq 2^{-q} \sum_{k=1}^{2^q} \frac{M_R(|e_k \cdot t|)}{|e_k \cdot t|}, \qquad (3.10.7)$$

where

$$M_R(|e_k \cdot t|) \leq \min(2, \, 2R|e_k \cdot t|). \qquad (3.10.8)$$

Hence, in order to prove (3.10.5), it suffices to show that

$$\int_\delta^\infty \cdots \int_\delta^\infty \frac{|f_0(t_1) \cdots f_0(t_q)|}{t_1 \cdots t_q} \frac{M_R(|e_k \cdot t|)}{|e_k \cdot t|} dt_1 \cdots dt_q = o(R) \qquad (3.10.9)$$

holds for every $k \in (1, \ldots, 2^q)$.

Denote the left side of the above equation by I_R. It follows

$$I_R = \sum_{n_1, \ldots, n_q = 1}^{\infty} \int_{n_1 \delta}^{(n_1+1)\delta} \cdots \int_{n_q \delta}^{(n_q+1)\delta} \frac{|f_0(t_1) \cdots f_0(t_q)|}{t_1 \cdots t_q}$$
$$\times \frac{M_R(|e_k \cdot t|)}{|e_k \cdot t|} dt_1 \cdots dt_q$$
$$\leq \sum_{n_1, \ldots, n_q = 1}^{\infty} \frac{1}{n_1 \cdots n_q} \delta^{-q} \int_{n_1 \delta}^{(n_1+1)\delta} \cdots \int_{n_q \delta}^{(n_q+1)\delta} |f_0(t_1) \cdots f_0(t_q)|$$
$$\times \frac{M_R(|e_k \cdot t|)}{|e_k \cdot t|} dt_1 \cdots dt_q.$$

Set

$$\tau_{n_1 \cdots n_q}(R) = \int_{n_1 \delta}^{(n_1+1)\delta} \cdots \int_{n_q \delta}^{(n_q+1)\delta} |f_0(t_1) \cdots f_0(t_q)| \cdot \frac{M_R(|e_k \cdot t|)}{|e_k \cdot t|} dt_1 \cdots dt_q.$$
$$(3.10.10)$$

We might as well take it for granted that

$$e_k = (\overbrace{1, \ldots, 1}^{a}, \overbrace{-1, \ldots, -1}^{q-a}).$$

Let us first prove that

I We uniformly have
$$\tau_{n_1\cdots n_q}(R) = o(R),$$
as $R \to \infty$, for all $(n_1 \cdots n_q) \in N_q$;

II When $|n_1 + \cdots + n_\alpha - (n_{\alpha+1} + \cdots + n_q)| > 2q\delta$,
$$\tau_{n_1\cdots n_q}(R) \leq C\big|n_1 + \cdots + n_\alpha - (n_{\alpha+1} + \cdots + n_q)\big|^{-1}$$
holds.

Making changes of variables as follows
$$\begin{cases} t_1 = s_1 - s_2 - \cdots - s_\alpha + s_{\alpha+1} + \cdots + s_q, \\ t_j = s_j, \qquad j = 2, \ldots, q, \end{cases}$$
we can get that
$$\tau_{n_1\cdots n_q}(R) = \int_{n_2\delta}^{(n_2+1)\delta} \cdots \int_{n_q\delta}^{(n_q+1)\delta} |f_0(s_2) \cdots f_0(s_q)|$$
$$\times h_{n_1}(s_2, \ldots, s_q; R)ds_2 \cdots ds_q,$$
where
$$h_{n_1}(s_2, \ldots, s_q; R) = \int_{E_{n_1}} |f_0(s_1 - \cdots - s_\alpha + \cdots + s_q)|\frac{M_R(|s_1|)}{|s_1|}ds_1.$$
Set
$$E_{n_1} = \Big\{s_1 : n_1\delta < s_1 - s_2 - \cdots - s_\alpha + s_{\alpha+1} + \cdots + s_q < (n_1+1)\delta\Big\}.$$
If we denote
$$u = s_1 + s_2 + \cdots + s_\alpha - (s_{\alpha+1} + \cdots + s_q),$$
then we have
$$E_{n_1} = \{s_1 : n_1\delta < s_1 - u < (n_1+1)\delta\}.$$
Thus, we have
$$h_{n_1}(s_2, \ldots, s_q; R) = \int_{n_1\delta+u}^{(n_1+1)\delta+u} |f_0(s_1 - u)|\frac{M_R(|s_1|)}{|s_1|}ds_1.$$

3.10 Strong summation

Since $f \in L(Q^n)$, associate to $\varepsilon > 0$, there exists $\xi > 0$ such that for any set $E \subset \mathbb{R}^n$, if $|E| < \xi$, then

$$\int_E |f(x)|dx < \varepsilon$$

holds.

Geometrically, there exists $\eta_0 > 0$, such that for every $m \in \mathbb{Z}^n$ and every $\gamma \geq 0$, the measure of the set

$$E_{m,\gamma} = \{x : x \in Q^n + 2\pi m, \ \gamma < |x| < \gamma + \eta_0\}$$

is always smaller than ξ. It follows that

$$\int_{E_{m,\gamma}} |f(x)|dx < \varepsilon. \tag{3.10.11}$$

Now we set $\eta_0 < \frac{1}{2}\delta$. We will prove that, for any $\gamma \geq 0$ and any $\eta > 0$, if $\eta \leq \eta_0$, then we have

$$\int_\gamma^{\gamma+\eta} |f_0(t)|dt \leq C\varepsilon, \tag{3.10.12}$$

where C is independedent of ε, γ and η.

Since f vanishes in $B(0,\delta)$, when $\gamma < \frac{1}{2}\delta$, then the left side of (3.10.12) became 0. Assume $\gamma \geq \frac{1}{2}\delta$, meanwhile, $2\gamma > \eta_0 + \gamma$. Thus we conclude that

$$\int_\gamma^{\gamma+\eta} |f_0(t)|dt \leq \frac{1}{\gamma^{n-1}} \int_\gamma^{\gamma+\eta_0} |f_0(t)|t^{n-1}dt$$

$$\leq \frac{2^{n-1}}{(\gamma+\eta_0)^{n-1}} \int_{\gamma<|x|<\gamma+\eta_0} |f(x)|dx$$

$$= \frac{2^{n-1}}{(\gamma+\eta)^{n-1}} \sum_m \int_{E_{m,\gamma}} |f(x)|dx.$$

Since the order of the number of the integer points falling in the spherical shell $\gamma < |x| < \gamma+\eta_0$ and taking the form of $2\pi m$ with $m \in \mathbb{Z}^n$ is the same as that of $(\gamma+\eta_0)^{n-1}$, we say that the number of m which guarantees $E_{m,\delta} \neq \emptyset$ shares the same order with $(\gamma+\eta_0)^{n-1}$. Hence, by (3.10.11), we can obtain (3.10.12).

In this way, combining (3.10.12) with (3.10.8) yields that

$$\int_{|s_1|<\frac{1}{2}\eta_0} |f_0(s_1-u)| \frac{M_R(|s_1|)}{|s_1|} ds_1 < CR\varepsilon. \tag{3.10.13}$$

Meanwhile, we also have

$$\int_{u+n_1\delta<s_1<(n_1+1)\delta+u, |s_1|>\frac{1}{2}\eta_0} |f_0(s_1-u)|\frac{M_R(|s_1|)}{|s_1|}ds_1$$

$$\leq \frac{4}{\eta_0}\int_{n_1\delta}^{(n_1+1)\delta}|f_0(t)|dt$$

$$\leq \frac{C}{\eta_0}\|f\|_1. \qquad (3.10.14)$$

Combining the two inequalities (3.10.13) with (3.10.14), we have that

$$h_{n_1}(s_2,\ldots,s_q; R) = o(R)$$

uniformly holds for all the parameters.

Consequently, the conclusion I holds.

When

$$\left|n_1+\cdots+n_\alpha-(n_{\alpha+1}+\cdots+n_q)\right| > 2q\delta,$$

for

$$(t_1,\ldots,t_q) \in \prod_{j=1}^{q}[n_j\delta,\ (n_j+1)\delta],$$

we must have

$$\left|t_1+\cdots+t_\alpha-(t_{\alpha+1}+\cdots+t_q)\right| > \frac{1}{2}\left|n_1+\cdots+n_\alpha-(n_{\alpha+1}+\cdots+n_q)\right|.$$

Therefore, the equality (3.10.10) directly implies the validity of the conclusion II.

The remainder is to estimate

$$I_R \leq \sum \delta^{-q}\frac{1}{n_1\cdots n_q}\tau_{n_1\cdots n_q}(R). \qquad (3.10.15)$$

We decompose the summation on the right side of the inequality 3.10.15 into two parts $\sum_1, \sum_2$ as

$$\sum_1 = \sum_{|n_1+\cdots+n_\alpha-(n_{\alpha+1}+\cdots+n_q)|\leq 2q\delta} \delta^{-q}\frac{1}{n_1\cdots n_q}\tau_{n_1\cdots n_q}(R)$$

and

$$\sum_2 = \sum_{|n_1+\cdots+n_\alpha-(n_{\alpha+1}+\cdots+n_q)|>2q\delta} \delta^{-q}\frac{1}{n_1\cdots n_q}\tau_{n_1\cdots n_q}(R).$$

3.10 Strong summation

Applying the estimate $\tau_{n_1\cdots n_q}(R) = o(R)$ in the conclusion I into $\sum_1$, we can easily get that

$$\sum_1 = o(R) \cdot \sum_{\mu=\alpha}^{\infty} \sum_{\mu_1=\alpha-1}^{\mu-1} \cdots \sum_{\mu_{\alpha-1}=1}^{\mu_{\alpha-2}-1} \sum_{\nu \geq q-\alpha, |\nu-\mu| \leq 2q\delta}$$

$$\sum_{\nu_1=q-\alpha-1}^{\nu-1} \cdots \sum_{\nu_{q-\alpha-1}=1}^{\nu_{q-\alpha-2}-1} \left(\frac{1}{\mu-\mu_1} \cdot \frac{1}{\mu_1-\mu_2} \cdots \frac{1}{\mu_{\alpha-2}-\mu_{\alpha-1}}\right) \frac{1}{\mu_{\alpha-1}}$$

$$\left(\frac{1}{\nu-\nu_1} \frac{1}{\nu_1-\nu_2} \cdots \frac{1}{\nu_{q-\alpha-2}-\nu_{q-\alpha-1}} \frac{1}{\nu_{q-\alpha-1}}\right).$$

When $\alpha > 1$ and $\mu \geq \alpha$, we have

$$\sum_{\mu_1=\alpha-1}^{\mu-1} \cdots \sum_{\mu_{\alpha-1}=1}^{\mu_{\alpha-2}=1} \frac{1}{\mu-\mu_1} \frac{1}{\mu_1-\mu_2} \cdots \frac{1}{\mu_{\alpha-2}-\mu_{\alpha-1}} \leq C \frac{\log^{\alpha-1}(1+\mu)}{\mu}. \tag{3.10.16}$$

Thus we have that

$$\sum_1 = o(R) \sum_{\mu=\alpha}^{\infty} \frac{\log^{\alpha-1}(1+\mu)}{\mu} \sum_{\nu \geq q-\alpha, |\nu-\mu| \leq 2q\delta} \frac{\log^{q-\alpha-1}(1+\nu)}{\nu}$$

$$= o(R) \sum_{\mu=1}^{\infty} \frac{\log^{q-2}(1+\mu)}{\mu^2}$$

$$= o(R). \tag{3.10.17}$$

On the other hand, using the conclusion II implies that

$$\sum_2 \leq C \sum_{|n_1+\cdots+n_\alpha-(n_{\alpha+1}+\cdots+n_q)|>2q\delta} \frac{1}{n_1\cdots n_q}$$

$$\times \frac{1}{|n_1+\cdots+n_\alpha-(n_{\alpha+1}+\cdots+n_q)|}$$

$$= C \sum_{\mu=\alpha}^{\infty} \sum_{\mu_1=\alpha-1}^{\mu-1} \cdots \sum_{\mu_{\alpha-1}=1}^{\mu_{\alpha-2}=1} \sum_{\nu \geq q-\alpha, |\nu-\mu|>2q\delta} \sum_{\nu_1=q-\alpha-1}^{\nu-1} \cdots \sum_{\nu_{q-\alpha-1}=1}^{\nu_{q-\alpha-2}-1}$$

$$\left(\frac{1}{\mu-\mu_1} \cdots \frac{1}{\mu_{\alpha-2}-\mu_{\alpha-1}} \frac{1}{\mu_{\alpha-1}}\right)$$

$$\times \left(\frac{1}{\nu-\nu_1} \cdots \frac{1}{\nu_{q-\alpha-2}-\nu_{q-\alpha-1}} \frac{1}{\nu_{q-\alpha-1}}\right) \frac{1}{|\mu-\nu|}.$$

Applying (3.10.16) into the above inequality, we can get that

$$\sum\nolimits_2 \le C \sum_{\mu=\alpha}^{\infty} \frac{\log^{\alpha-1}(1+\mu)}{\mu} \sum_{\nu \ge q-\alpha, |\nu-\mu| \ge 2q\delta} \frac{1}{|\nu-\mu|} \frac{\log^{q-\alpha-1}(1+\nu)}{\nu}$$

$$\le C \sum_{\mu=\alpha}^{\infty} \frac{\log^{\alpha-1}(1+\mu)}{\mu} \left(\sum_{\nu \ge \mu+2q\delta} + \sum_{q-\alpha \le \nu < \mu-2q\delta} \right) \frac{1}{|\nu-\mu|}$$

$$\times \frac{\log^{q-\alpha-1}(1+\nu)}{\nu}.$$

For $\mu \ge 1$, since we have

$$\sum_{\nu > \mu+2q\delta} \frac{1}{|\nu-\mu|} \frac{\log^{q-\alpha-1}(1+\nu)}{\nu}$$

$$\le \left(\sum_{\nu=\mu+1}^{2\mu} + \sum_{\nu=2\mu+1}^{\infty} \right) \frac{1}{\nu-\mu} \frac{\log^{q-\alpha-1}(1+\nu)}{\nu}$$

$$\le C \frac{1}{\mu} \log^{q-\alpha}(1+\mu) \sum_{q-\alpha \le \nu < \mu-2q\delta} \frac{1}{|\nu-\mu|} \frac{\log^{q-\alpha-1}(1+\nu)}{\nu}$$

$$\le \log^{q-\alpha-1}(1+\mu) \sum_{\nu=1}^{\mu-1} \frac{1}{\nu(\mu-\nu)}$$

$$\le C \frac{1}{\mu} \log^{q-\alpha}(1+\mu),$$

it follows that

$$\sum\nolimits_2 \le C \sum_{\mu=\alpha}^{\infty} \frac{\log^{q-1}(1+\mu)}{\mu^2} < +\infty. \tag{3.10.18}$$

Thus Both (3.10.17) and (3.10.18) yield that (3.10.9) holds. This completes the proof of Theorem 3.10.1. ∎

Theorem 3.10.2 Let $f \in L(Q^n)$ with $n \ge 2$. If there exists $\delta > 0$ and $p > 1$, such that, for $f \in L^p(B(x_0, \delta))$ with some $\delta > 0$,

$$\int_0^t \tau^{n-1} \{f_{x_0}(\tau) - f(x_0)\} d\tau = o(t^n),$$

3.10 Strong summation

as $t \to 0^+$, and
$$\int_0^t |f_{x_0}(\tau) - f(x_0)|^p d\tau = O(t),$$

then
$$\lim_{R \to \infty} \frac{1}{R} \int_0^R \left| S_r^{\frac{n-1}{2}}(f; x_0) - f(x_0) \right|^q dr = 0$$

holds for every $q > 0$.

By Theorem 3.2.3, the proof of Theorem 3.10.2 can be ascribed to that of the theorem about the Fourier integral.

Theorem 3.10.3 Let $g \in L(\mathbb{R}^n)$ with $n \geq 2$ and $g \in L^p(B(x_0, \delta))$ for some $\delta > 0$ and $p > 1$. If
$$\Phi_1(t) := \int_0^t \tau^{n-1} \{g_{x_0}(\tau) - g(x_0)\} d\tau = o(t^n),$$

as $t \to 0^+$, and
$$\Phi_2(t) := \int_0^t |g_{x_0}(\tau) - g(x_0)|^p d\tau = O(t),$$

then
$$\lim_{R \to \infty} \frac{1}{R} \int_0^R \left| B_r^{\frac{n-1}{2}}(g; x_0) - g(x_0) \right|^q dr = 0$$

holds for every $q > 0$, where $B_r^{\frac{n-1}{2}}$ is the Bochner-Riesz means of g with an order of $\frac{n-1}{2}$.

Proof. We might as well take it for granted that $1 < p < 2$, $\frac{1}{p} + \frac{1}{q} = 1$ and $x_0 = 0$, $g(x_0) = 0$.

According to the discussion in Chapter 2 about the Fourier integral, we have that
$$\lim_{r \to \infty} \left| B_r^{\frac{n-1}{2}}(g; 0) - C \int_{\frac{1}{r}}^\infty g_0(t) \frac{\cos\left(rt - \frac{n\pi}{2}\right)}{t} dt \right| = 0.$$

Thus it suffices to show that
$$\lim_{R \to \infty} \frac{1}{R} \int_1^R \left| \int_{\frac{1}{r}}^\infty g_0(t) \frac{\cos\left(rt - \frac{n\pi}{2}\right)}{t} dt \right|^q dr = 0.$$

Since
$$\cos\left(rt - \frac{n\pi}{2}\right) = \mathrm{Re}\left(e^{-i\frac{n\pi}{2}} e^{irt}\right),$$
we merely need to prove
$$\lim_{R\to\infty} \frac{1}{R} \int_1^R \left| \int_{\frac{1}{r}}^{\infty} g_0(t) \frac{e^{irt}}{t} dt \right|^q dr = 0. \qquad (3.10.19)$$

For every $\varepsilon \in (0,1)$, we have
$$\int_{\frac{1}{r}}^{\frac{1}{\varepsilon r}} \frac{g_0(t)}{t} e^{irt} dt = \varphi(t) \frac{1}{t} e^{irt} \bigg|_{\frac{1}{r}}^{\frac{1}{\varepsilon r}} + \int_{\frac{1}{r}}^{\frac{1}{\varepsilon r}} \varphi(t) \left(\frac{1}{t^2} e^{irt} - \frac{ir}{t} e^{irt} \right) dt,$$
where
$$\varphi(t) = \int_0^t g_0(\tau) d\tau = \Phi_1(t) \frac{1}{t^{n-1}} + (n-1) \int_0^t \frac{\Phi_1(\tau)}{\tau^n} d\tau = o(t),$$
as $t \to 0$.

Thus it follows that
$$\int_{\frac{1}{r}}^{\frac{1}{\varepsilon r}} \frac{g_0(t)}{t} e^{irt} dt = o(1) + o(1) \left(\int_{\frac{1}{r}}^{\frac{1}{\varepsilon r}} t\, dt + r \int_{\frac{1}{r}}^{\frac{1}{\varepsilon r}} dt \right)$$
$$= o(1) \left(1 + \log\frac{1}{\varepsilon} + \frac{1}{\varepsilon} \right) \qquad (3.10.20)$$
$$= o(1),$$
as $r \to +\infty$.

Therefore, it follows from (3.10.20) that
$$\alpha := \limsup_{R\to\infty} \frac{1}{R} \int_1^R \left| \int_{\frac{1}{r}}^{\infty} g_0(t) \frac{1}{t} e^{irt} dt \right|^q dr$$
$$\leq 2^q \limsup_{R\to\infty} \frac{1}{R} \int_1^R \left| \int_{\frac{1}{r}}^{\frac{1}{\varepsilon r}} g_0(t) \frac{1}{t} e^{irt} dt \right|^q dr$$
$$+ 2^q \limsup_{R\to\infty} \frac{1}{R} \int_1^R \left| \int_{\frac{1}{\varepsilon r}}^{\infty} g_0(t) \frac{1}{t} e^{irt} dt \right|^q dt$$
$$= 2^q \limsup_{R\to\infty} \frac{1}{R} \int_1^R \left| \int_{\frac{1}{\varepsilon r}}^{\infty} g_0(t) \frac{1}{t} e^{irt} dt \right|^q dr. \qquad (3.10.21)$$

3.10 Strong summation

Denote

$$\Phi(t,r) = \int_0^t g_0(\tau)e^{ir\tau}d\tau = \varphi(t)e^{ir\tau} - ir\int_0^t \varphi(\tau)e^{ir\tau}d\tau.$$

Since $\varphi(t) = o(t)$, it follows that

$$\Phi(t,r) = o(t) + ir \cdot o(t^2),$$

as $t \to 0$.

Meanwhile, we have

$$|\Phi(t,r)| \le \int_0^t |g_0(\tau)|d\tau \le M,$$

for any $r > 0$ and $t > 0$.

Thus we have

$$\int_{\frac{1}{\varepsilon r}}^\infty g_0(t)e^{irt}\frac{1}{t}dt = o(1) + \int_{\frac{1}{\varepsilon r}}^\infty \Phi(t,r)\frac{1}{t^2}dt.$$

Therefore, we conclude that

$$\alpha \le 2^q \limsup_{R\to\infty} \frac{1}{R}\int_1^R \left|\int_{\frac{1}{\varepsilon r}}^\infty \frac{\Phi(t,r)}{t^2}dt\right|^q dr$$

$$\le C_q \limsup_{R\to\infty} \frac{1}{R}\int_1^R \left|\int_{\frac{1}{\varepsilon R}}^\delta \frac{\Phi(t,r)}{t^2}dt\right|^q dr$$

$$+ C_q \limsup_{R\to\infty} \frac{1}{R}\int_1^R \left|\int_\delta^\infty \frac{\Phi(t,r)}{t^2}dt\right|^q dr.$$

The integral of the second term on the right side of the above inequality can be rewritten as

$$\frac{1}{R}\int_1^R h(r)dr,$$

where

$$h(r) = \left|\int_\delta^\infty \frac{\Phi(t,r)}{t^2}dt\right|^q.$$

Notice that $\Phi(t,r) \to 0$, as $r \to \infty$, and $|\Phi(t,r)| \le M$. It follows from the control convergence theorem that $h(r) \to 0$, as $r \to \infty$. Thus we have

$$\frac{1}{R}\int_1^R h(r)dr \to 0,$$

as $R \to \infty$.

Hence, we have that

$$\alpha \le C_q \limsup_{R \to \infty} \frac{1}{R} \int_{\frac{1}{\varepsilon\delta}}^{R} \left| \int_{\frac{1}{\varepsilon R}}^{\delta} \frac{\Phi(t,r)}{t^2} dt \right|^q dr$$

$$\le C_q \limsup_{R \to \infty} \frac{1}{R} \left\{ \int_{\frac{1}{\varepsilon R}}^{\delta} \frac{1}{t^2} \left(\int_{\frac{1}{\varepsilon\delta}}^{R} |\Phi(t,r)|^q dr \right)^{\frac{1}{q}} dt \right\}^q.$$

Applying Hausdorff-Young's inequality to the function $g_0 \chi_{[0,t)}$ with $0 < t < \delta$ in $L^p((0,\delta))$, we can get that

$$\left(\int_{\mathbb{R}} |\Phi(t,r)|^q dr \right)^{\frac{1}{q}} \le \left(\int_0^t |g_0(\tau)|^p d\tau \right)^{\frac{1}{p}} = O\left(t^{\frac{1}{p}}\right),$$

for $t \in (0,\delta)$.

Consequently, we have

$$\alpha \le C_q \limsup_{R \to \infty} \frac{1}{R} \left(\int_{\frac{1}{\varepsilon R}}^{\delta} t^{\frac{1}{p}-2} dt \right)^q \le C_q \limsup_{R \to \infty} \frac{1}{R} \varepsilon R = C_q \varepsilon.$$

Since ε is arbitrary, this leads to $\alpha = 0$, and (3.10.19) is proven. This completes the proof of Theorem 3.10.3. ∎

Chapter 4

THE CONJUGATE FOURIER INTEGRAL AND SERIES

4.1 The conjugate integral and the estimate of the kernel

The conjugate integral considered here is based on the singular integral theory by Calderón and Zygmund [CZ1].

We use normal notations. Let n denote the number of dimensions and $P(\cdot) \in \mathscr{A}_k^{(n)}$ for $k \geq 1$. Here $\mathscr{A}_k^{(n)}$ denotes the collection of all the homogeneous harmonic polynomials of degree k with n variables. We consider the kernel $K(x) = P(x)|x|^{-n-k}$, $x \neq \mathbf{0}$, whose principal value Fourier transform $\widehat{K}$ can be obtained as

$$\widehat{K}(y) = (-i)^k \frac{\Gamma\left(\frac{k}{2}\right)}{\pi^{\frac{n}{2}} 2^n \Gamma\left(\frac{n+k}{2}\right)} \cdot \frac{P(y)}{|y|^k},$$

for $y \neq 0$, and $\widehat{K}(0) = 0$. For any $f \in L(\mathbb{R}^n)$, we define, $x \in \mathbb{R}^n$,

$$\widetilde{f}_\varepsilon(x) = \frac{1}{(2\pi)^n} \int_{|y|>\varepsilon} f(x-y) K(y) dy, \qquad (4.1.1)$$

for $\varepsilon > 0$, and

$$\widetilde{f}(x) = \lim_{\varepsilon \to 0^+} \widetilde{f}_\varepsilon(x). \qquad (4.1.2)$$

The function $\widetilde{f}$ is the extension of the Hilbert transform of univariate function, which is called the multiple Hilbert transform by Calderón and

Zygmund. Here, we prefer to call $\widetilde{f}$ the Calderón-Zygmund transform of f. It is well known that the Calderón-Zygmund transform is of type (p,p) for $1 < p < \infty$, as well as weak type $(1,1)$ (see Stein [St4] or [St5]).

The conjugate Fourier integral of f is the following integral

$$\int_{\mathbb{R}^n} \hat{f}(y)\widehat{K}(y)e^{ix\cdot y}dy. \qquad (4.1.3)$$

From (4.1.1) and (4.1.2), it is easy to see that $\widetilde{f}$ is the convolution of f and K in the sense of principal value. Formally, $\hat{f}\widehat{K}$ should be $\widehat{(f*K)} = \widehat{\widetilde{f}}$ and then (4.1.3) is the inversion of the Fourier transform of $\widetilde{f}$. So we should turn back to $\widetilde{f}$. However, these discussions are formal. In fact, we can only discuss the relation between some kind of linear means of (4.1.3) and $\widetilde{f}$.

We define the Bochner-Riesz means of the conjugate Fourier integral (4.1.3) as

$$\widetilde{\sigma}_R^\alpha(f;x) = \int_{|y|<R} \hat{f}(y)\widehat{K}(y)\left(1 - \frac{|y|^2}{R^2}\right)^\alpha e^{ix\cdot y}dy, \qquad (4.1.4)$$

where the index α can be taken as some complex whose real part is bigger than -1 and $R > 0$.

Lemma 4.1.1 Let $P \in \mathscr{A}_k^{(n)}$ with $k \geq 1$, $n \geq 2$ and $K(x) = \frac{P(x)}{|x|^{n+k}}$. Then we have

$$\int_{|y|<1} P\left(\frac{y}{|y|}\right)(1-|y|^2)^{\alpha_0+\beta}e^{ix\cdot y}dy = i^k P(x)E(n,\beta,k,|x|), \qquad (4.1.5)$$

where

$$E(n,\beta,k,u) = \frac{\pi^{\frac{n}{2}}}{2^k}\sum_{j=0}^{\infty}\frac{(-1)^j\Gamma\left(\frac{n+k}{2}+j\right)\Gamma(\alpha_0+\beta+1)}{j!\Gamma\left(\frac{n}{2}+k+j\right)\Gamma\left(\frac{n+k}{2}+\alpha_0+\beta+1+j\right)}\left(\frac{u}{2}\right)^{2j},$$
$$(4.1.6)$$

$\alpha_0 = \frac{n-1}{2}$ and β is a complex number whose real part is bigger than $-\frac{n+1}{2}$.

This lemma was proven by Chang [Cha1]. For the sake of integrality, we give the proof in the following.

4.1 The conjugate integral and the estimate of the kernel

Proof. We have that

$$\int_{|y|<1} P\left(\frac{y}{|y|}\right)(1-|y|^2)^{\alpha_0+\beta}e^{ix\cdot y}dy$$

$$= \int_0^1 \left[\int_{|\xi|=t} P\left(\frac{\xi}{|\xi|}\right)(1-t^2)^{\alpha_0+\beta}e^{ix\cdot \xi}d\sigma(\xi)\right]dt$$

$$= \int_0^1 (1-t^2)^{\alpha_0+\beta}t^{n-1}\int_{|\xi|=1} P(\xi)e^{itx\cdot \xi}d\sigma(\xi)dt.$$

By the property of transforms of radial functions (see [SW1]), we can get that

$$\int_{|y|<1} P\left(\frac{y}{|y|}\right)(1-|y|^2)^{\alpha_0+\beta}e^{ix\cdot y}dy$$

$$= \int_0^1 (1-t^2)^{\alpha_0+\beta}t^{n-1}(i)^k(2\pi)^{\frac{n}{2}}\frac{J_{\frac{n}{2}+k-1}(t|x|)}{(t|x|)^{\frac{n}{2}-1}}P\left(\frac{x}{|x|}\right)dt. \qquad (4.1.7)$$

By the series expression of the Bessel functions and changing the order of integration and summation, we have that

$$i^k(2\pi)^{\frac{n}{2}}P\left(\frac{x}{|x|}\right)\sum_{j=0}^{\infty}\frac{(-1)^j}{j!\,\Gamma\left(j+\frac{n}{2}+k\right)}$$

$$\times \int_0^1 (1-t^2)^{\alpha_0+\beta}t^{n-1}\left(\frac{t|x|}{2}\right)^{2j+\frac{n}{2}+k-1}\frac{1}{(t|x|)^{\frac{n}{2}-1}}dt$$

$$= i^k(2\pi)^{\frac{n}{2}}P(x)\sum_{j=0}^{\infty}\frac{(-1)^j}{j!\,\Gamma\left(j+\frac{n}{2}+k\right)}$$

$$\times \int_0^1 (1-t^2)^{\alpha_0+\beta}t^{n+k-1}\left(\frac{t|x|}{2}\right)^{2j}dt\frac{1}{2^{\frac{n}{2}+k-1}}$$

$$= i^k\pi^{\frac{n}{2}}2^{-k}\sum_{j=0}^{\infty}\frac{(-1)^j}{j!\,\Gamma\left(j+\frac{n}{2}+k-1\right)}\left(\frac{|x|}{2}\right)^{2j}$$

$$\times \int_0^1 (1-t^2)^{\alpha_0+\beta}t^{n+k-1+2j}2dt.$$

Noticing the fact that

$$\int_0^1 (1-t^2)^{\alpha_0+\beta}t^{n+k-1+2j}2dt = \int_0^1 (1-t)^{\alpha_0+\beta}t^{\frac{n+k}{2}+j-1}dt$$

$$= B\left(\alpha_0+\beta+1,\frac{n+k}{2}+j\right),$$

we can get the lemma proven. ∎

In addition, for $u \geq 1$, Chang [Cha1] also gave an asymptotic estimate of the function $E(n, \beta, k, u)$. That is,

$$E(n, \beta, k, u) = \frac{c(n,k)}{u^{n+k}} + \frac{R_1(\beta, u)}{u^{n+k+1}} + \frac{R_2(\beta, u)}{u^{n+k+1+\beta}}$$
$$+ A(n, \beta) \frac{\cos\left(u - \frac{\pi}{2}(n+k+\beta)\right)}{u^{n+k+\beta}}, \quad (4.1.8)$$

where both R_1 and R_2 are bounded about u. The estimate is enough for the case when $|\text{Re}\beta| \leq 1$. However, for further need, it is necessary to give a more accurate estimate.

Lemma 4.1.2 *Let $\beta = \sigma + i\tau$ and $-\frac{n+1}{2} + h \leq \sigma \leq G$ with $0 < h < G < \infty$. For $u \geq 1$, we have*

$$E(n, \beta, k, u) = \frac{c(n,k)}{u^{n+k}} + \frac{G(n, \beta, k, u)}{u^{n+k+2}} + \frac{H(n, \beta, k, u)}{u^{n+k+2+\beta}}$$
$$+ \frac{B(n, \beta)}{u^{n+k+\beta}} \left(\sqrt{u} J_{n+k+\beta-\frac{1}{2}}(u) + k \frac{\frac{n+1}{2}+\beta}{\sqrt{u}} J_{n+k+\beta+\frac{1}{2}}(u) \right), \quad (4.1.9)$$

where G and H are both infinitely differentiable with respect to u,

$$c(n,k) = \pi^{\frac{n}{2}} 2^n \frac{\Gamma\left(\frac{n+k}{2}\right)}{\Gamma(k/2)},$$

$$B(n, \beta) = \pi^{\frac{n}{2}} 2^{n-\frac{1}{2}+\beta} \Gamma\left(\frac{n+1}{2} + \beta\right),$$

and

$$|G(n, \beta, k, u)| \leq A_{n,k,h,G} e^{\frac{3}{2}\pi|\tau|}, \quad |H(n, \beta, k, u)| \leq A_{n,k,h,G} e^{\frac{3\pi}{2}|\tau|}.$$

Proof. For $\xi > -(n+k)$, let

$$\varphi_\xi(x) = \int_0^\infty e^{-xt} t^{\frac{n}{2}+\xi} J_{\frac{n}{2}+k-1}(t) dt,$$

for $0 < x \leq 1$, and

$$f_\xi(u) = \int_0^\infty e^{-t} t^{\frac{n}{2}+\xi} J_{\frac{n}{2}+k-1}(ut) dt.$$

4.1 The conjugate integral and the estimate of the kernel

Clearly we have

$$f_\xi(u) = \frac{1}{u^{\frac{n}{2}+1+\xi}} \varphi_\xi\left(\frac{1}{u}\right).$$

From a formula in Bateman's book (see [Ba2], page 29), we have that

$$f_\xi(u) = \left(\frac{1}{1+u^2}\right)^{\frac{n}{2}+\xi+1} \Gamma(n+k+\xi) P_{\frac{n}{2}+\xi}^{-(\frac{n}{2}+k-1)}\left(\frac{1}{\sqrt{1+u^2}}\right),$$

where P_μ^ν is the Legendre function (see [Ba1]). It follows from the above equation that

$$\lim_{u\to\infty} u^{\frac{n}{2}+\xi+1} f_\xi(u) = \Gamma(n+k+\xi) P_{\frac{n}{2}+\xi}^{-(\frac{n}{2}+k-1)}(0) = \varphi_\xi(0^+). \quad (4.1.10)$$

For $x > 0$, it is obvious that

$$\varphi_\xi'(x) = -\int_0^\infty e^{-xt} t^{\frac{n}{2}+\xi+1} J_{\frac{n}{2}+k-1}(t) dt = -\varphi_{\xi+1}(x),$$

which leads to

$$\varphi_\xi'(0^+) = -\varphi_{\xi+1}(0^+). \quad (4.1.11)$$

Hence, by the L'Hospital rule, we have

$$f_\xi(u) = \varphi_\xi(0^+) u^{-(\frac{n}{2}+\xi+1)} + \varphi_\xi'(0^+) u^{-(\frac{n}{2}+\xi+2)} g_\xi(u) u^{-(\frac{n}{2}+\xi+3)}, \quad (4.1.12)$$

where $|g_\xi(u)| \leq A_{n,k,\xi}$.

Choose $\psi \in C_{[0,\infty)}^\infty$ such that

$$\psi(t) = \begin{cases} 1, & 0 \leq t \leq \frac{1}{3}, \\ 0, & t \geq \frac{2}{3}. \end{cases}$$

We construct a polynomial

$$Q(t) = \sum_{j=0}^{2n+3k+6} a_j t^j$$

satisfying

$$\frac{d^m}{dt^m}\left\{e^{-t} Q(t) - (1-t^2)^{\frac{n-1}{2}+\beta} \psi(t)\right\}\bigg|_{t=0} = 0, \quad (4.1.13)$$

for $m = 0, 1, \ldots, q$.

Obviously we have $a_0 = a_1 = 1$.

Let
$$I_1(u) = \int_0^\infty \psi(t)\left(1-t^2\right)^{\frac{n-1}{2}+\beta} t^{\frac{n}{2}} J_{\frac{n}{2}+k-1}(ut)dt,$$

$$I_1^*(u) = \int_0^\infty e^{-t} Q(t) t^{\frac{n}{2}} J_{\frac{n}{2}+k-1}(ut)dt,$$

and set
$$\Delta_1(u) = I_1^*(u) - I_1(u)$$
$$= \int_0^\infty \left[e^{-t}Q(t) - \left(1-t^2\right)^{\frac{n-1}{2}+\beta}\psi(t)\right] t^{-k} t^{\frac{n}{2}+k} J_{\frac{n}{2}+k-1}(ut)dt.$$
(4.1.14)

We denote by
$$\varepsilon(t) = \left[e^{-t}Q(t) - \left(1-t^2\right)^{\frac{n-1}{2}+\beta}\psi(t)\right] t^{-k}.$$

By the condition (4.1.13), it is clear that $t = 0$ is the zero point of $\varepsilon(t)$ of degree
$$q - k = 2(n+k+3),$$
which implies $\varepsilon^{(m)}(0) = 0$, for $m = 0, 1, \ldots, 2(n+k+3)$.

For $\varepsilon(t) \in C^{n+k+3}[0,\infty)$, we define an operator A by
$$A(\varepsilon)(t) = \frac{\varepsilon'(t)}{t}$$

and
$$A^{m+1}(\varepsilon)(t) = \frac{(A^m(\varepsilon))'(t)}{t},$$

for $0 \le m < n+k+3$.

Obviously, for the index $\lambda = n+k+3$, we have
$$A^\lambda(\varepsilon)(t) = \varepsilon^{(\lambda)}(t)t^{-\lambda} + a_1 \varepsilon^{(\lambda-1)}(t)t^{-\lambda-1} + \cdots + a_{-2\lambda+1}\varepsilon'(t)t^{-2\lambda+1}.$$

Therefore, $A^\lambda(\varepsilon)(t)$ is bounded in the neighborhood of the origin and for $t \ge 1$, $A^\lambda(\varepsilon)(t)$ is not bigger than the product of a polynomial about t and e^{-t}. Moreover, for $m \in [0, \lambda) \cap \mathbb{Z}$, $A^m(\varepsilon)(t)$ also satisfies
$$A^m(\varepsilon)(t)\Big|_{t=0} = 0.$$

4.1 The conjugate integral and the estimate of the kernel

Hence, by the formula

$$\frac{d}{dt}t^\nu J_\nu(ut) = ut^\nu J_{\nu-1}(ut),$$

and λ times the integration by parts, we get, for $u > 0$,

$$\Delta_1(u) = (-1)^\lambda \frac{1}{u^\lambda} \int_0^\infty A^\lambda(\varepsilon)(t) t^{\frac{n}{2}+k+\lambda} J_{\frac{n}{2}+k+\lambda-1}(ut)dt.$$

According to the definition of $\varepsilon(t)$, we know that there exists a positive number $M = M(n,k,h,G)$ such that

$$\int_0^\infty \left| A^\lambda(\varepsilon)(t) \, t^{\frac{n}{2}+k+\lambda} J_{\frac{n}{2}+k+\lambda-1}(ut) \right| dt \leq M(1+|\tau|)^M.$$

Consequently, we have

$$|\Delta_1(u)| \leq M(1+|\tau|)^M u^{-(n+k+3)}, \qquad (4.1.15)$$

for $u \geq 1$. We note

$$I_1^*(u) = f_0(u) + f_1(u) + \sum_{j=2}^q a_j f_j(u).$$

By (4.1.12), we have

$$f_0(u) = \varphi_0(0^+)u^{-(\frac{n}{2}+1)} + \varphi_0'(0^+)u^{-(\frac{n}{2}+2)} + g_0(u)u^{-(\frac{n}{2}+3)},$$

$$f_1(u) = \varphi_1(0^+)u^{-(\frac{n}{2}+2)} + \varphi_1'(0^+)u^{-(\frac{n}{2}+3)} + g_1(u)u^{-(\frac{n}{2}+4)},$$

and

$$\sum_{j=2}^q a_j f_j(u) = G(u) \, u^{-(\frac{n}{2}+3)}$$

with

$$|G(u)| \leq A_{n,k,\beta} \leq A_{n,k,h,G} e^{\frac{3\pi}{2}|\tau|}.$$

Thus we have

$$I_1^*(u) = \varphi_0(0^+)u^{-(\frac{n}{2}+1)} + (\varphi_0'(0^+) + \varphi_1(0^+))\, u^{-(\frac{n}{2}+2)} + G_1(u)u^{-(\frac{n}{2}+3)},$$

and

$$|G_1(u)| \leq A_{n,k,h,G} e^{\frac{3\pi}{2}|\tau|}. \qquad (4.1.16)$$

Combining (4.1.11) with the fact $\varphi_0'(0^+) + \varphi_1(0^+) = 0$, we get

$$I_1^*(u) = \varphi_0(0^+)u^{-(\frac{n}{2}+1)} + G_1(u)u^{-(\frac{n}{2}+3)}. \tag{4.1.17}$$

Now we again construct a polynomial

$$\overline{Q}(t) = \sum_{j=0}^{q} b_j t^j$$

such that the function

$$\eta(t) := \overline{Q}(1-t^2) - t^{-k}(1 - \psi(t))$$

satisfies $\eta^{(m)}(1) = 0$, for $m = 0, 1, \ldots, q$, where $q = 2n + 3k + 6$. It is easy to calculate that $b_0 = 1$ and $b_1 = \frac{1}{2}k$.

Set

$$I_2(u) = \int_0^1 (1 - \psi(t))(1 - t^2)^{\frac{n-1}{2}+\beta} t^{\frac{n}{2}} J_{\frac{n}{2}+k-1}(ut)dt,$$

$$I_2^*(u) = \int_0^1 \overline{Q}(1-t^2) t^{\frac{n}{2}+k} (1-t^2)^{\frac{n-1}{2}+\beta} J_{\frac{n}{2}+k-1}(ut)dt$$

and

$$\Delta_2(u) = I_2^*(u) - I_2(u)$$
$$= \int_0^1 \eta(t)(1-t^2)^{\frac{n-1}{2}+\beta} t^{\frac{n}{2}+k} J_{\frac{n}{2}+k-1}(ut)dt.$$

According to the property of $\eta(t)$, we have that

$$|\Delta_2(u)| \leq M(1+|\tau|)^M u^{-(n+k+3)}, \tag{4.1.18}$$

for $u \geq 1$.

We clearly have that

$$I_2^*(u) = \sum_{j=0}^{q} b_j \int_0^1 (1-t^2)^{\frac{n-1}{2}+\beta+j} t^{\frac{n}{2}+k} J_{\frac{n}{2}+k-1}(ut)dt.$$

By a formula in Bateman's book (see [Ba2], page 26), we have

$$\int_0^1 (1-t^2)^{\frac{n-1}{2}+\beta+j} t^{\frac{n}{2}+k} J_{\frac{n}{2}+k-1}(ut)dt$$
$$= 2^{\frac{n-1}{2}+\beta+j} \Gamma\left(\frac{n+1}{2} + \beta + j\right) J_{n-\frac{1}{2}+k+\beta+j}(u) u^{-(\frac{n-1}{2}+\beta+j)}.$$

4.1 The conjugate integral and the estimate of the kernel

By the asymptotic formula of the Bessel functions, we have

$$I_2^*(u) = 2^{\frac{n-1}{2}+\beta} \Gamma\left(\frac{n+1}{2}+\beta\right) u^{-\left(\frac{n+1}{2}+\beta\right)}$$

$$\left\{ J_{n-\frac{1}{2}+k+\beta}(u) + k \frac{\frac{n+1}{2}+\beta}{u} J_{n+\frac{1}{2}+k+\beta}(u) \right\} + \frac{H(u)}{u^{\frac{n}{2}+3+\beta}},$$

and

$$|H(u)| \leq A_{n,k,h,G} e^{\frac{3\pi}{2}|\tau|}.$$

Therefore, we have

$$I_2(u) = 2^{\frac{n-1}{2}+\beta} \Gamma\left(\frac{n+1}{2}+\beta\right) u^{-\left(\frac{n}{2}+1+\beta\right)}$$

$$\left[\sqrt{u}\, J_{n-\frac{1}{2}+k+\beta}(u) + k \frac{\frac{n+1}{2}+\beta}{\sqrt{u}} J_{n+\frac{1}{2}+k+\beta}(u)\right] + \frac{H(u)}{u^{\frac{n}{2}+3+\beta}} - \Delta_2(u).$$

(4.1.19)

From an equation in Bateman's book (see [Ba1], page 145), it follows that

$$P_{\frac{n}{2}}^{-\left(\frac{n}{2}+k-1\right)}(0) = 2^{-\left(\frac{n}{2}+k-1\right)} \pi^{\frac{1}{2}} \cos\left(\frac{k-1}{2}\pi\right) \frac{\Gamma\left(1-\frac{k}{2}\right)}{\Gamma\left(\frac{n+k+1}{2}\right)}.$$

Substituting

$$\cos\frac{k-1}{2}\pi = \sin\frac{k}{2}\pi = \frac{\pi}{\Gamma\left(1-\frac{k}{2}\right)\Gamma\left(\frac{k}{2}\right)}$$

into the above equation, we have

$$P_{\frac{n}{2}}^{-\left(\frac{n}{2}+k-1\right)}(0) = 2^{-\left(\frac{n}{2}+k-1\right)}\sqrt{\pi}\left(\Gamma\left(\frac{k}{2}\right)\Gamma\left(\frac{n+k+1}{2}\right)\right)^{-1}. \quad (4.1.20)$$

Again substituting this into (4.1.9), we obtain

$$\varphi_0(0^+) = \frac{\sqrt{\pi}}{2^{\frac{n}{2}+k-1}} \frac{\Gamma(n+k)}{\Gamma\left(\frac{n+k+1}{2}\right)\Gamma\left(\frac{k}{2}\right)}.$$

Applying the formula

$$\Gamma(n+k) = 2^{n+k-1}\pi^{-\frac{1}{2}}\Gamma\left(\frac{n+k}{2}\right)\Gamma\left(\frac{n+k+1}{2}\right)$$

in Bateman's book (see [Ba1], page 5), we can obtain that

$$\varphi_0(0^+) = \frac{2^{\frac{n}{2}}\Gamma\left(\frac{n+k}{2}\right)}{\Gamma\left(\frac{k}{2}\right)}. \tag{4.1.21}$$

From (4.1.5) and (4.1.7), we can see that

$$E(n,\beta,k,u) = \frac{(2\pi)^{\frac{n}{2}}}{u^{\frac{n}{2}+k-1}}\Big(I_1(u) + I_2(u)\Big).$$

By combining (4.1.15) with (4.1.19) together and replacing them into (4.1.21), we get Lemma 4.1.2. ∎

Substituting the expression of $\hat{f}$ into (4.1.4) and changing the order of the integration, we have that

$$\widetilde{\sigma}_R^\alpha(f;x) = \frac{1}{(2\pi)^n}\int_{\mathbb{R}^n} f(u)\int_{|y|<R} \widehat{K}(y)\left(1 - \frac{|y|^2}{R^2}\right)^\alpha e^{i(x-u)\cdot y}dydu$$

$$= (-1)^k \frac{\Gamma(\frac{k}{2})}{(2\pi)^n \pi^{\frac{n}{2}} 2^n \Gamma\left(\frac{n+k}{2}\right)}\int_{\mathbb{R}^n} f(x-u)$$

$$\int_{|y|<R} P\left(\frac{y}{|y|}\right)\left(1 - \frac{|y|}{R^2}\right)^\alpha e^{iu\cdot y}dydu.$$

Again substituting (4.1.5) into the above equation and let $\beta = \alpha - \alpha_0$, we have

$$\widetilde{\sigma}_R^{\alpha_0+\beta}(f;x) = \frac{\Gamma(k/2)R^{n+k}}{2^{2n}\pi^{\frac{3\pi}{2}}\Gamma\left(\frac{n+k}{2}\right)}\int_{\mathbb{R}^n} f(x-u)P(u)E(n,\beta,k,R|u|)du.$$

Set

$$b(n,k) = \frac{\Gamma\left(\frac{k}{2}\right)}{2^{2n}\pi^{\frac{3\pi}{2}}\Gamma\left(\frac{n+k}{2}\right)}, \tag{4.1.22}$$

and

$$\psi_x(t) = \int_{\mathbb{S}^{n-1}} f(x - t\xi)P(\xi)d\sigma(\xi), \tag{4.1.23}$$

for $t > 0$.

Thus, it follows that

$$\widetilde{\sigma}_R^{\alpha_0+\beta}(f;x) = b(n,k)R^{n+k}\int_0^\infty \psi_x(t)E(n,\beta,k,Rt)\, t^{n+k-1}dt. \tag{4.1.24}$$

We call the function $E(n,\beta,k,Rt)$ the kernel of the operator $\widetilde{\sigma}_R^{\alpha_0+\beta}$.

4.2 Convergence of Bochner-Riesz means for conjugate Fourier integral

Let us first prove a proposition parallel to the case of non-conjugate, which was proven by Lu [Lu3].

Theorem 4.2.1 Let $f \in L(\mathbb{R}^n)$ with $n \geq 2$, and $\alpha = \frac{n-1}{2} - \beta$ with $0 \leq \beta < 1$. If $P \in \mathscr{A}_k^{(n)}$ with $k \geq 1$ and $\psi_x(t)$ defined as in (4.1.23) satisfies that

$$\int_0^t |\psi_{x_0}(\tau)| \tau^{n-1} d\tau = o(t^n), \qquad (4.2.1)$$

as $t \to 0^+$, then, the equation

$$\lim_{R \to \infty} \left\{ \widetilde{\sigma}_R^\alpha(f; x_0) - \widetilde{f}_{\frac{1}{R}}(x_0) \right\} = 0 \qquad (4.2.2)$$

holds if and only if

$$\lim_{R \to \infty} R^\beta \int_{\frac{1}{R}}^{+\infty} \psi_{x_0}(t) \frac{\cos\left(Rt - \frac{\pi}{2}(n+k-\beta)\right)}{t^{1-\beta}} dt = 0. \qquad (4.2.3)$$

Proof. It follows from (4.1.1) and (4.1.23) that

$$\widetilde{f}_{\frac{1}{R}}(x) = \frac{1}{(2\pi)^n} \int_{\frac{1}{R}}^\infty \frac{1}{t} \psi_x(t) dt. \qquad (4.2.4)$$

By (4.1.24), we have

$$\widetilde{\sigma}_R^{\alpha_0 - \beta}(f; x_0) = b(n,k) R^{n+k} \int_0^\infty \psi_{x_0}(t) E(n, -\beta, k, Rt) \, t^{n+k-1} dt$$

$$= b(n,k) R^{n+k} \left(\int_0^{\frac{1}{R}} + \int_{\frac{1}{R}}^\infty \right) \psi_{x_0}(t) E(n, -\beta, k, Rt) \, t^{n+k-1} dt$$

$$= I_1 + I_2. \qquad (4.2.5)$$

Since

$$|E(n, -\beta, k, Rt)| \leq M_{n,k,\beta} < +\infty,$$

for $t \geq 0$, by condition (4.2.1), we have

$$I_1 = b(n,k) R^{n+k} \int_0^{\frac{1}{R}} \psi_{x_0}(t) E(n, -\beta, k, Rt) t^{n+k-1} dt$$
$$= o(1),$$

as $R \to +\infty$.

To estimate I_2, we denote by $1 - \beta = \delta > 0$. By (4.1.9) and the asymptotic formula of the Bessel functions

$$J_\nu(u) = \sqrt{\frac{2}{\pi u}} \cos\left(u - \frac{\pi}{2}\nu - \frac{\pi}{4}\right) + O\left(\frac{1}{u^{\frac{3}{2}}}\right),$$

for $u \geq 1$, we have

$$E(n, -\beta, k, Rt) = \frac{c(n,k)}{(Rt)^{n+k}} + \frac{B(n,-\beta)}{(Rt)^{n+k-\beta}} \cos\left(Rt - \frac{\pi}{2}(n+R-\beta)\right)$$
$$+ O\left(\frac{1}{(Rt)^{n+k+\delta}}\right).$$

Thus, we can obtain the estimate

$$I_2 = b(n,k)c(n,k) \int_{\frac{1}{R}}^{\infty} \psi_{x_0}(t) t^{-1} dt + b(n,k) B(n,-\beta) R^\beta \int_{\frac{1}{R}}^{\infty} \psi_{x_0}(t)$$
$$\frac{\cos\left(Rt - \frac{\pi}{2}(n+k-\beta)\right)}{t^{1-\beta}} dt + O\left(\frac{1}{R^\delta} \int_{\frac{1}{R}}^{\infty} \psi_{x_0}(t) \frac{1}{t^{1+\delta}} dt\right).$$
(4.2.6)

By the condition (4.2.1), we have

$$O\left(\frac{1}{R^\delta} \int_{\frac{1}{R}}^{\infty} \psi_{x_0}(t) \frac{1}{t^{1+\delta}} dt\right) = o(1)$$

as $R \to \infty$.

Since

$$b(n,k) = \frac{\Gamma(k/2)}{2^{2n} \pi^{\frac{3n}{2}} \Gamma\left(\frac{n+k}{2}\right)}$$

and

$$c(n,k) = \pi^{\frac{n}{2}} 2^n \frac{\Gamma\left(\frac{n+k}{2}\right)}{\Gamma\left(\frac{k}{2}\right)},$$

we have

$$b(n,k)c(n,k) = \frac{1}{(2\pi)^n}.$$

Therefore, it follows from (4.2.4)–(4.2.6) that

$$\widetilde{\sigma}_R^{\alpha_0 - \beta}(f; x_0) - \widetilde{f}_{\frac{1}{R}}(x_0) = b(n,k) B(n,-\beta) R^\beta \int_{\frac{1}{R}}^{\infty} \psi_{x_0}(t)$$
$$\frac{\cos\left(Rt - \frac{\pi}{2}(n+k-\beta)\right)}{t^{1-\beta}} dt + o(1), \quad (4.2.7)$$

as $R \to +\infty$.

4.2 Convergence of Bochner-Riesz means for conjugate Fourier integral

Together with the fact that $b(n,k)B(n,-\beta) \neq 0$, this leads to the conclusion of Theorem 4.2.1. ∎

Theorem 4.2.1 above improves the corresponding results by Lippman [Li1] and Golubov [Go2].

Remark 4.2.1 *The condition (4.2.1) of Theorem 4.2.1 can be replaced by the following weaker one*

$$\int_0^t \psi_{x_0}(\tau)\tau^{n-1}d\tau = o(t^n),$$

as $\to 0^+$.

Theorem 4.2.2 *Let* $f \in L(\mathbb{R}^n)$ *with* $n \geq 2$ *satisfies the condition (4.2.1) and*

$$\int_\eta^\infty \frac{|\psi_{x_0}(\tau+\eta) - \psi_{x_0}(\tau)|}{\tau}d\tau = o(1), \quad (4.2.8)$$

as $\eta \to 0$. *If* $\beta = 0$, *then (4.2.2) holds.*

Proof. Let $\theta = -\frac{\pi}{2}(n+k)$. By Theorem 4.2.1, we merely need to verify that if $\beta = 0$, then (4.2.3) holds. That is,

$$I(R) := \int_{\frac{1}{R}}^\infty \psi_{x_0}(t)\frac{\cos(Rt+\theta)}{t}dt = o(1), \quad (4.2.9)$$

as $R \to \infty$. By the periodicity of cosine functions and the condition (4.2.1), we have that

$$I(R) = \frac{1}{2}\int_{\frac{1}{R}}^{\frac{1+\pi}{R}} \psi_{x_0}(t)\frac{\cos(Rt+\theta)}{t}dt$$

$$+ \frac{1}{2}\int_{\frac{1}{R}}^\infty \left[\psi_{x_0}\left(t+\frac{\pi}{R}\right)\frac{-1}{t+\frac{\pi}{R}} + \psi_{x_0}(t)\frac{1}{t}\right]\cos(Rt+\theta)dt$$

$$= o(1) + \frac{1}{2}\int_{\frac{1}{R}}^\infty \left\{\psi_{x_0}\left(t+\frac{\pi}{R}\right)\left(\frac{1}{t} - \frac{1}{t+\frac{\pi}{R}}\right)\right.$$

$$\left. - \frac{\psi_{x_0}\left(t+\frac{\pi}{R}\right) - \psi_{x_0}(t)}{t}\right\}\cos(Rt+\theta)dt$$

$$= o(1) + \frac{\pi}{2R}\int_{\frac{1}{R}}^\infty \frac{\psi_{x_0}\left(t+\frac{\pi}{R}\right)}{t\left(t+\frac{\pi}{R}\right)}\cos(Rt+\theta)dt$$

$$- \frac{1}{2}\int_{\frac{1}{R}}^\infty \frac{\psi_{x_0}\left(t+\frac{\pi}{R}\right) - \psi_{x_0}(t)}{t}\cos(Rt+\theta)dt. \quad (4.2.10)$$

It follows from the condition (4.2.1) that

$$\frac{1}{R}\int_{\frac{1}{R}}^{\infty}\frac{|\psi_{x_0}(t+\frac{\pi}{R})|}{t(t+\frac{\pi}{R})}dt \leq \frac{2\pi}{R}\int_{\frac{1}{R}}^{\infty}\frac{|\psi_{x_0}(t)|}{t^2}dt$$

$$\leq \frac{2\pi}{R}\int_{\frac{1}{R}}^{\infty}(n+1)\frac{\int_0^t \tau^{n-1}|\psi_{x_0}(\tau)|d\tau}{t^{n+2}}dt$$

$$= o(1).$$

It follows from the condition that

$$\int_{\frac{1}{R}}^{\infty}\frac{\psi_{x_0}(t+\frac{\pi}{R})-\psi_{x_0}(t)}{t}\cos(Rt+\theta)dt = o(1).$$

Then (4.2.9) is proven and thus we have proven this theorem. ∎

Corollary 4.2.1 *Let $f \in L(\mathbb{R}^n)$ with $n \geq 2$. If there exists $\eta > 0$ such that*

$$\int_0^\eta t^{-1}\int_{\mathbb{S}^{n-1}}|f(x_0+t\xi)-f(x_0)|d\sigma(\xi)dt < \infty, \tag{4.2.11}$$

then

$$\lim_{R\to\infty}\left(\widetilde{\sigma_R}^{\frac{n-1}{2}}(f;x_0) - \widetilde{f}_{\frac{1}{R}}(x_0)\right) = 0 \tag{4.2.12}$$

holds.

Proof. It is obvious that (4.2.1) can be deduced from (4.2.11). For any $\varepsilon > 0$, by (4.2.11), there exists $\delta_1 > 0$ such that

$$\int_0^{2\delta_1} t^{-1}\int_{\mathbb{S}^{n-1}}|f(x_0+t\xi)-f(x_0)|d\sigma(\xi)dt < \varepsilon. \tag{4.2.13}$$

Since $f \in L(\mathbb{R}^n)$, $\psi_{x_0}(t)$, as a function of t, is in $L((\delta_1,+\infty))$. Hence, by the continuous property of integral, we have that there exists $\delta > 0$ with $\delta \leq \delta_1$, such that when $0 < \eta < \delta$, we have

$$\delta_1^{-1}\int_{\delta_1}^{\infty}|\psi_{x_0}(t+\eta)-\psi_{x_0}(t)|dt < \varepsilon.$$

Thus it follows that

$$\int_{\delta_1}^{\infty}\frac{|\psi_{x_0}(t+\eta)-\psi_{x_0}(t)|}{t}dt < \varepsilon. \tag{4.2.14}$$

On the other hand, when $0 < \eta < \delta$, by the condition (4.2.13), we conclude that

$$\int_\eta^{\delta_1} \frac{|\psi_{x_0}(t+\eta) - \psi_{x_0}(t)|}{t} dt$$

$$\leq \int_\eta^{\delta_1} \frac{|\psi_{x_0}(t+\eta)|}{t} dt + \int_\eta^{\delta_1} \frac{|\psi_{x_0}(t)|}{t} dt$$

$$\leq 2 \int_{2\eta}^{2\delta_1} \frac{|\psi_{x_0}(t)|}{t} dt + \int_\eta^{\delta_1} \frac{|\psi_{x_0}(t)|}{t} dt$$

$$\leq 3 \int_0^{2\delta_1} \frac{|\psi_{x_0}(t)|}{t} dt$$

$$\leq 3 \max_{\xi \in \mathbb{S}^{n-1}} |P(\xi)| \int_0^{2\delta_1} t^{-1} \int_{\mathbb{S}^{n-1}} |f(\chi_{x_0} + t\xi) - f(x_0)| d\sigma(\xi) dt$$

$$\leq M\varepsilon, \qquad (4.2.15)$$

where M is a constant.

Combining (4.2.14) with (4.2.15) yields

$$\int_\eta^\infty \frac{|\psi_{x_0}(t+\eta) - \psi_{x_0}(t)|}{t} dt \leq (M+1)\varepsilon.$$

This implies the validity of (4.2.8).

This completes the proof of Corollary 4.2.1. ∎

Corollary 4.2.1 gives the Dini type criteria for the convergence of the conjugate Bochner-Riesz means, which was first established by Lippman [Li1].

Other literatures in the research of the Bochner-Riesz means of the conjugate Fourier integral include the works by Wu [Wu1].

For the conjugate means of the critical index $\widetilde{\sigma}_R^{\frac{n-1}{2}}(f;x)$, we can regard it as the multiple extension of the one variable means with an order of zero, that is,

$$\int_{|y|<R} \hat{f}(y)\widehat{k}(y)e^{ixy} dy.$$

Thus we call

$$\frac{1}{R} \int_0^R \widetilde{\sigma}_u^{\frac{n-1}{2}}(f;x) du$$

the $(C,1)$ means of the conjugate Fourier integral. Lu established the following theorem in [Lu4].

Theorem 4.2.3 Let $f \in L(\mathbb{R}^n)$ with $n \geq 2$. If the condition (4.2.1) holds at x_0, then we have

$$\lim_{R \to \infty} \left\{ \frac{1}{R} \int_0^R \widetilde{\sigma}_u^{\frac{n-1}{2}}(f; x_0) du - \widetilde{f}_{\frac{1}{R}}(x_0) \right\} = 0. \tag{4.2.16}$$

Proof. By setting $\theta = -\frac{\pi}{2}(n+k)$ and (4.2.7), we have that

$$\frac{1}{R} \int_0^R \widetilde{\sigma}_u^{\frac{n-1}{2}}(f; x_0) du - \widetilde{f}_{\frac{1}{R}}(x_0)$$

$$= \frac{1}{R} \int_0^R b(n,k) B(n,0) \int_{\frac{1}{u}}^{\infty} \psi_{x_0}(t)$$

$$\times \frac{\cos(ut+\theta)}{t} dt du + \frac{1}{R} \int_0^R \widetilde{f}_{\frac{1}{u}}(x_0) du - \widetilde{f}_{\frac{1}{R}}(x_0) + o(1). \tag{4.2.17}$$

By (4.2.4), we conclude that

$$\frac{1}{R} \int_0^R \widetilde{f}_{\frac{1}{u}}(x_0) du - \widetilde{f}_{\frac{1}{R}}(x_0)$$

$$= \frac{1}{R} \int_0^R \left(\frac{1}{(2\pi)^n} \int_{\frac{1}{u}}^{\infty} \frac{\psi_{x_0}(t)}{t} dt \right) du - \frac{1}{(2\pi)^n} \int_{\frac{1}{R}}^{\infty} \frac{\psi_{x_0}(t)}{t} dt$$

$$= \frac{1}{(2\pi)^n} \left\{ \frac{1}{R} \int_{\frac{1}{R}}^{\infty} \left(R - \frac{1}{t} \right) \frac{\psi_{x_0}(t)}{t} dt - \int_{\frac{1}{R}}^{\infty} \frac{\psi_{x_0}(t)}{t} dt \right\}$$

$$= -\frac{1}{(2\pi)^n} \frac{1}{R} \int_{\frac{1}{R}}^{\infty} \frac{\psi_{x_0}(t)}{t^2} dt. \tag{4.2.18}$$

Let

$$F(t) = \int_0^t |\psi_{x_0}(\tau)| \tau^{n-1} d\tau.$$

Integration by parts implies that

$$\int_{\frac{1}{R}}^{\infty} \frac{|\psi_{x_0}(t)|}{t^2} dt = F(t) t^{-(n+1)} \Big|_{\frac{1}{R}}^{\infty} + (n+1) \int_{\frac{1}{R}}^{\infty} \frac{F(t)}{t^{n+2}} dt.$$

By the condition (4.2.1), that is, $F(t) = o(t^n)$ as $t \to 0^+$, we have that

$$\frac{1}{R} \int_{\frac{1}{R}}^{\infty} \frac{|\psi_{x_0}(t)|}{t^2} dt = o(1), \tag{4.2.19}$$

as $R \to \infty$.

4.3 The conjugate Fourier series

In addition, it follows that

$$\frac{1}{R}\int_0^R \left(\int_{\frac{1}{u}}^\infty \psi_{x_0}(t)\frac{\cos(ut+\theta)}{t}dt\right)du$$

$$= \frac{1}{R}\int_{\frac{1}{R}}^\infty \left(\int_{\frac{1}{t}}^R \cos(ut+\theta)du\right)\frac{\psi_{x_0}(t)}{t}dt$$

$$= \frac{1}{R}\int_{\frac{1}{R}}^\infty \frac{\sin(Rt+\theta)-\sin(1+\theta)}{t}\frac{\psi_{x_0}(t)}{t}dt. \qquad (4.2.20)$$

Combining (4.2.17)–(4.2.20) yields that Theorem 4.2.3 holds. ∎

It is easy to see that if $0 \leq \beta < 1$, then, under the condition of Theorem 4.2.3, we have

$$\lim_{R\to\infty}\left\{\frac{1}{R}\int_0^R \widetilde{\sigma_u}^{\frac{n-1}{2}-\beta}(f;x_0)du - \widetilde{f}_{\frac{1}{R}}(x_0)\right\} = 0. \qquad (4.2.21)$$

Since the condition (4.2.1) is obviously valid at the Lebesgue points, we can deduce from the weak $(1,1)$ type of the Calderón-Zygmund transform that the limit in (4.2.1) exists almost everywhere finite. Therefore, as a corollary, we can get the following theorem.

Theorem 4.2.4 *If $f \in L(\mathbb{R}^n)$, then we have*

$$\lim_{R\to\infty}\frac{1}{R}\int_0^R \widetilde{\sigma_u}^{\frac{n-1}{2}}(f;x)du = \widetilde{f}(x), \qquad (4.2.22)$$

for a.e. $x \in \mathbb{R}^n$.

Of course, if we replace $\widetilde{\sigma_u}^{\frac{n-1}{2}}$ by $\widetilde{\sigma_u}^{\frac{n-1}{2}-\beta}$ with $0 \leq \beta < 1$, then the conclusion still holds.

At last, we have to point out that, from Theorem 4.2.1, it is easy to see that for the critical index α_0, the convergence of $\widetilde{\sigma_R}^{\alpha_0}(f)$ only depends on the local property of the functions, while things are different for the order below the critical index.

4.3 The conjugate Fourier series

Suppose that $f \in L(Q^n)$ and its Fourier series is

$$\sigma(f)(x) = \sum C_m(f)e^{im\cdot x}. \qquad (4.3.1)$$

For $n=1$, it is well known that

$$\tilde{\sigma}(f)(x) = \sum_{m\neq 0}(-i\operatorname{sgn} m)C_m(f)e^{im\cdot x}, \qquad (4.3.2)$$

is called the conjugate series of $\sigma(f)$. By iterated superposition, this notion can be extended from the case of one variable to the case of multiple variables. For a function of n variables, there are $2^n - 1$ conjugate series and the status of these conjugate series are quite asymmetric, or we say their extent of the conjugation are different. Thus, the above factors complicate the research of the conjugate series. Especially, when we investigate the series by the spherical limit, the notion of conjugation is not quite adequate for the case of multiple variables. Hence, we will study the conjugate in the sense of Calderón-Zygmund.

Here, we only consider the conjugate series in correspondence with the spherical harmonic kernel which is parallel to the concept of the conjugate integral in the previous two sections.

Suppose that $K(x) = P(x)|x|^{-n-k}$ and $\widehat{K}$ is the Fourier transform of the principal value of K. We define by

$$\tilde{\sigma}(f)(x) = \sum \widehat{K}(m)C_m(f)e^{im\cdot x} \qquad (4.3.3)$$

and call it the conjugate series of $\sigma(f)$ about the kernel K, or the conjugate Fourier series of f for brief. If $n=1$, then $P(x)$ is a homogenous function of degree 1. That is, $P(x) = cx$, where c is a nonzero constant. Then by the expression of $\widehat{K}$ in Section 4.1, we have

$$\widehat{K}(y) = (-i)\frac{c}{2}\operatorname{sgn}(y). \qquad (4.3.4)$$

If we choose $c=2$, then (4.3.3) coincides with (4.3.2). Since we are familiar with (4.3.2), we only discuss the case when $n \geq 2$ in the following.

Let $m \in \mathbb{Z}^n$. We define

$$I_m = \frac{1}{(2\pi)^n}\int_{Q^n} K(x+2\pi m)dx. \qquad (4.3.5)$$

If $m=0$, the above integral should be put in the sense of principal value. That is to say,

$$I_0 = \frac{1}{(2\pi)^n}\lim_{\varepsilon\to 0^+}\int_{Q^n\setminus B(0,\varepsilon)} K(x)dx = \frac{1}{(2\pi)^n}\int_{Q^n\setminus B(0,1)} K(x)dx, \qquad (4.3.6)$$

4.3 The conjugate Fourier series

where we use
$$\int_{\mathbb{S}^{n-1}} P(\xi) d\sigma(\xi) = 0.$$

Let $x, y \in Q := Q^n$ and $m \in \mathbb{Z}^n$, $m \neq \mathbf{0}$. We set
$$\xi_m = \frac{x + 2\pi m}{x + 2\pi m}, \quad \eta_m = \frac{y + 2\pi m}{y + 2\pi m}.$$

Then, we have
$$K(x + 2\pi m) - I_m = \frac{1}{|Q|} \int_Q \left(\frac{P(\xi_m)}{|\xi_m|^n} - \frac{P(\eta_m)}{|\eta_m|^n} \right) dy.$$

Thus it follows that
$$|K(x + 2\pi m) - I_m|$$
$$\leq \frac{1}{|Q|} \int_Q \frac{|P(\xi_m) - P(\eta_m)|}{|\xi_m|^n} dy + \frac{1}{|Q|} \int_Q \left| \frac{1}{|\xi_m|^n} - \frac{1}{|\eta_m|^n} \right| |P(\eta_m)| dy.$$

Since we have
$$|P(\xi_m) - P(\eta_m)| \leq |\nabla P(\theta)| |\xi_m - \eta_m| = O\left(\frac{1}{|m|}\right)$$

and
$$\left| \frac{1}{|\xi_m|^n} - \frac{1}{|\eta_m|^n} \right| = O\left(\frac{1}{|m|^{n+1}}\right),$$

it follows that
$$|K(x + 2\pi m) - I_m| = O\left(\frac{1}{|m|^{n+1}}\right), \tag{4.3.7}$$

for $m \neq \mathbf{0}$. We define the periodization of K by
$$K^*(x) = \sum_{m \in \mathbb{Z}^n} [K(x + 2\pi m) - I_m],$$

for $x \in Q$ and $x \neq \mathbf{0}$.

By (4.3.7), if we get rid of the term of $m = \mathbf{0}$ in the series on the right side of the above equation, then the series is absolutely and uniformly convergent. Therefore,
$$K^*(x) = K(x) - I_0 + \sum_{m \neq 0} [K(x + 2\pi m) - I_m]$$

is continuous everywhere on Q except the origin.

We define f_ε^* by

$$f_\varepsilon^*(x) = \begin{cases} \dfrac{1}{(2\pi)^n} \displaystyle\int_{y \in Q, |y| > \varepsilon} f(x-y)K^*(y)dy, & \text{if } 0 < \varepsilon < \sqrt{n}\pi \\ 0, & \text{if } \varepsilon \geq \sqrt{n}\pi. \end{cases}$$

We call f_ε^* the truncated conjugate function of f relative to the kernel K. The limit of f_ε^* is defined by

$$f^*(x) = \lim_{\varepsilon \to 0^+} f_\varepsilon^*(x),$$

which is called the conjugate function of f. Let $\varepsilon\sqrt{n}\pi > 0$. Thus we have

$$f_\varepsilon^*(x) = \frac{1}{(2\pi)^n} \int_{y \in Q, |y| > \varepsilon} f(x-y)(K(y) - I_0)dy$$

$$+ \sum_{m \neq 0} \frac{1}{(2\pi)^n} \int_{y \in Q} [K(y + 2\pi m) - I_m] f(x-y)dy$$

$$+ \sum_{m \neq 0} \frac{-1}{(2\pi)^n} \int_{|y| \leq \varepsilon} f(x-y)[K(y + 2\pi m) - I_m]dy$$

$$= \frac{1}{(2\pi)^n} \int_{y \in Q, |y| > \varepsilon} f(x-y)(K(y) - I_0)dy + \sum\nolimits_1 + \sum\nolimits_2.$$

For the item $\sum_1$ in the above equation, let

$$S_j = \{x \in \mathbb{Z}^n : j \leq |x| < j+1\},$$

for $j = 0, 1, 2, \ldots$, and by the spherical summation, we have

$$\sum\nolimits_1 = \sum_{j=1}^{\infty} \sum_{m \in S_j} \left\{ \frac{1}{(2\pi)^n} \int_{Q + 2\pi m} f(x-y)K(y)dy - I_m C_0(f) \right\}.$$

Let $N \in \mathbb{N}$ and $N > 10n$. We set

$$E_N = \bigcup_{j=1}^{N} \bigcup_{m \in S_j} (Q + 2\pi m)$$

and

$$D_N = \left\{ y : y \notin Q, |y| \leq 2\pi(N - \sqrt{n}) \right\}.$$

For any $y \in D_N$, there exists $m \in \mathbb{Z}^n$ such that $y \in Q + 2\pi m$. However, since $y \notin Q$, $m \neq 0$, by $|y - 2\pi m| \leq \sqrt{n}\pi$, we have that

$$2\pi |m| \leq |y| + \sqrt{n}\pi < 2N\pi, \quad 0 < |m| < N.$$

4.3 The conjugate Fourier series

Therefore, we have $y \in E_N$, which implies $D_N \subset E_N$.

On the other hand, for any $y \in E_N$, we have $|y| \leq 2\pi N + \sqrt{n}\pi$. We denote by $r_N = 2\pi(N - \sqrt{n})$ and $\rho_N = 2\pi(N + \sqrt{n})$. Then we have

$$E_N \subset B(0, \rho_N), \quad D_N = B(0, r_N) \setminus Q$$

and

$$\{E_N \setminus D_n\} \subset \{B(0, \rho_N) \setminus B(0, r_N)\}.$$

From the above discussion, we obtain that

$$\left| \sum_{j=1}^{N} \sum_{m \in S_j} \int_{Q+2\pi m} f(x-y)K(y)dy - \int_{D_N} f(x-y)K(y)dy \right|$$

$$\leq \int_{r_N < |y| < \rho_N} |f(x-y)K(y)|dy. \qquad (4.3.8)$$

Since for $|y| \in (r_N, \rho_N)$, we have $|K(y)| = O\left(\frac{1}{N^n}\right)$ and

$$\int_{r_N < |y| < \rho_N} |f(x-y)|dy = O(N^{n-1})\|f\|_{L(Q)}.$$

Then the item on the right side of (4.3.8) is $O\left(\frac{1}{N}\right)$.

Meanwhile, we have

$$\sum_{j=1}^{N} \sum_{m \in S_j} I_m = \frac{1}{|Q|} \sum_{j=1}^{N} \sum_{m \in S_j} \int_{Q+2\pi m} K(y)dy$$

$$= \frac{1}{|Q|} \int_{E_N} K(y)dy$$

$$= \frac{1}{|Q|} \int_{D_N} K(y)dy + \frac{1}{|Q|} \int_{E_N \setminus D_N} K(y)dy, \qquad (4.3.9)$$

where

$$\left| \int_{E_N \setminus D_N} K(y)dy \right| \leq \int_{r_N < |y| < \rho_N} \frac{C}{|y|^n} dy = O\left(\frac{1}{N}\right).$$

By the harmonicity, we have

$$\int_{D_N} K(y)dy$$
$$= \int_{B(o,\sqrt{n\pi})\backslash Q} K(y)dy + \int_{\sqrt{n\pi}<|y|<r_N} K(y)dy$$
$$= \int_{B(o,\sqrt{n\pi})\backslash Q} K(y)dy \int_{\sqrt{n\pi}}^{r_N} \frac{1}{t^{n+k}} \int_{|\xi|=t} P(\xi)d\sigma(\xi)dt$$
$$= \int_{B(o,\sqrt{n\pi})\backslash Q} K(y)dy. \qquad (4.3.10)$$

It follows from (4.3.8), (4.3.9) and (4.3.10) that

$$\sum\nolimits_1 = \lim_{N\to\infty} \left\{ \int_{D_N} f(x-y)K(y)dy - C_0(f)\int_{B(o,\sqrt{n\pi})\backslash Q} K(y)dy \right\} \frac{1}{|Q|}$$
$$= \frac{1}{(2\pi)^n} \int_{\sqrt{n\pi}}^{\infty} \left\{ \frac{1}{t} \int_{\mathbb{S}^{n-1}} f(x-t\xi)P(\xi)d\sigma(\xi) \right\} dt$$
$$+ \frac{1}{(2\pi)^n} \int_{B(o,\sqrt{n\pi})\backslash Q} [f(x-y) - C_0(f)]K(y)dy. \qquad (4.3.11)$$

According to (4.3.7), we have

$$\left|\sum\nolimits_2\right| \le c \sum_{m\neq 0} \frac{1}{|m|^{n+1}} \int_{|y|<\varepsilon} |f(x-y)|dy.$$

Combining the fact that $f \in L(Q)$ with the absolute continuity of integration, we have

$$\sum\nolimits_2 = o(1), \qquad (4.3.12)$$

uniformly holds about x, as $\varepsilon \to 0^+$. Consequently, we have

$$f_\varepsilon^*(x) = \frac{1}{(2\pi)^n} \int_\varepsilon^\infty \frac{1}{t} \int_{\mathbb{S}^{n-1}} f(x-t\xi)P(\xi)d\sigma(\xi)dt$$
$$- \frac{1}{(2\pi)^n} \int_{B(o,\sqrt{n\pi})\backslash B(o,\varepsilon)} C_0(f)K(y)dy + o(1).$$

That is,

$$f_\varepsilon^*(x) = \frac{1}{(2\pi)^n} \int_\varepsilon^\infty \frac{1}{t} \int_{\mathbb{S}^{n-1}} f(x-t\xi)P(\xi)d\sigma(\xi)dt + o(1), \qquad (4.3.13)$$

4.3 The conjugate Fourier series

holds uniformly about x, as $\varepsilon \to 0^+$. Thus, we obtain the expression of the conjugate function

$$f^*(x) = \frac{1}{(2\pi)^n} \lim_{\varepsilon \to 0^+, \rho \to +\infty} \int_\varepsilon^\rho \frac{1}{t} \int_{\mathbb{S}^{n-1}} f(x - t\xi) P(\xi) d\sigma(\xi) dt$$

$$= \frac{1}{(2\pi)^n} \lim_{\varepsilon \to 0^+, \rho \to +\infty} \int_{\varepsilon < |y| < \rho} f(x - y) K(y) dy. \quad (4.3.14)$$

We can notice that $f_\varepsilon^*(x)$ is the convolution of an integrable function and a continuous function on Q^n and so is continuous. Hence, the upper limit of the integral on the right side of (4.3.13) tends to ∞, which is a limit procedure of everywhere converging. And thus (4.3.14), which is worth being discussed, is the limit procedure when $\varepsilon \to 0^+$.

The existence of the conjugate functions and related problems were solved by Calderón and Zygmund [CZ1], [CZ2]. Let us now formulate the following theorem.

Theorem 4.3.1 *The truncated conjugate function is defined as above. We also define the conjugate maximal operator φ by*

$$\varphi(f)(x) = \sup_{\varepsilon > 0} |f_\varepsilon^*(x)|,$$

for $f \in L(Q^n)$. Then, we have

$$\|\varphi(f)\|_p \leq c \frac{p^2}{p-1} \|f\|_p,$$

for $1 < p < \infty$, and φ is weak-type $(1,1)$ and satisfies

$$\|\varphi(f)\|_1 \leq c \left(\int_Q |f| \log^+ |f| dx + 1 \right).$$

According to (4.3.14), for convenience, we sometimes replace f_ε^* by

$$\overline{f}_\varepsilon(x) = \frac{1}{(2\pi)^n} \int_\varepsilon^\infty \frac{1}{t} \int_{\mathbb{S}^{n-1}} f(x - t\xi) P(\xi) d\sigma(\xi) dt. \quad (4.3.15)$$

In the following, we give another basic conclusion.

Theorem 4.3.2 *For $f^* \in L(Q)$, the series (4.3.2) is the Fourier series of f^*, that is,*

$$C_m(f^*) = \widehat{K}(m) C_m(f),$$

for $m \in \mathbb{Z}^n$.

This theorem can be proven by the method for the case of one variable (see [Zy1]) and we omit the proof here.

We will consider the convergence and approximation of the Bochner-Riesz means of the series (4.3.2)

$$\widetilde{S}_R^\alpha(f;x) = \sum_{|m|<R} \left(1 - \frac{|m|^2}{R^2}\right)^\alpha \widehat{K}(m) C_m(f) e^{im \cdot x}, \qquad (4.3.16)$$

where α is a complex number with real part being bigger than -1.

4.4 Kernel of Bochner-Riesz means of conjugate Fourier series

By (4.3.16), we define $\widetilde{S}_R^\alpha(f;x)$ as

$$\widetilde{S}_R^\alpha(f;x) = \frac{1}{|Q|} \int_Q f(x-y) \widetilde{D}_R^\alpha(y) dy, \qquad (4.4.1)$$

where the kernel $\widetilde{D}_R^\alpha(\cdot)$ has the following expression

$$\widetilde{D}_R^\alpha(y) = \sum_{|m|<R} \widehat{K}(m) \left(1 - \frac{|m|^2}{R^2}\right)^\alpha e^{im \cdot x}. \qquad (4.4.2)$$

In Section 4.1, we calculate that (see Lemma 4.1.1 and 4.1.2)

$$\int_{|y|<1} P\left(\frac{y}{|y|}\right) (1-|y|^2)^{\alpha_0 + \beta} e^{ix \cdot y} dy = i^{-k} P(x) \cdot E(n, \beta, k, |x|), \qquad (4.4.3)$$

where $n \geq 2$, $k \geq 1$, $\alpha_0 = \frac{n-1}{2}$, Re $\beta > -\frac{n+1}{2}$ and

$$E(n, \beta, k, u) = \frac{C(n,k)}{u^{n+k}} + \frac{G(n, \beta, k, u)}{u^{n+k+2}} + \frac{H(n, \beta, k, u)}{u^{n+k+2+\beta}}$$

$$+ \frac{B(n, \beta)}{u^{n+k+\beta}} \left(\sqrt{u} J_{n+k+\beta-\frac{1}{2}}(u) + k \frac{\frac{n+1}{2} + \beta}{\sqrt{u}} J_{n+k+\beta+\frac{1}{2}}(u)\right). \qquad (4.4.4)$$

Let us now turn to verifying that for Re $\beta > 0$, the function

$$f(y) = P\left(\frac{y}{|y|}\right) \left(1 - \frac{|y|^2}{R^2}\right)^{\alpha_0 + \beta} \chi_{(0,R)}(|y|)$$

satisfies the condition of Theorem 1.2.3 in Chapter 1 for $y \neq 0$.

4.4 Kernel of Bochner-Riesz means of conjugate Fourier series

Firstly, since Re $\beta > 0$, f is continuous except at the origin and

$$(i) \quad \sum_{m \neq 0} |f(m)| = \sum_{0 < |m| < R} |f(m)| < +\infty.$$

By (4.4.3), we have that

$$g(y) := \int_{\mathbb{R}^n} f(x) e^{ix \cdot y} dx = i^k R^{n+k} P(y) E(n, \beta, k, R|y|). \tag{4.4.5}$$

According to the harmonicity of P, we get that

$$\int_{|y|<\rho} g(y) dy = \int_0^\rho i^k R^{n+k} E(n, \beta, k, Rt) \int_{|\xi|=t} P(\xi) d\sigma(\xi) dt = 0, \tag{4.4.6}$$

$\rho > 0$, which yields

$$\frac{1}{(2\pi)^n} \lim_{\rho \to \infty} \int_{|y|<\rho} g(y) dy = 0. \tag{4.4.7}$$

So due to the asymptotic formula (4.4.4), we obtain

(ii) $|g(y)| = O(|y|^{-n})$, as $|y| \to \infty$.

Finally, define

$$A_m = \frac{1}{|Q|} \int_Q g(x + 2\pi m) dx,$$

for $m \in \mathbb{Z}^n$.

For $N \in \mathbb{N}$ with $N > 10n$, let $\rho = 2\pi N$. Combining (4.4.6) with (ii), we have that

$$\left| \sum_{j=0}^N \sum_{m \in S_j} A_m \right| = \left| \sum_{j=0}^N \sum_{m \in S_j} A_m - \frac{1}{|Q|} \int_{|x|<\rho} g(x) dx \right|$$

$$\leq \frac{1}{|Q|} \int_{\rho - 2\pi n < |x| < \rho + 2\pi n} |g(x)| dx$$

$$= O\left(\frac{1}{N^n}\right) N^{n-1}$$

$$= O\left(\frac{1}{N}\right),$$

which implies

$$\sum_{j=0}^\infty \sum_{m \in S_j} A_m = 0. \tag{4.4.8}$$

Since we have

$$\sum_{j=0}^{N}\sum_{m\in S_j} g(x+2\pi m) = \sum_{j=0}^{N}\sum_{m\in S_j}\left(g(x+2\pi m)-A_m\right) + \sum_{j=0}^{N}\sum_{m\in S_j} A_m,$$

to prove the locally uniform convergence of the left side of the above equation about x as $N\to\infty$, it suffices to show that

$$\sum_{j=0}^{N}\sum_{m\in S_j}\left(g(x+2\pi m)-A_m\right) \qquad (4.4.9)$$

is locally and uniformly convergent.

Let $|x|\le M$. When $|m|\ge 2(M+1)$, we have

$$|g(x+2\pi m)-A_m| \le \frac{1}{|Q|}\int_{Q+2\pi m}|g(x)-g(y)|dy.$$

By (4.4.4), we get that

$$g(y) = i^k C(n,k)\frac{P(y)}{|y|^{n+k}} + O\left(|y|^{-n-\delta}\right),$$

for $|y|>1$, where $\delta = \min(1,\operatorname{Re}\beta)>0$. Thus, for $x,y\in Q+2\pi m$, we have

$$|g(x)-g(y)| = O\left(\frac{1}{|y|^{n+\delta}}\right) = O\left(\frac{1}{|m|^{n+\delta}}\right).$$

Therefore, (4.4.9) is uniformly convergent for $|x|\le M$.

Since the conditions for the Poisson summation formula are all satisfied, we have, for $\operatorname{Re}\beta>0$, that

$$\sum_{j=0}^{\infty}\sum_{m\in S_j} i^k R^{n+k} P(x+2\pi m) E(n,\beta,k,R|x+2\pi m|)$$

$$= \sum_{|m|<R} P\left(\frac{m}{|m|}\right)\left(1-\frac{|m|^2}{R^2}\right)^{\alpha_0+\beta} e^{im\cdot x}. \qquad (4.4.10)$$

Multiplying both sides of the above equation by the following constant

$$(-i)^k \frac{\Gamma(\frac{k}{2})}{\pi^{\frac{n}{2}} 2^n \Gamma\left(\frac{n+k}{2}\right)} = (-i)^k \frac{1}{C(n,k)},$$

we can obtain the following result.

Theorem 4.4.1 *For* Re $\beta > 0$, *we have*

$$\widetilde{D}_R^{\alpha_0+\beta}(y) = \frac{\Gamma(\frac{k}{2})R^{n+k}}{\pi^{\frac{n}{2}}2^n\Gamma\left(\frac{n+k}{2}\right)} \sum_{j=0}^{\infty} \sum_{m \in S_j} P(y+2\pi m)E(n,\beta,k,R|y+2\pi m|). \quad (4.4.11)$$

Here, the dimension n is bigger than 1. The above expression was established by Chang [Cha1].

By the uniform convergence of the series on the right side of (4.4.11), we directly obtain the following result from Theorem 4.4.1.

Proposition 4.4.1 *For* Re $\beta > 0$, *we have*

$$\widetilde{S}_R^{\alpha_0+\beta}(f;x) = \frac{1}{|Q|} \frac{1}{C(n,k)} \int_0^{\to\infty} \psi_x\left(\frac{t}{R}\right) t^{n+k-1} E(n,\beta,k,t) dt, \quad (4.4.12)$$

where we use the notation $\int_0^{\to\infty} = \lim_{\rho\to\infty} \int_0^{\rho}$ *and*

$$\psi_x(s) = \int_{|\xi|=1} f(x-s\xi)P(\xi)d\sigma(\xi).$$

Remark 4.4.1 *When* Re $\beta \leq 0$, *the above equation does not hold since we cannot prove the locally uniform convergence of the series (4.4.9). From this, we can see that the index* $\alpha_0 = \frac{n-1}{2}$ *also has the critical properties for the conjugate Bochner-Riesz means. Later on, a series of facts will be supplemented to illustrate it.*

4.5 The maximal operator of the conjugate partial sum

Our purpose is to give an estimate of the type of the maximal operator at the critical index

$$\widetilde{S}_*^{\frac{n-1}{2}}(f)(x) := \sup_{R>0}\left|\widetilde{S}_R^{\frac{n-1}{2}}(f;x)\right|. \quad (4.5.1)$$

Lemma 4.5.1 *Let* $\beta = \sigma + i\tau$, $\sigma > 0$ *and* $\tau \in \mathbb{R}$. *If we define*

$$M^{\beta}(f)(x) = \sup_{R>0}\left|\widetilde{S}_R^{\frac{n-1}{2}+\beta}(f;x) - \overline{f}_{\frac{1}{R}}(x)\right|,$$

then
$$M^\beta(f)(x) \le A_n(P)e^{2\pi|\tau|}\frac{2^\sigma(1+\sigma)\Gamma\left(\frac{n+1}{2}+\sigma\right)}{\sigma}M(f)(x) \quad (4.5.2)$$
holds.

Here, we replace the truncated conjugate function introduced at the beginning of Section 4.3 by $\overline{f}_\varepsilon$ defined as in (4.3.15). We denote by M the Hardy-Littlewood maximal operator and by $A_n(P)$ a constant only related to n and the function P.

Proof. By (4.3.15), we have that
$$\overline{f}_{\frac{1}{R}}(x) = \frac{1}{|Q|}\int_1^\infty \frac{1}{t}\psi_x\left(\frac{t}{R}\right)dt. \quad (4.5.3)$$

According to (4.4.12), we get that
$$\widetilde{S}_R^{\alpha_0+\beta}(f;x) = \frac{1}{|Q|}\frac{1}{C(n,k)}\int_0^1 \psi_x\left(\frac{t}{R}\right)t^{n+k-1}E(n,\beta,k,t)dt$$
$$+ \frac{1}{|Q|}\frac{1}{C(n,k)}\int_1^\infty \psi_x\left(\frac{t}{R}\right)t^{n+k-1}E(n,\beta,k,t)dt.$$

We denote the first term on the right side of the above equation by $A_R(x)$. Obviously, it follows that
$$|A_R(x)| \le A_n(P)M(f)(x). \quad (4.5.4)$$

Thus, we have that
$$\widetilde{S}_R^{\alpha_0+\beta}(f;x) - \overline{f}_{\frac{1}{R}}(x) = A_R(x) + \frac{1}{|Q|}\int_1^\infty \psi_x\left(\frac{t}{R}\right)t^{-1}$$
$$\left\{\frac{t^{n+k}}{C(n,k)}E(n,\beta,k,t) - 1\right\}dt. \quad (4.5.5)$$

Due to (4.4.4), we obtain that
$$\left|\frac{t^{n+k}}{C(n,k)}E(n,\beta,k,t) - 1\right| \le A_{n,k}e^{2\pi|\tau|}\left(\frac{1}{t^2}+\frac{1}{t^\sigma}\right). \quad (4.5.6)$$

4.5 The maximal operator of the conjugate partial sum

It is easy to calculate that

$$\int_1^\infty \left|\psi_x\left(\frac{t}{R}\right)\right| \frac{dt}{t^{1+\sigma}} = \sum_{j=0}^\infty \int_{2^j}^{2^{j+1}} \left|\psi_x\left(\frac{t}{R}\right)\right| \frac{dt}{t^{1+\sigma}}$$

$$\leq A_n(P) \sum_{j=0}^\infty \frac{1}{2^{\sigma j}} Mf(x) \qquad (4.5.7)$$

$$\leq A_n(P) \frac{1+\sigma}{\sigma} Mf(x).$$

It follows from (4.5.4), (4.5.5), (4.5.6) and (4.5.7) that (4.5.2) naturally holds.

This completes the proof of Lemma 4.5.1. ∎

Lemma 4.5.2 *Let $\beta = \sigma + i\tau$, σ and $\tau \in \mathbb{R}$. If $\frac{n+1}{2} > \sigma > -\frac{n-1}{2}$, then we have*

$$\left\|\widetilde{S}_*^{\frac{n-1}{2}+\beta}(f)\right\|_2 \leq A_n(P) \frac{e^{2\pi|\tau|}}{\sigma + \frac{n-1}{2}} \|f\|_2. \qquad (4.5.8)$$

Proof. According to Theorem 4.3.1, when $f \in L^2(Q^n)$, we have $f^* \in L^2(Q^n)$ and

$$\|f^*\|_2 \leq A_n(P)\|f\|_2.$$

By Theorem 4.3.2, if $f^* \in L^2(Q^n)$, we have

$$\widetilde{S}_R^{\frac{n-1}{2}+\beta}(f) = S_R^{\frac{n-1}{2}+\beta}(f^*).$$

Thus we have that

$$\left\|\widetilde{S}_*^{\frac{n-1}{2}+\beta}(f)\right\|_2 = \left\|S_*^{\frac{n-1}{2}+\beta}(f^*)\right\|_2$$

$$\leq A_n e^{2\pi|\tau|} \frac{1}{\sigma + \frac{n-1}{2}} \|f^*\|_2$$

$$\leq A_n(P) e^{2\pi|\tau|} \frac{1}{\sigma + \frac{n-1}{2}} \|f\|_2.$$

This completes the proof of Lemma 4.5.2. ∎

Since $A_n(P)$ does not play any role in our discussion, we can rewrite it as A for short.

C4. THE CONJUGATE FOURIER INTEGRAL AND SERIES

According to Lemma 4.5.1 and 4.5.2, and by Stein's interpolation theorem, we can imitate the ways of dealing with the Bochner-Riesz operator in Section 3.5.

Theorem 4.5.1 *If $1 < p < \infty$, then we have*

$$\left\| \widetilde{S}_*^{\frac{n-1}{2}}(f) \right\|_p \leq A \frac{p^3}{(p-1)^2} \|f\|_p. \tag{4.5.9}$$

In the following, we will start from the estimate of the type of the maximal operator on the partial sum and then deduce another applicable type of operators.

Definition 4.5.1 *Suppose that P is a homogeneous harmonic polynomial of degree k, where k is a nonnegative integer. For $f \in L(Q)$, let*

$$P(f)(x;t) = \int_{\mathbb{S}^{n-1}} f(x - t\xi) P(\xi) d\sigma(\xi), \tag{4.5.10}$$

for $t \geq 0$. We define an operator TP by

$$TP(f)(x) = \sup_{R>1} \left\{ \limsup_{\beta \to 0^+} \left| \int_1^\infty P(f)\left(x; \frac{t}{R}\right) \frac{\cos\left(t - \frac{\pi}{2}(n+k+\beta)\right)}{t^{1+\beta}} dt \right| \right\}. \tag{4.5.11}$$

Lemma 4.5.3 *Let $1 < p < \infty$. We have*

$$\|TP(f)\|_p \leq A \frac{p^3}{(p-1)^2} \|f\|_p. \tag{4.5.12}$$

Proof. Firstly, we consider the case $k > 0$. By choosing $\beta \in (0,1)$ and (4.5.5), we have that

$$\widetilde{S}_R^{\alpha_0+\beta}(f;x) = \overline{f}_{\frac{1}{R}}(x) + A_R(x) + \frac{1}{|Q|} \int_1^\infty \psi_x\left(\frac{t}{R}\right) t^{-1}$$

$$\times \left(\frac{t^{n+k}}{c(n,k)} E(n,\beta,k,t) - 1\right) dt.$$

It follows from (4.4.4) that

$$\frac{t^{n+k}}{C(n,k)} E(n,\beta,k,t) - 1 = O\left(\frac{1}{t^2}\right) + \frac{B(n,\beta)}{C(n,k)} \sqrt{\frac{2}{\pi}}$$

$$\times \frac{\cos\left(t - \frac{\pi}{2}(n+k+\beta)\right)}{t^\beta} + O\left(\frac{1}{t^{1+\beta}}\right).$$

4.5 The maximal operator of the conjugate partial sum

Therefore, we get that

$$\frac{1}{|Q|}\frac{B(n,\beta)}{C(n,k)}\sqrt{\frac{2}{\pi}}\int_1^\infty \psi_x\left(\frac{t}{R}\right)\frac{\cos\left(t-\frac{\pi}{2}(n+k+\beta)\right)}{t^{1+\beta}}dt$$
$$= \widetilde{S}_R^{\alpha_0+\beta}(f;x) - \overline{f}_{\frac{1}{R}}(x) - A_R(x) + \int_1^\infty \psi_x\left(\frac{t}{R}\right) O\left(\frac{1}{t^2}\right) dt, \quad (4.5.13)$$

where

$$B(n,\beta) = \pi^{\frac{n}{2}} 2^{n-\frac{1}{2}+\beta} \Gamma\left(\frac{n+1}{2}+\beta\right).$$

Combining (4.5.4) with (4.5.7), we obtain that

$$|A_R(x)| + \left|\int_1^\infty \psi_x\left(\frac{t}{R}\right)\frac{1}{t^2}dt\right| \le AMf(x). \quad (4.5.14)$$

Set

$$N(f)(x) = \sup_{R>0}\left|\overline{f}_{\frac{1}{R}}(x)\right|.$$

Then, it follows from (4.5.13) that

$$\lim_{\beta\to 0^+}\left|\int_1^\infty P(f)\left(x;\frac{t}{R}\right)\frac{\cos\left(t-\frac{\pi}{2}(n+k+\beta)\right)}{t^{1+\beta}}dt\right|$$
$$\le A\left\{\widetilde{S}_*^{\alpha_0}(f)(x) + N(f)(x) + M(f)(x)\right\}.$$

Obviously taking sup for R on the left side of the above inequality equals to $TP(f)(x)$. By making use of the estimate about the type of $\widetilde{S}_*^{\alpha_0}$ in Theorem 4.5.1 and that of the operator φ in Theorem 4.3.1, we have the equivalence of φ and N (cf. (4.3.13)). In addition, we use the estimate about the type of the operator M to get (4.5.12).

For the case $k = 0$, we replace $\widetilde{S}_R^{\frac{n-1}{2}+\beta}$ with $S_R^{\frac{n-1}{2}+\beta}$ in the above argument and utilize the estimate about the type of $S_*^{\alpha_0}$ (in the same form as Theorem 4.5.1) to carry out the similar discussions. We note that for the case when $k = 0$, $P(\cdot)$ is a constant and then we have $P(f)(x;t) = Cf_x(t)$. This is the end of the proof of Lemma 4.5.3. ∎

Lemma 4.5.4 *If the operator T defined on $L(Q)$ satisfies*

$$\|T(f)\|_p \le A_0 p\|f\|_p,$$

for $2 \leq p < \infty$, then there exists a constant A only depending on A_0 such that
$$|\{x \in Q : T(f)(x) > \lambda\}| \leq A e^{-\frac{\lambda}{A\|f\|_\infty}}, \qquad (4.5.15)$$
for any $\lambda > 0$.

Proof. Let $f \in L^\infty(Q)$ and set
$$E_\lambda = \{x \in Q : T(f)(x) > \lambda\}.$$
Then, we have
$$|E_\lambda| \leq \int_{E_\lambda} \left|\frac{T(f)(x)}{\lambda}\right|^p dx \leq \left(\frac{A_0 p \|f\|_p}{\lambda}\right)^p,$$
for $p \geq 2$.

If $\lambda > 2A_0 \|f\|_\infty e$, by choosing $p = \frac{\lambda}{A_0 \|f\|_\infty e}$, then we immediately have
$$|E_\lambda| \leq e^{-\frac{\lambda}{A_0 \|f\|_\infty}} e.$$
Hence, we merely need to take $A = |Q|e^2 + A_0 e$ and then (4.5.15) holds.
This completes the proof of Lemma 4.5.4. ∎

4.6 The relations between the conjugate series and integral

For the non-conjugate case, by the Stein's theorem (see Section 3.2) about the relation between the series and the integral, we can transfer the problem of series into that of integral in many cases. This facilitates out research of the series. For the conjugate case, we have mentioned the conjugate integral in the previous two sections of this chapter. It is easy to see that the research of the conjugate integral is more convenient than that of the conjugate series. As far as the case at the critical index is concerned, the Bochner-Riesz means of the series has no integral expression. While for the integration, the formula (4.1.24) is valid in the large domain of $\mathrm{Re}\,\beta > -\frac{n+1}{2}$. Therefore, it is necessary to extend Stein's theorem to the conjugate case. Chang [Cha1] just did such a work in this field.

Let $P(x) \in \mathscr{A}_k^{(n)}$, $k \geq 1$ and $K(x) = P(x)|x|^{-n-k}$. In Section 4.4, we have discussed the kernel $\widetilde{D}_R^\alpha(y)$ (see (4.4.2)). Now, we denote by
$$D_R^S(y) = \sum_{0 < |m| < R} \frac{P(m)}{|m|^k} \left(1 - \frac{|m|^2}{R^2}\right)^{\alpha_0 + s} e^{im\cdot x}, \qquad (4.6.1)$$

4.6 The relations between the conjugate series and integral

where $\alpha_0 = \frac{n-1}{2}$ and s is a complex with $\mathrm{Re}\, s > -\frac{n+1}{2}$. We also denote the kernel of the conjugate integral by

$$H_R^s(y) = \int_{|x|<k} \frac{P(x)}{|x|^k} \left(1 - \frac{|x|^2}{R^2}\right)^{\alpha_0+s} e^{ix\cdot y} dx. \qquad (4.6.2)$$

We easily have

$$\widetilde{D}^{\alpha_0+s}(y) = (-i)^k \frac{\Gamma(\frac{k}{2})}{\pi^{\frac{n}{2}} 2^n \cdot \Gamma\left(\frac{n+k}{2}\right)} D_R^s(y)$$

$$= \sum_{|m|<R} \widehat{k}(m) \left(1 - \frac{|m|^2}{R^2}\right)^{\alpha_0+s} e^{im\cdot y},$$

and

$$\widetilde{H}_R^{\alpha_0+s}(y) := \int_{|x|<R} \widehat{K}(x) \left(1 - \frac{|x|^2}{R^2}\right)^{\alpha_0+s} e^{ix\cdot y} dx$$

$$= (-i)^k \frac{\Gamma(\frac{k}{2})}{\pi^{\frac{n}{2}} 2^n \Gamma\left(\frac{n+k}{2}\right)} H_R^s(y).$$

Set

$$C'_{n,k} = (-i)^k \frac{\Gamma(k/2)}{\pi^{\frac{n}{2}} 2^n \Gamma\left(\frac{n+k}{2}\right)} = (-i)^k \frac{1}{C(n,k)},$$

where $C(n,k)$ was introduced in Section 4.1 (see Lemma 4.1.2).

For $f \in L(Q^n)$, we have

$$\widetilde{S}_R^{\alpha_0+s}(f;x) = \frac{1}{|Q|} \int_Q f(x-y) \widetilde{D}_R^{\alpha_0+s}(y) dy,$$

while for $g \in L(\mathbb{R}^n)$, we also have

$$\widetilde{\sigma}_R^{\alpha_0+s}(g;x) = \frac{1}{|Q|} \int_{\mathbb{R}^n} g(x-y) \widetilde{H}_R^{\alpha_0+s}(y) dy.$$

Set

$$\triangle_R^s(x) = D_R^s(x) - H_R^s(x). \qquad (4.6.3)$$

Similar to the non-conjugate case, we aim at proving the following theorem.

Theorem 4.6.1 *There exists a constant A such that*

$$\sup \|\triangle_R^0\|_p \leq Ap, \qquad (4.6.4)$$

for $1 \leq p < +\infty$.

From (4.6.4), we can deduce that there exists a constant $\alpha > 0$ such that

$$\sup_{R>0} \int_Q e^{\alpha|\Delta_R(x)|} dx \leq A. \qquad (4.6.5)$$

Furthermore, the inequality (4.6.5) yields the next desired result.

Theorem 4.6.2 *Suppose that $f \in L\log^+ L(Q^n)$ with $n > 1$, $g = f\chi_Q$ and G is a closed subset of Q. Then we have*

$$\lim_{R \to \infty} \sup_{x \in G} \left| \widetilde{S}_R^{\alpha_0}(f; x) - \widetilde{\sigma}_R^{\alpha_0}(g; x) - C(x) \right| = 0, \qquad (4.6.6)$$

where

$$C(x) = \frac{1}{(2\pi)^n} \int_{Q^n} f(y) \left[K^*(x-y) - K(x-y) \right] dy. \qquad (4.6.7)$$

In the following, we will briefly formulate the steps of the proof, without giving any detailed proof like the non-conjugate case.

Step I. Let $\sigma = \operatorname{Re} s \in (0, 1]$. By (4.4.11), we have

$$D_R^s(x) = i^k \sum_{j=0}^{\infty} \sum_{m \in S_j} P(x + 2\pi m) E(n, s, k, R|x + 2\pi m|) R^{n+k}. \qquad (4.6.8)$$

Combining (4.6.2) with (4.1.5), we have that

$$H_R^s(x) = i^k P(x) E(n, s, k, R|x|) R^{n+k}. \qquad (4.6.9)$$

Subtracting (4.6.9) from (4.6.8), we get that

$$\triangle_R^s(x) = i^k \sum_{j=1}^{\infty} \sum_{m \in S_j} P(x + 2\pi m) E(n, s, k, R|x + 2\pi m|) R^{n+k}. \qquad (4.6.10)$$

Let

$$F(x) = i^k C(n, k) \sum_{j=1}^{\infty} \sum_{m \in S_j} \frac{P(x + 2\pi m)}{|x + 2\pi m|^{n+k}}, \qquad (4.6.11)$$

for $x \in Q$.

In (4.4.8) and (4.4.9), we replace $g(x)$ by $P(x)|x|^{-n-k} = K(x)$ and the summation by $\sum_{j=1}^{\infty} \sum_{m \in S_j}$. Similar arguments show that the series in (4.6.11) is locally and uniformly convergent and thus $F \in C(Q)$.

4.6 The relations between the conjugate series and integral

Lemma 4.6.1 *Let $s = \sigma + i\tau$, $0 < \sigma \le 1$ and $\tau \in \mathbb{R}$. Then we have*

$$\sup_{x \in Q} |\triangle_R^s(x) - F(x)| \le \frac{Ae^{2\pi|\tau|}}{\sigma R^\sigma}. \tag{4.6.12}$$

Proof. By (4.4.4) we have that

$$|\triangle_R^s(x) - F(x)| \le Ae^{2\pi|\tau|} \sum_{|m| \ne 0} \frac{1}{R^\sigma |m|^{n+\sigma}},$$

for $x \in Q$.

This completes the proof of Lemma 4.6.1. ∎

Step II. Let $-\frac{1}{2} \le \operatorname{Re} s = \sigma \le 0$. We choose $\psi \in C^1([0,1])$ satisfying

$$\int_0^1 \psi(u) u^j \, du = 0,$$

for $j = 0, 1, \ldots, n-1$.

We define some modified quantities as follows

$$\widetilde{\triangle}_R^s(x) = \int_0^1 \left(1 + \frac{u}{R}\right)^{n-1+2s} \psi(u) \triangle_{R+u}^s(x) \, du,$$

$$\widetilde{D}_R^s(x) = \int_0^1 \left(1 + \frac{u}{R}\right)^{n-1+2s} \psi(u) H_{R+u}^s(x) \, du,$$

and

$$\widetilde{H}_R^s(x) = \int_0^1 \left(1 + \frac{u}{R}\right)^{n-1+2s} \psi(u) H_{R+u}^s(x) \, du.$$

Lemma 4.6.2 *Let $S = \sigma + i\tau$, $-\frac{1}{2} \le \sigma \le 0$ and $\tau \in \mathbb{R}$. Then we have*

$$\sup_{x \in Q} \left|\widetilde{\triangle}_R^s(x)\right| \le Ae^{2\pi|\tau|} R^{-\sigma}. \tag{4.6.13}$$

Proof. By making use of the uniform convergence of (4.6.10) for $\sigma > 0$, we have

$$\widetilde{\triangle}_R^s(x) = \int_0^1 \triangle_{R+u}^s(x) \left(1 + \frac{u}{R}\right)^{n-1+2s} \psi(u) \, du$$

$$= i^k \sum_{j=1}^\infty \sum_{m \in S_j} P(x + 2\pi m) \int_0^1 E(n, s, k, (R+u)|x + 2\pi m|)$$

$$\times (R+u)^{n+k} \psi(u) \, du. \tag{4.6.14}$$

Then we will verify the fact that the series on the right side of the above equation is internally closed uniformly convergent about s when $\sigma \in (-1, 1)$. And thus the series are analytic with respect to s. Since the analyticity of $\widetilde{\triangle}_R^s(x)$ about s is evident, the equation (4.6.13) ia valid when $\sigma > -1$.

The key point lies in the fact that when the series which execute summation about m with $m \neq 0$ is added in some terms and makes the integral about some parameter, $u \in [0, 1]$, for one time, then its order characterizing the increase about $|m|$ will decrease by 1. This is equivalent to multiplication by the factor $|m|^{-1}$. So does the case of the series $\sum_{m \neq 0} \frac{\cos u|m|}{|m|^n}$.

Thus we immediately obtain (4.6.13) from (4.6.14), because it has a fine property of convergence on the right side.

This completes the proof of Lemma 4.6.2. ∎

Lemma 4.6.3 *Let $s = \sigma + i\tau$, $-\frac{1}{2} \leq \sigma \leq 0$ and $\tau \in \mathbb{R}$. Then we have*

$$\sup_{x \in Q} \left| \widetilde{H}_R^s(x) - H_R^s(x) \right| \leq AR^{-\sigma} e^{2\pi|\tau|}. \tag{4.6.15}$$

Proof. By the definition, we have that

$$\widetilde{H}_R^s(x) - H_R^s(x) = R^{-(n-1+2s)} \int_0^1 \left\{ (R+u)^{n-1+2s} H_{R+u}^s(x) - R^{n-1+2s} H_R^s(x) \right\}$$
$$\times \psi(u) du.$$

By the Taylor formula, we have that

$$(R+u)^{n-1+2s} H_{R+u}^s(x) - R^{n-1+2s} H_R^s(x) = \sum_{j=1}^{n-1} a_j u^j + E(n),$$

where the remainder $E(n)$ satisfies

$$|E(n)| \leq \frac{1}{n!} \sup_{0 \leq u \leq 1} \left| \frac{d^n}{du^n} \{ (R+u)^{n-1+2s} H_{R+u}^s(x) \} \right|. \tag{4.6.16}$$

It follows from the orthogonal condition of ψ that

$$\widetilde{H}_R^s(x) - H_R^s(x) = R^{-(n-1+2s)} \int_0^1 E(n) \psi(u) du,$$

which implies

$$\left| \widetilde{H}_R^s(x) - H_R^s(x) \right| \leq \frac{A}{R^{n-1+2\sigma}} \int_0^1 |E(n)| du. \tag{4.6.17}$$

4.6 The relations between the conjugate series and integral

Set
$$f(u) = (R+u)^{n-1+2s} H^s_{R+u}(x) = i^k P(x)(R+u)^{2n+k-1+2s} E(n,s,k,(R+u)|x|).$$

Applying the Leibniz formula, we have
$$f^{(n)}(u) = i^k P(x) \sum_{j=0}^{n} A_j (R+u)^{2n+k-1+2s-j} |x|^{n-j} E^{(n-j)}(n,s,k,(R+u)|x|),$$

where
$$A_j = C_n^j (2n+k-1+2s) \cdots (2n+k-1+2s-j+1).$$

Thus we obtain
$$\left| f^{(n)}(u) \right| \le \sum_{j=0}^{n} A e^{\pi|\tau|/2} (R+u)^{2n+k-1+2\sigma-j} |x|^{n+k-j}$$
$$\times \left| E^{(n-j)}(n,s,k,(R+u)|x|) \right|,$$

for $|\sigma| < 1$.

Since
$$E(n,\beta,k,t) = \frac{\pi^{n/2}}{2^k} \sum_{j=0}^{\infty} \frac{(-1)^j \, \Gamma\left(\frac{n+k}{2}+j\right) \Gamma\left(\frac{n+1}{2}+\beta\right)}{j! \, \Gamma\left(\frac{n}{2}+k+j\right) \Gamma\left(\frac{n+k}{2}+\frac{n+1}{2}+\beta+j\right)} \left(\frac{t}{2}\right)^{2j},$$

E is analytic about t. If we take a derivative with respect to t, then we have
$$E'(n,\beta,k,t) = -\frac{1}{2\pi} t E(n+2, \beta-1, k, t).$$

By induction, we get that, for $\nu = 1, 2, \ldots$,
$$E^{(2\nu-1)}(n,\beta,k,t) = \sum_{j=0}^{\nu-1} a_{\nu,j} \, t^{2j+1} E(n+2(\nu+j), \beta-(\nu+j), k, t),$$

and
$$E^{(2\nu)}(n,\beta,k,t) = \sum_{j=0}^{\nu} b_{\nu,j} t^{2j} \, E(n+2(\nu+j), \beta-(\nu+j), k, t), \quad (4.6.18)$$

where $|a_{\nu,j}| + |b_{\nu,j}| \le A_\nu$.

For $\sigma \in (-1, 0]$, it follows from (4.6.18) and the asymptotic formula (4.4.4) that
$$\left|E^{(\nu)}(n, s, k, t)\right| \leq A_\nu \frac{1}{t^{n+k+\sigma}},$$
for $t > 0$ and $\nu = 0, 1, 2, \ldots$.

Since $x \in Q$, we have
$$\left|f^{(n)}(u)\right| \leq A e^{2\pi|\tau|} R^{n-1+\sigma},$$
for $u \in [0, 1]$.

By (4.6.16), we have
$$|E(n)| \leq A e^{2\pi|\tau|} R^{n-1+\sigma}.$$

By substituting this into (4.6.17), we get (4.6.15).
This finishes the proof of Lemma 4.6.3. ■

Lemma 4.6.4 Let $s = \sigma + i\tau$, $-\frac{1}{2} \leq \sigma \leq 0$ and $\tau \in \mathbb{R}$. Then we have
$$\left\|\widetilde{D}_R^s - D_R^s\right\|_2 \leq A e^{2\pi|\tau|} R^{-\sigma}, \tag{4.6.19}$$
for any $R > 0$. Here we denote by $\|\cdot\|_2$ the L^2 norm on Q^n.

Proof. The proof of this lemma can be ascribed to the fact in the non-conjugate case, which has already been obtained.

Let
$$\lambda_m = \left(1 - \frac{|m|^2}{R^2}\right)^{\alpha_0+s} \chi_{[0,R)}(|m|)$$
and
$$\widetilde{\lambda}_m = \int_0^1 \left[1 - \frac{|m|^2}{(R+u)^2}\right]^{\alpha_0+s} \chi_{[0,R+u)}(|m|) \psi(u) \left(1 + \frac{u}{R}\right)^{n-1+2s} du.$$

For the non-conjugate case, we have
$$\left(\sum_m \left|\widetilde{\lambda}_m - \lambda_m\right|^2\right)^{\frac{1}{2}} \leq A e^{2\pi|\tau|} R^{-\sigma}.$$

The Fourier coefficients of $D_R^s(x)$ and $\widetilde{D}_R^s$ are $P\left(\frac{m}{|m|}\right) \lambda_m$ and $P\left(\frac{m}{|m|}\right) \widetilde{\lambda}_m$, $m \neq 0$, respectively, and their coefficients of order zero are both zero. Thus

4.6 The relations between the conjugate series and integral

we immediately get (4.6.19) by the boundedness of $P\left(\frac{m}{|m|}\right)$ with $m \neq 0$. ∎

Combining Lemma 4.6.2, 4.6.3 and 4.6.4, we obtain the L^2 estimate
$$\|\triangle_R^s(x) - F(x)\|_2 \leq A e^{2\pi|\tau|} R^{-\sigma}, \tag{4.6.20}$$
for $-\frac{1}{2} \leq \sigma \leq 0$.

Step III. Similar to the non-conjugate case, by the L^∞ estimate (4.6.12), L^2 estimate (4.6.20) and the method of the complex interpolation, we have
$$\sup_{R>0} \|\triangle_R^0(x) - F(x)\|_p \leq Ap,$$
for $2 \leq p < \infty$.

From the above discussion, we have proven Theorem 4.6.1.

Finally, we deduce the Lebesgue constant of the conjugate kernel $\widetilde{D}_R^{\frac{n-1}{2}}$. That is,
$$\widetilde{L}_R := \widetilde{L}_R^{\frac{n-1}{2}} := \int_Q \left|\widetilde{D}_R^{\frac{n-1}{2}}(y)\right| dy. \tag{4.6.21}$$

Theorem 4.6.3 *There exists a constant α', only depending on the dimension n and the spherical harmonic function P, such that*
$$L_R = \alpha' \log R + O(1), \tag{4.6.22}$$
as $R \to \infty$.

Proof. Since
$$\widetilde{D}_R^{\alpha_0}(y) = \widetilde{H}_R^{\alpha_0}(y) + (-i)^k \frac{1}{C(n,k)} \triangle_R^0(y),$$
we have
$$L_R = \int_Q \left|\widetilde{H}_R^{\alpha_0}(y)\right| dy + r,$$
where
$$|r| \leq A \int_Q |\triangle_R^0(y)| \, dy = O(1).$$

However, it follows that
$$\int_Q |\widetilde{H}_R^{\alpha_0}(y)| dy = A \int_0^1 |E(n,o,k,Rt)| R^{n+k} t^{n+k-1} dt + O(1)$$
$$= A \int_{\frac{1}{R}}^1 \left|\frac{1}{t}\left\{\alpha + \beta \cos\left(Rt - \frac{\pi}{2}(n+k)\right)\right\}\right| dt + O(1)$$
$$= \alpha' \log R + O(1). \tag{4.6.23}$$

By Lemma 4.1.2, we have

$$\alpha = C(n,k) = \frac{2^n \pi^{\frac{n}{2}} \Gamma\left(\frac{n+k}{2}\right)}{\Gamma\left(\frac{k}{2}\right)}$$

and

$$\beta = \sqrt{\frac{2}{\pi}} B(n,0) = \frac{2^n \pi^{\frac{n}{2}} \Gamma\left(\frac{n+1}{2}\right)}{\sqrt{\pi}}.$$

By the definition of β-function, we conclude that if $z \geq y > 0$ and $x > 0$, then $B(z,x) \leq B(y,x)$. On the other hand, by the result (see [Ba1]),

$$B(x,y)B(x+y,z) = B(z,x)B(x+z,y),$$

we have

$$B(x+y,z) = \frac{B(z,x)}{B(y,x)} B(x+z,y) \leq B(x+z,y),$$

for $z \geq y > 0$.

Consequently, it follows that

$$B\left(\frac{n+1}{2}, \frac{k}{2}\right) \leq B\left(\frac{n+k}{2}, \frac{1}{2}\right).$$

That is,

$$\frac{\Gamma\left(\frac{n+k}{2}\right)}{\Gamma\left(\frac{k}{2}\right)} \geq \frac{\Gamma\left(\frac{n+1}{2}\right)}{\Gamma\left(\frac{1}{2}\right)}$$

for $k \geq 1$. This implies $\alpha \geq \beta$. Here the equal sign holds only if $k = 1$. This completes the proof of Theorem 4.6.3. ∎

4.7 Convergence of Bochner-Riesz means of conjugate Fourier series

As an application of the theorems in Section 4.6, we will give, in this section, the results about the a.e.-convergence of the conjugate series due to Lu [Lu6], which is parallel to the results in Section 3.6.

Theorem 4.7.1 *Let $1 < q \leq \infty$. If $f \in B_q \bigcap L\log^+ L(Q^n)$ with $n > 1$, $P \in \mathscr{A}_k^{(n)}$, $k \geq 1$, then by taking the conjugation about P, we have*

$$\lim_{R \to \infty} \widetilde{S}_R^{\frac{n-1}{2}}(f;x) = f^*(x) \qquad (4.7.1)$$

for a.e. $x \in Q^n$.

4.7 Convergence of Bochner-Riesz means of conjugate Fourier series

Proof. Let $g = f\chi_Q$. Since $f \in L\log^+ L(Q)$, by Theorem 4.6.2, we have that
$$\lim_{R\to\infty} \left\{ \widetilde{S}_R^{\frac{n-1}{2}}(f;x) - \widetilde{\sigma}_R^{\frac{n-1}{2}}(g;x) - C(x) \right\} = 0,$$
for any $x \in Q^\circ$, where
$$C(x) = \frac{1}{|Q^n|} \int_{Q^n} (K^*(x-y) - K(x-y)) f(y) dy.$$

Since $K^*(u) - K(u)$ is internally and closed uniformly continuous on the set $2Q^n$, the function C is continuous on $(Q^n)^\circ$. In addition, we have
$$C(x) = f^*(x) - \widetilde{g}(x),$$
for almost everywhere $x \in Q^n$. It suffices to show that
$$\widetilde{\sigma}_R^{\frac{n-1}{2}}(g;x) \longrightarrow \widetilde{g}(x)$$
for a.e. $x \in Q^n$. And therefore we merely need to prove that
$$\widetilde{\sigma}_*(g)(x) := \sup_{R>0} \left| \widetilde{\sigma}_R^{\frac{n-1}{2}}(g;x) \right|$$
satisfies the weak-type estimate
$$|\{x : \widetilde{\sigma}_*(g)(x) > \lambda\}| < \frac{1}{\lambda} A N_q(g). \tag{4.7.2}$$

By the superposition principle, it suffices to prove this for each q-block.

Suppose that b is a q-block for $1 < q < \infty$. As what we have mentioned before in Theorem 4.3.1, it has been proven in the theory of the Calderón-Zygmund singular integral that the Calderón-Zygmund transform or Hilbert transform is of strong-type (q,q) with $1 < q < \infty$. And thus $\widetilde{b} \in L^q(\mathbb{R}^n)$. Here, we introduce the following facts without proof that, for $b \in L^q$ with $1 < q < \infty$, we have
$$\sigma_R^{\alpha_0}(\widetilde{b}) = \widetilde{\sigma}_R^{\alpha_0}(b).$$

Thus we have
$$\|\widetilde{\sigma}_*(b)\|_q = \left\| \sigma_* \left(\widetilde{b} \right) \right\|_q \leq A_q \left\| \widetilde{b} \right\|_q \leq A_q \|b\|_q. \tag{4.7.3}$$

By noticing the kernel
$$\widetilde{H}_R^{\alpha_0}(y) = O\left(\frac{1}{|y|^n}\right)$$

together with (4.7.3), we can deduce the validity of the equation (4.7.2) for the q-block b, just as we have done in the non-conjugate case. And therefore, we have finished the proof of the theorem. ∎

We have to point out that, similar to the non-conjugate case, the convergence of the $\widetilde{S}_R^{\frac{n-1}{2}}(f;x)$ also does not satisfy the localization principle, which was proven by Lippman [Li1]. For the similarity between its proof and that of the non-conjugate case, we omit the detailed proof here.

4.8 $(C,1)$ means in the conjugate case

The arithmetic mean of the conjugate Bochner-Riesz means $\widetilde{S}_R^{\frac{n-1}{2}}(f;x)$ of the critical order

$$\frac{1}{R}\int_0^R \widetilde{S}_u^{\frac{n-1}{2}}(f;x)du \tag{4.8.1}$$

is called the conjugate $(C,1)$ mean. It should have some good convergence properties.

Let $\beta \in (0,1)$. We first consider $S_R^{\alpha_0+\beta}$ and then turn to (4.8.1) by taking limit $\beta \to 0^+$.

Theorem 4.8.1 Let $f \in L(Q^n)$ with $n > 1$, and $P \in \mathscr{A}_k^{(n)}$ with $k \geq 1$. Set

$$\psi_x(t) = \int_{\mathbb{S}^{n-1}} f(x-ty)P(y)d\sigma(y). \tag{4.8.2}$$

If

$$\int_0^t |\psi_x(\tau)|d\tau = o(t), \tag{4.8.3}$$

holds, as $t \to 0^+$, then we have

$$\lim_{R \to \infty}\left(\widetilde{S}_R^{\alpha_0+\beta}(f;x) - \overline{f}_{\frac{1}{R}}(x)\right) = 0, \tag{4.8.4}$$

for $\beta > 0$.

Proof. By (4.3.15) and (4.4.12) (also in (4.5.5)), we conclude that

$$\widetilde{S}_R^{\alpha_0+\beta}(f;x) - \overline{f}_{\frac{1}{R}}(x)$$

$$= \frac{1}{|Q|}\frac{1}{C(n,k)}\int_0^1 \psi_x\left(\frac{t}{R}\right)t^{n+k-1}E(n,\beta,k,t)dt \tag{4.8.5}$$

$$+ \frac{1}{|Q|}\frac{1}{C(n,k)}\int_1^\infty \psi_x\left(\frac{t}{R}\right)t^{n+k-1}\left(E(n,\beta,k,t) - \frac{C(n,k)}{t^{n+k}}\right)dt.$$

4.8 $(C,1)$ means in the conjugate case

We denote the first term on the right side of the equation (4.8.5) by $A_R(x)$. From the expression of E in (4.1.6), we have that when $\beta > 0$, $|E(n,\beta,k,t)| \leq A$ uniformly holds about β and $t \in [0,1]$. And therefore, under the condition of (4.8.3), there uniformly holds about β

$$A_R(x) = o(1), \tag{4.8.6}$$

as $R \to \infty$.

By Lemma 4.1.2, we have that when $\beta \in (0,1)$, with respect to β,

$$E(n,\beta,k,t) - \frac{C(n,k)}{t^{n+k}} = O\left(\frac{1}{t^{n+k+1}}\right) + A(n,\beta)\frac{\cos\left(t - \frac{\pi}{2}(n+k+\beta)\right)}{t^{n+k+\beta}}$$

uniformly holds, where

$$A(n,\beta) = \pi^{\frac{n-1}{2}} 2^{n+\beta} \Gamma\left(\frac{n+1}{2} + \beta\right).$$

Thus, it follows from the condition (4.8.3) that

$$\int_1^\infty \psi_x\left(\frac{t}{R}\right) t^{n+k-1} \left[E(n,\beta,k,t) - \frac{C(n,k)}{t^{n+k}}\right] dt$$
$$= A(n,\beta) \int_1^\infty \psi_x\left(\frac{t}{R}\right) \frac{\cos\left(t - \frac{\pi}{2}(n+k+\beta)\right)}{t^{1+\beta}} dt + o(1), \tag{4.8.7}$$

as $R \to \infty$.

Let

$$A_\beta = \frac{1}{|Q|}\frac{1}{C(n,k)} A(n,\beta) = \frac{1}{(2\pi)^n} \frac{2^\beta}{\sqrt{\pi}} \frac{\Gamma\left(\frac{n+1}{2} + \beta\right) \Gamma\left(\frac{k}{2}\right)}{\Gamma\left(\frac{n+k}{2}\right)}.$$

By substituting (4.8.6) and (4.8.7) into (4.8.5), we obtain that

$$\widetilde{S}_R^{\alpha_0+\beta}(f;x) - \overline{f}_{\frac{1}{R}}(x)$$
$$= A_\beta \int_1^\infty \psi_x\left(\frac{t}{R}\right) \frac{\cos\left(t - \frac{\pi}{2}(n+k+\beta)\right)}{t^{1+\beta}} dt + o(1)$$
$$= \frac{A_\beta}{R^\beta} \int_{\frac{1}{R}}^\infty \psi_x(t) \frac{\cos\left(Rt - \frac{\pi}{2}(n+k+\beta)\right)}{t^{1+\beta}} dt + o(1), \tag{4.8.8}$$

holds uniformly about $\beta \in (0,1)$, as $R \to \infty$.

Since the first term on the right side of (4.8.8) is obviously $o(1)$, we have actually proved Theorem 4.8.1. ∎

Remark 4.8.1 *From the proof of Theorem 4.8.1, we can weaken the condition (4.8.3) into the following*

$$\int_0^t \psi_x(\tau) d\tau = o(t),$$

as $t \to 0^+$.

If f *vanishes in* $B(x; \delta)$ *with* $\delta > 0$, *then (4.8.3) is valid. And thus Theorem 4.8.1 contains the localization principle over the critical index.*

Our main results are formulated as follows.

Theorem 4.8.2 *Let* $f \in L(Q^n)$ *with* $n > 1$. *If (4.8.3) holds, then we have*

$$\lim_{R \to \infty} \frac{1}{R} \int_0^R \left(\widetilde{S}_u^{\frac{n-1}{2}}(f; x) - f_{\frac{1}{u}}^*(x) \right) du = 0. \qquad (4.8.9)$$

Proof. Let $\beta \in (0, 1)$. By (4.8.8), when $R \to \infty$, we have that

$$\frac{1}{R} \int_0^R \left(\widetilde{S}_u^{\alpha_0 + \beta}(f; x) - f_{\frac{1}{u}}^*(x) \right) du$$
$$= A_\beta \int_{\frac{1}{R}}^\infty \frac{\psi_x(t)}{t^{1+\beta}} \left\{ \frac{1}{R} \int_{\frac{1}{t}}^R \frac{\cos(ut - \theta)}{u^\beta} du \right\} dt + o(1), \qquad (4.8.10)$$

where $\theta = \frac{\pi}{2}(n + k + \beta)$ and $o(1)$ is uniformly valid for $\beta \in (0, 1)$.

Set

$$M_R(\beta, t, \tau) = t \int_\tau^R \frac{1}{(ut)^\beta} \cos(ut - \theta) du, \qquad (4.8.11)$$

for $\beta > 0$, $\tau \in (0, \infty)$ with $t\tau \geq 1$.

By the mean value theorem of integral, we have

$$M_R(\beta, t, \tau) = \frac{t}{(t\tau)^\beta} \int_\tau^\xi \cos(ut - \theta) du = \frac{1}{(\tau t)^\beta} \int_{t\tau}^{\tau\xi} \cos(u - \theta) du.$$

Since

$$|M_R(\beta, t, \tau)| \leq 2 \qquad (4.8.12)$$

holds for $\beta > 0$, $\tau \in (0, \infty)$ with $t\tau \geq 1$, we conclude that

$$\sup_{\beta \in (0,1)} \left| A_\beta \int_{\frac{1}{R}}^{\infty} \frac{\psi_x(t)}{t^{1+\beta}} \frac{1}{R} \int_{\frac{1}{t}}^{R} \frac{\cos(ut-\theta)}{u^\beta} du dt \right|$$

$$\leq A \int_{\frac{1}{R}}^{\infty} \frac{|\psi_x(t)|}{t^2} \frac{1}{R} \sup_{\beta \in (0,1)} \left| M_R\left(\beta, t, t^{-1}\right) \right| dt \qquad (4.8.13)$$

$$\leq A \frac{1}{R} \int_{\frac{1}{R}}^{\infty} \frac{|\psi_x(t)|}{t^2} dt$$

$$= o(1),$$

as $R \to \infty$.

Substituting (4.8.13) into (4.8.10) and taking $\beta \to 0^+$, we get (4.8.9). This completes the proof. ∎

Since the condition (4.8.3) is valid for a.e. $x \in \mathbb{R}^n$, then so is (4.8.9). In addition, since $f_\varepsilon^*(x) \to f^*(x)$, for a.e. $x \in \mathbb{R}^n$, then Theorem 4.8.2 shows that

$$\lim_{R \to \infty} \frac{1}{R} \int_0^R \widetilde{S}_u^{\frac{n-1}{2}}(f;x) du = f^*(x)$$

holds for a.e. $x \in \mathbb{R}^n$.

The above results were obtained by Wang [Wa2].

4.9 The strong summation of the conjugate Fourier series

Because (4.8.9) is valid for a.e. $x \in \mathbb{R}^n$, it is virtually natural to ask whether there is any possibility to strengthen it to the strong limit. That is the strong summation problem, which is far from being completely solved, like the non-conjugate case. Here we will give the sufficient condition for the strong summation at a fixed point. This parallels to the non-conjugate case. We shall prove this result in three steps. We first describe the localized results and then transform the problem to the conjugate integral by the condition of the locally p integral with $p > 1$ as well as the results in Section 4.6. Finally, we prove the conclusion about the strong summation for the conjugate integral.

Step I. The localization.

Theorem 4.9.1 Let $f \in L(Q^n)$ with $n > 1$. If f vanishes in $B(x,\delta)$ with $\delta > 0$ and $q > 0$, then we have

$$\lim_{R \to \infty} \frac{1}{R} \int_0^R \left| \widetilde{S}_u^{\frac{n-1}{2}}(f;x) - \overline{f}_{\frac{1}{u}}(x) \right|^q du = 0. \qquad (4.9.1)$$

Proof. Let $\beta \in (0,1)$. By (4.8.8), we merely need to prove that

$$\limsup_{R \to \infty} \sup_{\beta \in (0,1)} \frac{1}{R} \int_{\frac{1}{\delta}}^R |I^\beta(u)|^q du = 0, \qquad (4.9.2)$$

where

$$I^\beta(u) = \int_{\frac{1}{u}}^\infty \psi_x(t) \frac{\cos\left(ut - \frac{\pi}{2}(n+k+\beta)\right)}{u^\beta t^{1+\beta}} dt. \qquad (4.9.3)$$

For $u > \delta^{-1}$, this implies $u^{-1} < \delta$. At the same time, $\psi_x(t)$ equals to zero when $t < \delta$. Thus $I^\beta(u)$ in (4.9.3) can be rewritten as

$$I^\beta(u) = \int_\delta^\infty \psi_x(t) \frac{\cos\left(ut - \frac{\pi}{2}(n+k+\beta)\right)}{u^\beta t^{1+\beta}} dt, \qquad (4.9.4)$$

for $\delta^{-1} \leq u \leq R$.

In addition, by Hölder's inequality, it suffices to prove $q = 2^\lambda$ with $\lambda \in \mathbb{N}$. Since

$$\cos(ut - \theta) = \cos ut \cos \theta + \sin ut \sin \theta,$$

we merely need to show that

$$\limsup_{R \to \infty} \sup_{\beta \in (0,1)} \frac{1}{R} \int_{\frac{1}{\delta}}^R |J^\beta(u)|^q du = 0, \qquad (4.9.5)$$

where $q = 2^\lambda$ and J^β is taken as the following two functions

$$\int_\delta^\infty \psi_x(t) \frac{\cos ut}{u^\beta t^{1+\beta}} dt \qquad (4.9.6)$$

and

$$\int_\delta^\infty \psi_x(t) \frac{\sin ut}{u^\beta t^{1+\beta}} dt. \qquad (4.9.7)$$

Since the method of the proof for the two different cases are the same, we choose J^β as in (4.9.6). Since $q = 2^\lambda$ and $J^\beta(u)$ is a real function, we have

$$\left| J^\beta(u) \right|^q = \left(J^\beta(u) \right)^q.$$

4.9 The strong summation of the conjugate Fourier series

We have

$$\frac{1}{R}\int_{\frac{1}{\delta}}^{R}\left|J^{\beta}(u)\right|^{q}du$$

$$=\frac{1}{R}\int_{(\delta,\infty)^{q}}\frac{\psi_{x}(t_{1})\cdots\psi_{x}(t_{q})}{(t_{1}\cdots t_{q})^{1+\beta}}\int_{1/\delta}^{R}\left(\frac{1}{u^{q\beta}}\prod_{j=1}^{q}\cos ut_{j}\right)du\,dt_{1}\cdots dt_{q}.$$

Line up all the q-dimensional vectors whose components are taken as either 1 or -1. We denote these vectors by $\{e_l\}_{l=1}^{2^q}$, where

$$e_{l}=(e_{l,1},\ldots,e_{l,q}),\quad e_{l,j}=1\text{ or }-1,\ j=1,2,\ldots,q.$$

By the inner product

$$e_l \cdot t = e_{l,1}t_1 + \cdots + e_{l,q}t_q,$$

we have

$$\prod_{j=1}^{q}\cos ut_{j}=\frac{1}{t^{q}}\cdot\sum_{l=1}^{2q}\cos(e_{l}\cdot t)u,$$

where $t=(t_1,\ldots,t_q)\in\mathbb{R}^q$.

According to the mean value formula of integral, we have that, for $(e_l \cdot t)\neq 0$,

$$\int_{\frac{1}{\delta}}^{R}\frac{1}{u^{q\beta}}\cos u(e_{l}\cdot t)du$$

$$=\delta^{q\beta}\int_{\frac{1}{\delta}}^{\xi}\cos(e_{l}\cdot t)udu$$

$$=\delta^{q\beta}(e_{l}\cdot t)^{-1}\left[\sin\xi(e_{l}\cdot t)-\sin\delta^{-1}(e_{l}\cdot t)\right],$$

for $\xi\in(\delta^{-1},R)$.

Thus we have

$$\left|\int_{\frac{1}{\delta}}^{R}\frac{1}{u^{q\beta}}\cos u(e_{l}\cdot t)du\right|\leq\frac{A}{|e_{l}\cdot t|+R^{-1}},$$

for $\beta\in(0,1)$, $t\in\mathbb{R}^q$.

Set

$$h(t_j)=|\psi_x(t_j)|,$$

for $j = 1, 2, \ldots, q$. Then it suffices to show that

$$\lim_{R\to\infty} \frac{1}{R} \int_{(\delta, \infty)^q} \left(\prod_{j=1}^{q} \frac{h(t_j)}{t_j}\right) (|e_l \cdot t| + R^{-1})^{-1} dt = 0, \qquad (4.9.8)$$

for $l = 1, \ldots, 2^q$.
Let
$$e_l \cdot t = t_1 + \cdots + t_\alpha - (t_{\alpha+1} + \cdots + t_q).$$

We divide the integral into

$$\int_{(\delta, \infty)^q} = \sum_{m_1 \cdots m_q = 1}^{\infty} \int_{[m\delta, (m+1)\delta]},$$

where

$$[m\delta, (m+1)\delta] = \prod_{j=1}^{q} [m_j \delta, (m_j + 1)\delta] := \triangle_m.$$

Set
$$\tau_m(R) = \int_{\triangle_m} \frac{h(t_1) \cdots h(t_q) dt_1 \cdots dt_q}{|t_1 + \cdots + t_\alpha - (t_{\alpha+1} + \cdots + t_q)| + R^{-1}}.$$

We can ascribe the problem into proving

$$\sum_{m \in \mathbb{N}^q} \frac{1}{m_1 \cdots m_q} \tau_m(R) = o(R). \qquad (4.9.9)$$

Clearly we have that

$$\int_a^{a+\delta} h(s) ds \le A_\delta \|f\|_{L(Q)}, \qquad (4.9.10)$$

for any $a \ge \delta$, and for any $\varepsilon > 0$, there exists $\eta > 0$ such that

$$\int_a^{a+\eta} h(s) ds < A\varepsilon, \qquad (4.9.11)$$

for $a \ge \delta$.
Since f vanishes in $B(x; \delta)$, the above restriction $a \ge \delta$ can be canceled. These two things are based on the periodicity and the local integrability of f.
We first prove the following results.

4.9 The strong summation of the conjugate Fourier series

(a) $\tau_m(R) = o(R)$ uniformly holds for $m \in \mathbb{N}^q$.

(b) If $|m_1 + \cdots + m_\alpha - (m_{\alpha+1} + \cdots + m_q)| > 2q\delta$, we have

$$|t_1 + \cdots + t_\alpha - (t_{\alpha+1} + \cdots + t_q)| \geq \frac{\delta}{2}|m_1 + \cdots + m_\alpha - (m_{\alpha+1} + \cdots + m_q)|,$$

for $t \in \Delta_m$, which implies

$$\tau_m(R) \leq \frac{A}{|m_1 + \cdots + m_\alpha - (m_{\alpha+1} + \cdots + m_q)|}.$$

Since (b) is obvious, we will prove (a) in the following.
Let

$$s = -(t_2 + \cdots + t_\alpha) + t_{\alpha+1} + \cdots + t_q.$$

Then we have

$$\int_{[m_1\delta,(m_1+1)\delta]} \frac{h(t_1)}{|t_1 - s| + R^{-1}} dt_1$$

$$= \int_{\{t_1:|t_1-s|<\frac{\eta}{2}\} \cap [m_1\delta,(m_1+1)\delta]} \frac{h(t_1)}{|t_1 - s| + R^{-1}} dt_1$$

$$+ \int_{\{t_1:|t_1-s|\geq\frac{\eta}{2}\} \cap [m_1\delta,(m_1+1)\delta]} \frac{h(t_1)}{|t_1 - s| + R^{-1}} dt_1$$

$$:= I_1 + I_2.$$

It follows from (4.9.11) and (4.9.10) that

$$I_1 < AR\varepsilon$$

and

$$I_2 < A\frac{1}{\eta}\|f\|_{L(Q)}.$$

Thus it follows that

$$\tau_m(R) \leq A\left(R\varepsilon + \frac{1}{\eta}\right) \int_{m_2\delta}^{(m_2+1)\delta} h(t_2)dt_2 \cdots \int_{m_q\delta}^{(m_q+1)\delta} h(t_q)dt_q$$

$$\leq A\left(R\varepsilon + \frac{1}{\eta}\right)$$

holds uniformly for $m \in \mathbb{N}^q$. This implies (a).

Now we rewrite the sum on the left side of the equation (4.9.9) as

$$\sum_{m\in\mathbb{N}^q} = \sum_{|m_1+\cdots+m_\alpha-(m_{\alpha+1}+\cdots+m_q)|<2q\delta} + \sum_{\text{other } m} := \sigma_1 + \sigma_2.$$

Applying the estimate of (b) in σ_1, we get $\sigma_1 = O(1)$. And by the estimate of (a) in σ_2 we have $\sigma_2 = o(R)$. One can check the detailed steps in the computation of (3.10.17) and (3.10.18) in Chapter 3. Hence (4.9.9) holds.

This finishes the proof of Theorem 4.9.1. ∎

Step II. The strong summation of the conjugate Fourier integral.

Theorem 4.9.2 *Let $g \in L(\mathbb{R}^n)$ with $n > 1$. We have, at the point x,*

$$\int_0^t \left(\int_{|y|=1} g(x-\tau y) P(y) d\sigma(y) \right) d\tau = o(t), \qquad (4.9.12)$$

as $t \to 0$. In addition, if g is r integrable on $x + B(0;\delta)$ with $\delta > 0$ and $r > 1$, and

$$\int_0^t \left| \int_{|y|=1} g(x-\tau y) P(y) d\sigma(y) \right|^r d\tau = O(t), \qquad (4.9.13)$$

then we have

$$\lim_{R\to\infty} \frac{1}{R} \int_0^R \left| \tilde{\sigma}_u^{\alpha_0}(g;x) - \tilde{g}_{\frac{1}{u}}(x) \right|^q du = 0, \qquad (4.9.14)$$

for any $q > 0$.

Proof. We might as well take it for granted that $x = 0$ and $r < 2$. It suffices to prove for the conjugate number $q = \frac{r}{r-1}$ with the index of r.

Define

$$\psi(t) = \int_{|y|=1} g(-ty) P(y) d\sigma(y)$$

and

$$h(t) = \int_0^t \psi(\tau) d\tau.$$

Then the conditions (4.9.12) and (4.9.13) are represented as

$$h(t) = o(t)$$

4.9 The strong summation of the conjugate Fourier series

and
$$\int_0^t |\psi(\tau)|^r d\tau = O(t).$$

According to (4.1.24), we have
$$\tilde{\sigma}_R^{\alpha_0}(g;0) = \frac{1}{(2\pi)^n} R^{n+k} \int_0^\infty \psi(t) t^{n+k-1} \frac{E(n,0,k,Rt)}{C(n,k)} dt.$$

By the definition
$$\tilde{g}_{\frac{1}{R}}(0) = \frac{1}{(2\pi)^n} \int_{\frac{1}{R}}^\infty \psi(t) t^{-1} dt,$$

we also obtain
$$\tilde{\sigma}_R^{\alpha_0}(g;0) - \tilde{g}_{\frac{1}{R}}(0) = \frac{1}{(2\pi)^n} \int_{\frac{1}{R}}^\infty \psi(t) t^{n+k-1} \left(R^{n+k} \frac{E(n,0,k,Rt)}{C(n,k)} - \frac{1}{t^{n+k}} \right) dt$$
$$+ \frac{1}{(2\pi)^n} \int_0^{\frac{1}{R}} \psi(t) t^{n+k-1} \frac{1}{C(n,k)} R^{n+k} E(n,0,k,Rt) dt$$
$$:= I_1 + I_2. \tag{4.9.15}$$

By integration by parts, we have that
$$I_2 = \frac{1}{(2\pi)^n C(n,k)} \left\{ h(t) t^{n+k-1} R^{n+k} E(n,0,k,Rt) \Big|_0^{R^{-1}} \right.$$
$$\left. - \int_0^{1/R} h(t) \frac{d}{dt} \left[t^{n+k-1} R^{n+k} E(n,0,k,Rt) \right] dt \right\}.$$

By substituting $h(t) = o(t)$ into the above equation and (4.6.18), we have that
$$I_2 = o(1) + A \int_0^{1/R} h(t) \frac{d}{dt} (E(n,0,k,Rt)) t^{n+k-1} R^{n+k} dt$$
$$= o(1) + A \int_0^{1/R} h(t) t^{n+k} E(n+2,-1,k,Rt) R^{n+k+2} dt$$
$$= o(1). \tag{4.9.16}$$

Again, integration by parts for I_1, we have
$$I_1 = \frac{1}{(2\pi)^n} \left\{ h(t) t^{n+k-1} R^{n+k} \left(\frac{E(n,0,k,Rt)}{C(n,k)} - \frac{1}{(Rt)^{n+k}} \right) \Big|_{\frac{1}{R}}^\infty \right.$$
$$\left. + \int_{\frac{1}{R}}^\infty h(t) R^{n+k} \frac{d}{dt} \left[t^{n+k-1} \left(\frac{E(n,0,k,Rt)}{C(n,k)} - \frac{1}{(Rt)^{n+k}} \right) \right] dt \right\}.$$

By the facts
$$|h(t)| \leq M < +\infty, \quad h(t) = o(t)$$
as $t \to 0$, and
$$E(n, 0, k, Rt) = O\left(\frac{1}{(Rt)^{n+k}}\right),$$
for $t \geq \frac{1}{R}$, we have that the first term on the right side is o(1). According to the formula (4.6.18) and the expression (4.4.4) of the function E, we calculate that
$$\frac{d}{dt}\left[t^{n+k-1}\left(\frac{E(n,0,k,Rt)}{C(n,k)} - \frac{1}{(Rt)^{n+k}}\right)\right]R^{n+k}$$
$$= A_1\frac{\sin Rt}{t^2} + A_2\frac{\cos Rt}{t^2} + A_3\frac{R\sin Rt}{t} + A_4\frac{R\cos Rt}{t} + O\left(\frac{1}{Rt^3}\right).$$

From the above equation and the conditions $h(t) = o(t)$, as $t \to 0$, and $|h(t)| \leq M$, we get that
$$I_1 = B_1\int_{R^{-1}}^{\infty}\frac{\psi(t)}{t}\cos Rt\,dt + B_2\int_{R^{-1}}^{\infty}\frac{\psi(t)}{t}\sin Rt\,dt$$
$$+ B_3\int_{R^{-1}}^{\infty}\frac{\psi(t)}{Rt^2}\cos Rt\,dt + B_4\int_{R^{-1}}^{\infty}\frac{\psi(t)}{Rt^2}\sin Rt\,dt + o(1).$$

We denote the four integrals on the right side of the above equation by $J_1(R), J_2(R), J_3(R), J_4(R)$, respectively. It follows from (4.9.15) and (4.9.16) that
$$\widetilde{\sigma}_R^{\alpha_0}(g;0) - \widetilde{g}_{\frac{1}{R}}(0) = \sum_{\nu=1}^{4}B_\nu J_\nu(R) + o(1).$$

Thus the problem can be ascribed into proving
$$\lim_{R\to\infty}\frac{1}{R}\int_0^R|J_\nu(u)|^q du = 0, \tag{4.9.17}$$
for $\nu = 1, 2, 3, 4$. In the following, we only write down the proof for the case $\nu = 1$ and the proofs for $\nu = 2, 3, 4$ are similar.

Set
$$\alpha = \limsup_{R\to\infty}\frac{1}{R}\int_0^R|J_1(u)|^q du.$$

4.9 The strong summation of the conjugate Fourier series

For any $\varepsilon \in (0,1)$, we have that

$$\int_{\frac{1}{u}}^{\frac{1}{\varepsilon u}} \frac{\psi(t)}{t} \cos ut\, dt = h(t) \frac{\cos ut}{t}\Big|_{\frac{1}{u}}^{\frac{1}{\varepsilon u}} + \int_{\frac{1}{u}}^{\frac{1}{\varepsilon u}} h(t)\left(\frac{\cos ut}{t^2} + \frac{u\sin ut}{t}\right) dt$$

$$= o(1) + o(1)\log\frac{1}{\varepsilon} + o(1)\frac{1}{\varepsilon}$$

$$= o(1),$$

as $R \to \infty$. Hence, we have

$$\alpha \leq 2^q \limsup_{R\to\infty} \frac{1}{R} \int_0^R \left|\int_{\frac{1}{u}}^{\frac{1}{\varepsilon u}} \frac{\psi(t)}{t} \cos ut\, dt\right|^q du$$

$$+ 2^q \limsup_{R\to\infty} \frac{1}{R} \int_0^R \left|\int_{\frac{1}{\varepsilon u}}^{\infty} \frac{\psi(t)}{t} \cos ut\, dt\right|^q du$$

$$= 2^q \limsup_{R\to\infty} \frac{1}{R} \int_0^R \left|\int_{\frac{1}{\varepsilon u}}^{\infty} \frac{\psi(t)}{t} \cos ut\, dt\right|^q du. \tag{4.9.18}$$

Let

$$\Phi(u,t) = \int_0^t \psi(\tau)\cos u\tau\, d\tau.$$

Obviously we have that that

$$\Phi(u,t) = h(t)\cos ut + u\int_0^t h(\tau)\sin u\tau\, d\tau$$

$$= o(t) + uo(t^2), \tag{4.9.19}$$

as $t \to 0$, and

$$|\Phi(u,t)| \leq \int_0^t |\psi(\tau)|d\tau \leq M < +\infty, \tag{4.9.20}$$

for any u, t.
Therefore, we have

$$\int_{\frac{1}{\varepsilon u}}^{\infty} \frac{\psi(t)}{t} \cos ut\, dt = \Phi(u,t)\frac{1}{t}\Big|_{\frac{1}{\varepsilon u}}^{\infty} + \int_{\frac{1}{\varepsilon u}}^{\infty} \Phi(u,t) t^{-2} dt$$

$$= o(1) + \int_{\frac{1}{\varepsilon u}}^{\infty} \Phi(u,t) t^{-2} dt,$$

as $u \to \infty$. This implies

$$\alpha \leq A_q \limsup_{R\to\infty} \cdot \frac{1}{R} \int_0^R \left(\int_{\frac{1}{\varepsilon u}}^\infty \frac{|\Phi(u,t)|}{t^2} dt \right)^q du$$

$$\leq A_q \limsup_{R\to\infty} \frac{1}{R} \int_0^{\frac{1}{\varepsilon \delta}} \left(\int_{\frac{1}{\varepsilon u}}^\infty \frac{|\Phi(u,t)|}{t^2} dt \right)^q du$$

$$+ A_q \limsup_{R\to\infty} \frac{1}{R} \int_{\frac{1}{\varepsilon \delta}}^R \left(\int_{\frac{1}{\varepsilon u}}^\delta \frac{|\Phi(u,t)|}{t^2} dt \right)^q du$$

$$+ A_q \limsup_{R\to\infty} \frac{1}{R} \int_{\frac{1}{\varepsilon \delta}}^R \left(\int_\delta^\infty \frac{|\Phi(u,t)|}{t^2} dt \right)^q du.$$

Since $|\Phi(u,t)| \leq M$, the first term on the right side is obviously zero. For fixed t, by the Riemann-Lebesgue theorem, we get that $\lim_{u\to\infty} \Phi(u,t) = 0$, which leads to the third term is also zero. By the controlled convergence theorem, we have that

$$\lim_{u\to\infty} \left(\int_\delta^\infty \frac{|\Phi(u,t)|}{t^2} dt \right)^q = 0.$$

Then, according to the generalized Minkowski's inequality, we get that

$$\alpha \leq A^q \limsup_{R\to\infty} \frac{1}{R} \int_{\frac{1}{\varepsilon \delta}}^R \left(\int_{\frac{1}{\varepsilon u}}^\delta \frac{|\Phi(u,t)|}{t^2} dt \right)^q du$$

$$\leq A^q \limsup_{R\to\infty} \frac{1}{R} \left\{ \int_{\frac{1}{\varepsilon R}}^\delta \frac{1}{t^2} \left(\int_0^R |\Phi(u,t)|^q du \right)^{\frac{1}{q}} dt \right\}^q.$$

It follows from the Hausdorff-Young's inequality that

$$\left(\int_0^R |\Phi(u,t)|^q du \right)^{\frac{1}{q}} \leq A_q \left(\int_0^t |\psi(\tau)|^r d\tau \right)^{\frac{1}{r}} \leq A_q t^{\frac{1}{r}}, \qquad (4.9.12)'$$

which implies

$$\alpha \leq A_q \limsup_{R\to\infty} \frac{1}{R} \left\{ \int_{\frac{1}{\varepsilon R}}^\delta \frac{1}{t^{2-\frac{1}{r}}} \right\}^q \leq A_q \frac{1}{R} (\varepsilon R)^{(1-\frac{1}{r})q} = A_q \varepsilon.$$

Together with the arbitrary of ε, this yields $\alpha = 0$. This completes the proof. ∎

Step III. Combining Theorem 4.9.1, 4.9.2 with 4.6.2, we have the next result.

Theorem 4.9.3 *Let $f \in L(Q^n)$, $n > 1$, and f is r $(r > 1)$ integrable on $B(x, \delta)$ with $\delta > 0$. If the conditions (4.9.12) and (4.9.13) hold for f, then for any $q > 0$, we have*

$$\lim_{R \to \infty} \frac{1}{R} \int_0^R \left| \widetilde{S}_u^{\frac{n-1}{2}}(f;x) - \overline{f}_{\frac{1}{u}}(x) \right|^q du = 0.$$

The above results about the strong summation were done by Wang [Wa7].

4.10 Approximation of continuous functions

For the case of one dimension, the approximation of continuous functions and their conjugate functions through their partial sum in Fourier series defined on set of total measure, was solved by Oskolkov [O1]. Now we turn to discuss the case of higher dimension. We have such a result that let $f \in Lip\alpha$ ($0 < \alpha < 1$), if we consider uniform approximation by $S_R^{\frac{n-1}{2}}(f)$, then we only get the order of approximation as $\frac{\log R}{R^\alpha}$. If we restrict functions on set of total measure, then the order of the a.e. approximation can be $O(\log \log R / R^\alpha)$, and we can get the same results about the conjugate functions.

Recall the definitions of the classes of functions $W^1 L^\infty = Lip1$ and $W^2 L^\infty$ in Section 3.7. There we have proven that if $f \in W^2 L^\infty$, there holds

$$S_R^{\frac{n-1}{2}}(f;x) - f(x) = O\left(\frac{1}{R^2}\right) \tag{4.10.1}$$

a.e. $x \in \mathbb{R}^n$. Now we take further research on the problem of the approximation of some functions which include the conjugate functions.

Lemma 4.10.1 *If $f \in W^1 L^\infty$, then we have*

$$|S_R^{\alpha_0}(f;x) - f(x)| \leq \frac{2\Gamma(\frac{n+1}{2})}{\pi^{\frac{n+1}{2}}} \frac{1}{R} \left\{ \sum_{j=1}^n TP_j(f_j)(x) + \theta_R(f)(x) \right\}, \tag{4.10.2}$$

where $P_j(x) = x_j$, TP_j is the operator defined as in Definition 4.5.1 and the formulas (4.5.10) and (4.5.11), and $f_j = \frac{\partial f}{\partial x_j}$ for $(j = 1, 2, \ldots, n)$. The remainder term $\theta_R(f)(x)$ satisfies

(I) $0 \leq \theta_R(f)(x) \leq A \cdot \max\{\|f_j\|_\infty : j = 1, 2, \ldots, n\}$,

(II) $\lim_{R \to \infty} \theta_R(f)(x) = 0$, for a.e. $x \in Q^n$.

Proof. Let $\beta \in (0,1)$. Due to Bochner's formula, we get

$$S_R^{\frac{n-1}{2}+\beta}(f;x) - f(x) = C'_{n,\beta} \int_0^\infty \left(f_x\left(\frac{t}{R}\right) - f(x)\right) \frac{J_{n-\frac{1}{2}+\beta}(t)}{t^{\frac{1}{2}+\beta}} dt,$$

where

$$C'_{n,\beta} = \frac{2^{\frac{1}{2}+\beta}\Gamma(\frac{n+1}{2}+\beta)}{\Gamma(n/2)}.$$

Set

$$F_\beta(t) = \int_t^\infty s^{-\frac{1}{2}-\beta} J_{n-\frac{1}{2}+\beta}(s) ds.$$

Then, for $t \geq 1$, we conclude that

$$\begin{aligned} F_\beta(t) &= \sqrt{\frac{2}{\pi}} \int_2^\infty \Big\{ \frac{\cos(s - \frac{\pi}{2}(n+\beta))}{s^{1+\beta}} \\ &\quad - A_{n,\beta} \frac{\sin(s - \frac{\pi}{2}(n+\beta))}{s^{2+\beta}} + O\left(\frac{1}{s^3}\right) \Big\} ds \\ &= \sqrt{\frac{2}{\pi}} \frac{\cos(t - \frac{\pi}{2}(n+1+\beta))}{t^{1+\beta}} + O\left(\frac{1}{t^2}\right), \end{aligned} \quad (4.10.3)$$

as $t \to \infty$, where "O" holds uniformly about $\beta \in (0,1)$. For $0 \leq t < 1$, we have

$$F_\beta(t) = \int_t^1 s^{-\frac{1}{2}-\beta} J_{n-\frac{1}{2}+\beta}(s) ds + F_\beta(1) = O(1), \quad (4.10.4)$$

where "O" holds uniformly about $\beta \in (0,1)$. Then it follows from integration by parts that

$$\begin{aligned} \int_0^\infty \left(f_x\left(\frac{t}{R}\right) - f(x)\right) \frac{J_{n-\frac{1}{2}+\beta}(t)}{t^{\frac{1}{2}+\beta}} dt &= -\int_0^\infty F_\beta(t) \frac{d}{dt} f_x\left(\frac{t}{R}\right) dt \\ &= -\left(\int_0^1 + \int_0^\infty\right) F_\beta(t) \frac{d}{dt} f_x\left(\frac{t}{R}\right) dt \\ &= I_1 + I_2. \end{aligned}$$

Applying (4.10.4), we have

$$|I_1| \leq A \int_0^1 \left|\frac{d}{dt} f_x\left(\frac{t}{R}\right)\right| dt.$$

4.10 Approximation of continuous functions

It is easy to see that

$$\frac{d}{dt} f_x\left(\frac{t}{R}\right) = \frac{1}{R}\frac{1}{\omega_{n-1}} \int_{\mathbb{S}^{n-1}} \sum_{j=1}^{n} f_j\left(x - \frac{t}{R}\xi\right)(-\xi_j) d\sigma(\xi)$$

$$= -\frac{1}{\omega_{n-1}}\frac{1}{R} \sum_{i=1}^{n} P_j(f_j)\left(x; \frac{t}{R}\right). \tag{4.10.5}$$

Obviously, if $f_j(x)$ is finite, then we have

$$\left|P_j\left(f_j\left(x;\frac{t}{R}\right)\right)\right| = \left|\int_{\mathbb{S}^{n-1}} \left(f_j\left(x - \frac{t}{R}\xi\right) - f_j(x)\right)\xi_j d\sigma(\xi)\right|$$

$$\leq \int_{\mathbb{S}^{n-1}} \left|f_j\left(x - \frac{t}{R}\xi\right) - f_j(x)\right| d\sigma(\xi).$$

If x belongs to the intersection of the sets of the Lebesgue points of $f_j = \frac{\partial f}{\partial x_j}$, for $j = 1, 2, \ldots, n$, we denote by

$$\Phi_j(x; s) = \int_0^s \int_{|\xi|=t} |f_j(x - \xi) - f_j(x)| d\sigma(\xi) dt.$$

Then, we have $\Phi_j(x; s) = o(s^n)$, as $s \to 0^+$. Consequently, we have

$$\int_0^1 \left|P_j(f_j)\left(x;\frac{t}{R}\right)\right| dt \leq R \int_0^{\frac{1}{R}} \frac{1}{t^{n-1}} \int_{|\xi|=t} |f_j(x - \xi) - f_j(x)| d\sigma(\xi) dt$$

$$= R\left\{\frac{1}{t^{n-1}}\Phi_j(x;t)\Big|_0^{\frac{1}{R}} + (n-1) \int_0^{\frac{1}{R}} \frac{\Phi_j(x;t)}{t^n} dt\right\}$$

$$= o(1),$$

as $R \to \infty$.

Thus we have $I_1 = o(\frac{1}{R})$.

Combining (4.10.3) with (4.10.5), we have

$$\omega_{n-1} I_2 = -\sqrt{\frac{2}{\pi}}\frac{1}{R}\sum_{i=1}^{n} \int_1^{\infty} P_j(f_j)\left(x;\frac{t}{R}\right) \frac{\cos\left(t - \frac{\pi}{2}(k+1+\beta)\right)}{t^{1+\beta}} dt + I_2',$$

where
$$|I_2'| \le A \int_1^\infty \left|\frac{d}{dt} f_x\left(\frac{t}{R}\right)\right| \frac{1}{t^2} dt$$
$$\le A \frac{1}{R} \sum_{j=1}^n \int_{\frac{1}{R}}^\infty R^{-1} t^{-n-1} \int_{|\xi|=t} |f_j(x-\xi) - f_j(x)| d\sigma(\xi) dt$$
$$\le A \frac{1}{R} \sum_{j=1}^n \int_{\frac{1}{R}}^\infty R^{-1} t^{-n-2} \Phi_j(x;t) dt$$
$$= o\left(\frac{1}{R}\right).$$

Set
$$\theta_R(f;x) = \sqrt{\frac{\pi}{2}} R \overline{\lim_{\beta \to 0^+}} |I_1 + I_2'|.$$

We have
$$\left|S_R^{\frac{n-1}{2}}(f;x) - f(x)\right| \le \frac{1}{R} \sqrt{\frac{2}{\pi}} C_{n,0}' \left(\sum_{j=1}^n TP_j(f_j)(x) + \theta_R(f)(x)\right).$$

Here when x belongs to the intersection of the sets of the Lebesgue points of f_j, then $\theta_R(f)$ equals to o(1), as $R \to \infty$. Thus the lemma is proven. ∎

Suppose $P(x)$ is the homogeneous harmonic polynomial with order k, then for every $j \in \{1, \ldots, n\}$, $x_j P(x) \in \mathcal{P}_{k+1}$. Due to the decomposition theorem on harmonic polynomials (see [SW1]), $x_j P(x)$ can be uniquely decomposed into the following form
$$x_j P(x) = P_{j,k+1}(x) + P_{j,k-1}(x)|x|^2 + \cdots + P_{j,k-1-2\nu_k}(x)|x|^{2\nu_k}, \quad (4.10.6)$$
where $\nu_k = [\frac{k+1}{2}]$ and $P_{j,k+1-2l} \in \mathcal{A}_{k+1-2l}$, for $l = 0, 1, 2, \ldots, \nu_k$.

Lemma 4.10.2 *If $f \in W^1 L^\infty$, then we have*
$$\left|\widetilde{S}_R^{\frac{n-1}{2}}(f;x) - f^*(x)\right| \le A \frac{1}{R} \left(\sum_{j=1}^n \sum_{l=0}^{\nu_k} TP_{j,k+1-2l}(f_j)(x) + \widetilde{\theta}_R(f)(x)\right),$$
where $TP_{j,k+1-2l}$ is the operator defined (4.5.11) in Definition 4.5.1 to $P_{j,k+1-2l}$ which came from (4.10.6). Here $P(x)$ is the homogeneous harmonic polynomial about conjugate function with order k, $k \ge 1$, f_j is the j-th partial derivative on f, and the remainder term $\widetilde{\theta}(f)(x)$ satisfies
$$0 \le \widetilde{\theta}_R(f)(x) \le A \max\{\|f_j\|_\infty : j = 1, 2, \ldots, n\}.$$

4.10 Approximation of continuous functions

Proof. Let $\beta \in (0,1)$. Combining (4.3.14) with (4.4.12), we get

$$\tilde{S}_R^{\alpha_0+\beta}(f;x) - f^*(x) = \frac{1}{|Q|} \int_0^\infty P(f)\left(x; \frac{t}{R}\right) t^{n+h-1}$$
$$\times \left(\frac{E(n,\beta,k,t)}{C(n,k)} - \frac{1}{t^{n+k}}\right) dt. \quad (4.10.7)$$

Set

$$G_\beta(t) = \int_t^\infty s^{n+k-1}\left(\frac{E(n,\beta,k,s)}{C(n,k)} - \frac{1}{s^{n+k}}\right) ds,$$

for $t > 0$.

Due to the asymptotic formula (4.4.4), we have

$$\frac{E(n,\beta,k,s)}{C(n,k)} - s^{-n-k} = O\left(s^{-n-k-2}\right) + A_1(\beta)\frac{\cos\left(s - \frac{\pi}{2}(n+k+\beta)\right)}{s^{n+k+\beta}}$$
$$+ A_2(\beta)\frac{\sin(s - \frac{\pi}{2}(n+k+\beta))}{s^{n+k+1+\beta}}.$$

By making use of the integration by parts, we have that

$$\int_t^\infty \frac{\cos(s-\theta)}{s^{1+\beta}} ds = -\frac{\sin(t-\theta)}{t^{1+\beta}} + (1+\beta)\frac{\cos(t-\theta)}{t^{2+\beta}} + O\left(\frac{1}{t^2}\right),$$

and

$$\int_t^\infty \frac{\sin(s-\theta)}{s^{2+\beta}} ds = O\left(\frac{1}{t^2}\right)$$

hold uniformly about $\beta \in [0,1]$.

Consequently, we have

$$G_\beta(t) = -A_1(\beta)\frac{\sin(t-\theta)}{t^{1+\beta}} + O\left(\frac{1}{t^2}\right)$$

and

$$\theta = \frac{\pi}{2}(n+k+\beta), \quad (4.10.8)$$

for $t \geq 1$.

While, for $0 < t < 1$, we get

$$G_\beta(t) = G_\beta(1) + \int_t^1 s^{n+k-1}\left(\frac{E(n,\beta,k,s)}{C(n,k)} - \frac{1}{s^{n+k}}\right) ds$$
$$= O(1) + \log t \quad (4.10.9)$$

uniformly holds about $\beta \in [0,1]$. Applying the integration by parts to (4.10.7), we have that

$$\widetilde{S}_R^{\alpha_0+\beta}(f;x) - f^*(x) = \frac{1}{|Q|}\left(\int_0^1 + \int_1^\infty\right)\frac{d}{dt}\left(P(f)\left(x;\frac{t}{R}\right)\right)G_\beta(t)dt$$
$$= I_1 + I_2,$$

where

$$\frac{d}{dt}\left(P(f)\left(x;\frac{t}{R}\right)\right) = -\frac{1}{R}\sum_{j=1}^n \int_{\mathbb{S}^{n-1}} f_j\left(x-\frac{t}{R}\xi\right)\xi_j P(\xi)d\sigma(\xi).$$

By (4.10.6), we obtain that

$$\xi_j P(\xi) = \sum_{l=0}^{\nu_k} P_{j,k+1-2l}(\xi)$$

holds on $\mathbb{S}^{n-1}$. Then it follows that

$$\frac{d}{dt}\left(P(f)\left(x;\frac{t}{R}\right)\right) = -\frac{1}{R}\sum_{j=1}^n\sum_{l=0}^{\nu_k}\int_{\mathbb{S}^{n-1}} f_j\left(x-\frac{t}{R}\xi\right)P_{j,k+1-2l}(\xi)d\sigma(\xi)$$
$$= -\frac{1}{R}\sum_{j=1}^n\sum_{l=0}^{\nu_k} P_{j,k+1-2l}(f_j)\left(x;\frac{t}{R}\right). \quad (4.10.10)$$

It is easy to see that

$$\left|\frac{d}{dt}\left(P(f)\left(x;\frac{t}{R}\right)\right)\right| \leq \frac{1}{R}A\max\left\{\|f_j\|_\infty : j=1,2,\ldots,n\right\}. \quad (4.10.11)$$

Concisely, we write $M = \max\{\|f_j\|_\infty : j=1,2,\ldots,n\}$. Thus combining with (4.10.9), we get

$$|I_1| \leq \frac{A}{R}M\int_0^1 (1+|\log t|)dt = \frac{A}{R}M. \quad (4.10.12)$$

Due to (4.10.8) and (4.10.9), we obtain that

$$I_2 = \frac{A_1(\beta)}{R}\sum_{j=1}^n\sum_{l=0}^{\nu_k}\int_1^\infty P_{j,k+1-2l}(f_j)\left(x;\frac{t}{R}\right)\frac{\cos\left(t-\frac{\pi}{2}(n+k+1+\beta)\right)}{t^{1+\beta}}dt$$
$$+ \int_1^\infty O\left(\frac{M}{R}\frac{1}{t^2}\right)dt \quad (4.10.13)$$

4.10 Approximation of continuous functions

uniformly holds about β.

By Definition 5.1, we have

$$\limsup_{\beta \to 0^+} |I_2| \leq \frac{A}{R} \left(\sum_{j=1}^{n} \sum_{l=0}^{\nu_k} TP_{j,k+1-2^l}(f_j)(x) + M \right). \tag{4.10.14}$$

Thus combining (4.10.13) with (4.10.14), we get the proof Lemma 4.10.2. ∎

Lemma 4.10.3 *(A) If $f \in W^1 L^\infty$, then we have*

$$\omega(f^*; \delta) = O\left(\delta \log \frac{1}{\delta}\right).$$

(B) If $f \in W^2 L^\infty$, then we have

$$\omega_2(f^*; \delta) = O\left(\delta^2 \log \frac{1}{\delta}\right).$$

Here ω and ω_2 denote the modulus of continuity with the first and second order, respectively.

Proof. According to the definition, we have

$$f^*(x) = \frac{1}{|Q|} \int_Q f(x-y) K^*(y) dy$$
$$= \frac{1}{|Q|} \int_Q f(x-y) K(y) dy$$
$$+ \frac{1}{|Q|} \int_Q f(x-y) \left(\sum_{m \neq 0} (K(y+2\pi m) - I_m) - I_0 \right) dy$$
$$= g_1(x) + g_2(x).$$

It is easy to see that

$$\left| \sum_{m \neq 0} [K(y+2\pi m) - I_m] - I_0 \right| \leq A,$$

for $y \in Q$. Then we have

$$\omega(g_2; \delta) \leq A \omega(f; \delta)$$

and
$$\omega_2(g_2;\delta) \leq A\omega_2(f;\delta).$$

Set $h \in Q$ with $0 < |h| \leq 1/2$. Then we have

$$g_1(x) = \frac{1}{|Q|}\int_{0+}^{|h|} P(f)(x;t)\frac{1}{t}dt + \frac{1}{|Q|}\int_{|y|>|h|, y \in Q} f(x-y)K(y)dy$$
$$= g_{1,1}(x) + g_{1,2}(x).$$

Applying the integration by parts, we have

$$g_{1,1} = \frac{1}{|Q|}\left(P(f)(x;t)\log t\Big|_0^{|h|} - \int_0^{|h|} \log t \sum_{j=1}^n P_j(f_j)(x;t)dt\right),$$

where $P_j(x)$ is the polynomial $-x_j P(x)$. Then it follows that

$$g_{1,1}(x) = \frac{1}{|Q|}\left(P(f)(x;|h|)\log|h| - \sum_{j=1}^n \int_0^{|h|} \log t P_j(f_j)(x;t)dt\right).$$

For any $t > 0$, we have

$$|P(f)(x+h;t) - P(f)(x;t)| \leq A\omega(f;|h|)$$

and

$$|P(f)(x+h;t) + P(f)(x-h;t) - 2P(f)(x;t)| \leq A\omega_2(f;|h|).$$

Thus, we have that

$$\omega(g_{1,1};|h|) \leq A\omega(f;|h|)\log\frac{1}{|h|} + A\,M\int_0^{|h|}\log\frac{1}{t}dt$$
$$= O\left(|h|\log\frac{1}{|h|}\right),$$

for $f \in W^1 L^\infty$, and

$$\omega_2(g_{1,1};|h|) \leq A\omega_2(f;|h|)\log\frac{1}{|h|} + AM\int_0^{|h|}\log\frac{1}{t}dt \cdot |h|$$
$$= O\left(|h|^2 \log\frac{1}{|h|}\right),$$

for $f \in W^2 L^\infty$.

4.10 Approximation of continuous functions

For the term $g_{1,2}$, it is easy to get that

$$\omega(g_{1,2}; |h|) \leq \int_{|y|>|h|} \omega(f; |h|)|K(y)|dy$$
$$= O\left(|h| \log \frac{1}{|h|}\right),$$

and

$$\omega_2(g_{1,2}; |h|) \leq \int_{|y|>|h|} \omega_2(f; |h|)|K(y)|dy$$
$$= O\left(|h|^2 \log \frac{1}{|h|}\right).$$

Combining these results, we get the proof of Lemma 4.10.3. ∎

Remark 4.10.1 *From the above proof, we see that for the general kernel $K(x) = \frac{\Omega(x/|x|)}{|x|^n}$, which satisfies the conditions*

$$\int_{\mathbb{S}^{n-1}} \Omega(\xi) d\sigma(\xi) = 0$$

and

$$\int_0^1 \frac{\omega(\Omega; \delta)}{\delta} d\delta < \infty,$$

Lemma 4.10.3 is still true.

Theorem 4.10.1 *If $f \in W^2 L^\infty(Q^n)$, then we have*

$$\left| S_R^{\frac{n-1}{2}}(f; x) - f(x) \right| = O\left(\frac{1}{R^2}\right) \qquad (4.10.15)$$

for a.e. $x \in Q^n$,

$$\left\| S_R^{\frac{n-1}{2}}(f) - f \right\|_c = O\left(\frac{\log R}{R^2}\right) \qquad (4.10.16)$$

and

$$\left\| \widetilde{S}_R^{\frac{n-1}{2}}(f) - f^* \right\|_c = O\left(\frac{\log R}{R^2}\right). \qquad (4.10.17)$$

Proof. (4.10.15) and (4.10.16) have been proven in Theorem 3.7.3. For (4.10.17), it should point out that the Lebesgue constant of the conjugate function is also $O(\log R)$, then (4.10.17) can be obtained in the same way as (4.10.16). ∎

Theorem 4.10.2 *If $f \in W^1 L^\infty(Q^n)$, then both equalities*

$$\left| S_R^{\frac{n-1}{2}}(f;x) - f(x) \right| = o\left(\frac{1}{R}\right) \qquad (4.10.18)$$

and

$$\left| \widetilde{S}_R^{\frac{n-1}{2}}(f;x) - f^*(x) \right| = O\left(\frac{1}{R}\right) \qquad (4.10.19)$$

hold for a.e. $x \in Q^n$.

Proof. For any $\varepsilon > 0$, we denote the Steklov function of f by

$$f_\varepsilon(x) = \frac{1}{(2\varepsilon)^n} \int_{(-\varepsilon,\varepsilon)^n} f(x+y)dy.$$

It easily follows that

$$\frac{\partial f_\varepsilon(x)}{\partial x_j} = \frac{1}{(2\varepsilon)^n} \int_{(-\varepsilon,\varepsilon)^n} \frac{\partial}{\partial x_j} f(x+y)dy$$

$$= \frac{1}{(2\varepsilon)^n} \int_{(-\varepsilon,\varepsilon)^{n-1}} (f(x+\bar{y}+\varepsilon e_j) - f(x+\bar{y}-\varepsilon e_j))d\bar{y},$$

where

$$\bar{y} = (y_1, \ldots, y_{j-1}, 0, y_{j+1}, \ldots, y_n),$$
$$e_j = (0, \ldots, 0, 1, 0, \ldots, 0),$$

and

$$d\bar{y} = dy_1 \cdots dy_{j-1} dy_{j+1} \cdots dy_n.$$

Obviously, we have

$$\frac{\partial f_\varepsilon}{\partial x_j} \in W^1 L^\infty$$

for $j = 1, 2, \ldots, n$, and

$$\left\| \frac{\partial^2 f_\varepsilon}{\partial x_j \partial x_l} \right\|_\infty \leq \frac{1}{\varepsilon} \max \left\{ \left\| \frac{\partial f}{\partial x_j} \right\|_\infty : j = 1, 2, \ldots, n \right\}.$$

4.10 Approximation of continuous functions

It evidently follows from Theorem 4.10.1 that

$$R\left\|S_R^{\frac{n-1}{2}}(f_\varepsilon) - f_\varepsilon\right\|_c = O\left(\frac{\log R}{R}\right). \tag{4.10.20}$$

We set $g = f - f_\varepsilon$ and define the operator γ by

$$\gamma(f)(x) = \limsup_{R \to \infty} R\left|S_R^{\frac{n-1}{2}}(f;x) - f(x)\right|.$$

By the inequality

$$R\left|S_R^{\frac{n-1}{2}}(f;x) - f(x)\right| \leq R\left|S_R^{\frac{n-1}{2}}(g;x) - g(x)\right| + R\left|S_R^{\frac{n-1}{2}}(f_\varepsilon;x) - f_\varepsilon(x)\right|$$

and (4.10.20), we have

$$\gamma(f)(x) \leq \gamma(g)(x).$$

By the fact $g \in W^1L^\infty$ and conclusion (II) in Lemma 8.1, we obtain

$$\gamma(g)(x) \leq A \sum_{j=1}^{n} TP_j(g_j)(x)$$

for a.e. $x \in Q^n$.

Using the estimate about the type of the operator TP in Section 4.5, we have

$$\|TP_j(g_j)\|_2 \leq A\|g_j\|_2.$$

Since

$$g_j(x) = \frac{\partial f(x)}{\partial x_j} - \frac{\partial f_\varepsilon(x)}{\partial x_j} \frac{1}{(2\varepsilon)^n} \int_{(-\varepsilon,\varepsilon)^n} \left(\frac{\partial f(x)}{\partial x_j} - \frac{\partial f(x+y)}{\partial x_j}\right) dy,$$

we have

$$\|g_j\|_2 \leq \frac{1}{(2\varepsilon)^n} \int_{(-\varepsilon,\varepsilon)^n} \sqrt{\int_{Q^n} |f_j(x) - f_j(x+y)|^2 dx} \cdot dy$$

$$\leq \omega\left(f_j; \sqrt{n}\varepsilon\right)_{L^2(Q^n)}.$$

Finally, we get

$$\|\gamma(f)\|_2 \leq A \sum_{j=1}^{n} \omega\left(f_j; \sqrt{n}\varepsilon\right)_{L^2(Q^n)}.$$

Taking $\varepsilon \to 0^+$, we get $\gamma(f)(x) = 0$ for a.e. $x \in Q^n$. Thus, we get (4.10.18) proven. As (4.10.19) is the direct result from Lemma 4.10.2. Finally, we get the Theorem proven. ∎

The following Theorem is our main goal.

Theorem 4.10.3 *Suppose that $f \in C(Q^n)$ and $f \notin W^1 L^\infty$. Let*

$$\omega(\delta) = \omega(f; \delta)$$

and

$$\bar{\omega}(\delta) = \frac{\delta}{\omega(\delta)} \omega(1), \quad 0 < \delta \le 1.$$

We define $\delta_0 = 1$ and

$$\delta_{m+1} = \min\left\{\delta : \max\left(\frac{\omega(\delta)}{\omega(\delta_m)}, \frac{\bar{\omega}(\delta)}{\bar{\omega}(\delta_m)}\right) = \frac{1}{6}\right\}$$

and the function $\Omega(\delta)$ by

$$\Omega(\delta) = 6^{-m}, \delta \in (\delta_{m+1}, \delta_m],$$

for $m = 0, 1, 2, \ldots$. We also denote

$$\rho_R(f; x) = \max\left\{\left|S_R^{\frac{n-1}{2}}(f; x) - f(x)\right|, \left|\widetilde{S}_R^{\frac{n-1}{2}}(f; x) - f^*(x)\right|\right\}.$$

Then we have

(I) $\rho_R(f; x) \le C(x) \omega\left(\dfrac{1}{R}\right) \log\log \dfrac{9}{\Omega(1/R)}$, *for $R \ge 1$, where $C(x)$ is a.e. finite, and*

$$\left|\{x \in Q^n : C(x) > \lambda\}\right| \le A e^{-\frac{\lambda}{A}}$$

for $\lambda > 0$.

(II) $\limsup\limits_{R \to \infty} \rho_R(f; x) \left\{\omega\left(\dfrac{1}{R}\right) \log\log \dfrac{9}{\Omega(1/R)}\right\}^{-1} \le A$ *for a.e. $x \in Q^n$.*

Proof. Firstly, it follows from the fact $f \notin W^1 L^\infty$ that

$$\lim_{\delta \to 0^+} \omega(\delta) \delta^{-1} = +\infty. \tag{4.10.21}$$

4.10 Approximation of continuous functions

Thus the above definitions make sense. Set

$$Q_f = \left\{ x \in Q : \lim_{R \to \infty} \rho_R(f;x) = 0 \right\}.$$

It is easy to see that $|Q_f| = |Q|$.

For any $\lambda > 0$, we define

$$G_R(\lambda) = \left\{ x \in Q_f : \rho_R(f;x) > \lambda \omega\left(\frac{1}{R}\right) \log \log \frac{9}{\Omega(1/R)} \right\},$$

for $R \geq 1$, and

$$\Delta_m(\lambda) = \bigcup_{R \in [\delta_m^{-1}, \delta_{m+1}^{-1})} G_R(\lambda),$$

for $m = 0, 1, 2, \ldots$. Then, for any $m \in \mathbb{Z}_+$, one of the following two equations

$$\omega(\delta_{m+1}) = \frac{1}{6} \omega(\delta_m)$$

and

$$\frac{\delta_{m+1}}{\omega(\delta_{m+1})} = \frac{1}{6} \frac{\delta_m}{\omega(\delta_m)}$$

must holds at least.

We divide this problem into two cases.

Case 1. Let $\omega(\delta_{m+1}) = \frac{1}{6}\omega(\delta_m)$. We choose g as the optimal approximate trigonometric polynomial for f with the order δ_m^{-1}, and set $f - g = h$. Let $R \in [\delta_m^{-1}, \delta_{m+1}^{-1})$. Then we have

$$\|h\|_c = E_{\delta_m^{-1}}(f) \leq A\omega(\delta_m) \leq A\omega\left(\frac{1}{R}\right).$$

It obviously follows that

$$\rho_R(f;x) \leq \rho_R(g;x) + \rho_R(h;x)$$

and

$$\rho_R(h;x) \leq \|h\|_c \rho_R\left(\frac{h}{\|h\|_c}; x\right)$$

$$\leq A\omega\left(\frac{1}{R}\right)\left(S_*^{\frac{n-1}{2}}\left(\frac{h}{\|h\|_c}; x\right) + \tilde{S}_*^{\frac{n-1}{2}}\left(\frac{h}{\|h\|_c}; x\right)\right).$$

In addition, we have that

$$\left|S_R^{\frac{n-1}{2}}(g;x) - g(x)\right| \leq \left|S_R^{\frac{n-1}{2}}(g;x) - S_R^{\frac{n+1}{2}}(g;x)\right| + \left|S_R^{\frac{n+1}{2}}(g;x) - g(x)\right|$$

$$\leq \frac{1}{R^2}\left|S_R^{\frac{n-1}{2}}(-\Delta g;x)\right| + A\omega\left(g;\frac{1}{R}\right)$$

$$\leq \frac{1}{R^2}\|\Delta g\|_c S_*^{\frac{n-1}{2}}\left(\frac{\Delta g}{\|\Delta g\|_c};x\right) + A\omega\left(\frac{1}{R}\right).$$

Since

$$\|\Delta g\|_c \leq A\delta_m^{-2}\omega_2(g;\delta_m) \leq AR^2\omega_2\left(f;\frac{1}{R}\right),$$

we have

$$\left|S_R^{\frac{n-1}{2}}(g;x) - g(x)\right| \leq A\omega\left(\frac{1}{R}\right)\left(S_*^{\frac{n-1}{2}}\left(\frac{\Delta g}{\|\Delta g\|_c};x\right) + 1\right). \qquad (4.10.22)$$

For the conjugate case, similarly we get that

$$\left|\widetilde{S}_R^{\frac{n-1}{2}}(g;x) - g^*(x)\right| \leq A\omega\left(\frac{1}{R}\right)\widetilde{S}_*^{\frac{n-1}{2}}\left(\frac{\Delta g}{\|\Delta g\|_c};x\right)$$

$$+ \left|\widetilde{S}_R^{\frac{n-1}{2}}(g;x) - g^*(x)\right|.$$

We conclude that

$$\widetilde{S}_R^{\frac{n-1}{2}}(g;x) - g^*(x) = \frac{1}{|Q|}\int_0^\infty P(g)\left(x;\frac{t}{R}\right)t^{n+k-1}\left(\frac{E(n,1,k,t)}{C(n,k)} - \frac{1}{t^{n+k}}\right)dt$$

$$= \frac{1}{|Q|}\int_0^1 P(g)\left(x;\frac{t}{R}\right)t^{n+k-1}\left(\frac{E(n,1,k,t)}{C(n,k)} - \frac{1}{t^{n+k}}\right)dt$$

$$= \frac{1}{|Q|}\int_1^\infty P(g)\left(x;\frac{t}{R}\right)t^{n+k-1}\left(\frac{E(n,1,k,t)}{C(n,k)} - \frac{1}{t^{n+k}}\right)dt$$

$$:= I_1 + I_2. \qquad (4.10.23)$$

By choosing

$$M = \max\{\|g_j\| : j = 1,2,\ldots,n\},$$

we have that

$$|P(g)|\left(x;\frac{t}{R}\right) = \left|\int_{|\xi|=1}\left(g\left(x - \frac{t}{R}\xi\right) - g(x)\right)P(\xi)d\sigma(\xi)\right|$$

$$\leq AM\frac{t}{R}.$$

4.10 Approximation of continuous functions

Applying the condition $\omega(\delta_m) = 6\omega(\delta_{m+1})$, we obtain

$$|I_1| \leq A\frac{1}{R}M \leq A\frac{1}{R}\delta_m^{-1}\omega(g;\delta_m) \leq A\omega\left(\frac{1}{R}\right). \qquad (4.10.24)$$

According to the expansion formula

$$\frac{E(n,1,k,t)}{C(n,k)} - \frac{1}{t^{n+k}} = O\left(\frac{1}{t^{n+k+2}}\right) + At^{-(n+k+1)}\cos\left(t - \frac{\pi}{2}(n+k+1)\right),$$

we have

$$I_2 = A\int_1^\infty P\left(g;\frac{t}{R}\right)t^{-2}\cos\left(t - \frac{\pi}{2}(n+k+1)\right)dt$$
$$+ \int_1^\infty P\left(g;\frac{t}{R}\right)O\left(\frac{1}{t^3}\right)dt$$
$$= J + \tau.$$

It easily follows that

$$\tau = \int_1^\infty P\left(g;\frac{t}{R}\right)O\left(\frac{1}{t^3}\right)dt = O\left(\omega(g;\frac{1}{R})\right),$$

and

$$|\tau| \leq A\omega\left(g;\frac{1}{R}\right) \leq A\omega(g;\delta_m) \leq 6A\omega(\delta_{m+1}) \leq A\omega\left(\frac{1}{R}\right). \qquad (4.10.25)$$

Here we denote by $\theta = \frac{\pi}{2}(n+k+1)$. By the property of cosine function, we have

$$J = A\frac{1}{2}\int_1^\infty \left(P(g)\left(x;\frac{t}{R}\right) - P(g)\left(x;\frac{t+\pi}{R}\right)\right)\frac{\cos(t-\theta)}{t^2}dt$$
$$+ A\frac{1}{2}\int_1^\infty P(g)\left(x;\frac{t+\pi}{R}\right)\left(\frac{1}{t^2} - \frac{1}{(t+\pi)^2}\right)\cos(t-\theta)dt$$
$$- A\frac{1}{2}\int_{1-\pi}^1 P(g)\left(x;\frac{t+\pi}{R}\right)\frac{\cos(t-\theta)}{(t+\pi)^2}dt,$$

which implies

$$|J| \leq A\omega\left(g;\frac{1}{R}\right) \leq A\omega\left(\frac{1}{R}\right). \qquad (4.10.26)$$

Combining (4.10.23), (4.10.24), (4.10.25) with (4.10.26), we get that

$$\left\|\widetilde{S}_R^{\frac{n+1}{2}}(g) - g^*\right\|_c \leq A\omega\left(\frac{1}{R}\right).$$

Consequently, we have

$$\left|\widetilde{S}_R^{\frac{n+1}{2}}(g;x) - g^*(x)\right| \le A\omega(\frac{1}{R})\left\{\widetilde{S}_*^{\frac{n+1}{2}}\left(\frac{\Delta g}{\|\Delta g\|_c};x\right) + 1\right\}. \quad (4.10.27)$$

Concisely, for any $h \in C(Q)$, we denote by $h/\|h\|_c = \bar{h}$. Due to (4.10.22) and (4.10.27), we have

$$\rho_R(g;x) \le A\omega\left(\frac{1}{R}\right)\left\{S_*^{\frac{n-1}{2}}(\overline{\Delta g};x) + \widetilde{S}_*^{\frac{n-1}{2}}(\overline{\Delta g};x) + 1\right\}.$$

Thus, if we write

$$S_* = S_*^{\frac{n-1}{2}} + \widetilde{S}_*^{\frac{n-1}{2}}$$

for short, then we have

$$\rho_R(f;x) \le A\omega\left(\frac{1}{R}\right)\left\{S_*(\bar{h};x) + S_*(\overline{\Delta g};x) + 1\right\}. \quad (4.10.28)$$

Because of the estimate of $\widetilde{S}_*^{\frac{n-1}{2}}$ in Section 4.5 and the estimate of $S_*^{\frac{n-1}{2}}$ in Section 3.4, we get the similar estimate of S_*. Using the Theorem 4.5.1, for any $f \in L^\infty(Q^n)$, we have that

$$\left|\{x \in Q : S_*(f)(x) > \lambda\log(m+2)\}\right| \le A(m+2)^{-\frac{\lambda}{A\|f\|_\infty}}. \quad (4.10.29)$$

Notice that when $R \in [\delta_m^{-1}, \delta_{m+1}^{-1})$, there holds $\Omega(\frac{1}{R}) = 6^{-m}$. Then we have

$$G_R(\lambda) \subset \left\{x \in Q_f : \rho_R(f;x) > \lambda\omega\left(\frac{1}{R}\right)\log(2+m)\right\} := E_m(\lambda),$$

which yields

$$\Delta_m(\lambda) \subset E_m(\lambda).$$

Then due to (8.29), we obtain

$$|\Delta_m(\lambda)| \le A(m+2)^{-\frac{\lambda}{A}}. \quad (4.10.30)$$

Case 2. If

$$\frac{\delta_{m+1}}{\omega(\delta_{m+1})} = \frac{1}{6}\frac{\delta_m}{\omega(\delta_m)},$$

by the property of the modulus of continuity, when $R \in [\delta_m^{-1}, \delta_{m+1}^{-1})$, we have

$$\frac{\omega(\delta_{m+1})}{\delta_{m+1}} = 6\frac{\omega(\delta_m)}{\delta_m} \le 12R\omega\left(\frac{1}{R}\right). \quad (4.10.31)$$

4.10 Approximation of continuous functions

Now we let $g(x)$ be the best approximate trigonometric polynomial for $f(x)$ with the order δ_{m+1}^{-1}, and set $h = f - g$. Similar to Case 1, when $x \in Q_f$, we have

$$\left|S_R^{\frac{n-1}{2}}(f;x) - f(x)\right| \leq A\omega\left(\frac{1}{R}\right) S_*^{\frac{n-1}{2}}(\bar{h};x) + \left|S_R^{\frac{n-1}{2}}(g;x) - g(x)\right|.$$

Applying Lemma 10.1, we get

$$\left|S_R^{\frac{n-1}{2}}(g;x) - g(x)\right| \leq \frac{A}{R}\left\{\sum_{j=1}^{n} TP_j(g_j)(x) + \theta_R(g)(x)\right\},$$

where θ_R satisfies

$$|\theta_R(g)(x)| \leq M \leq A\delta_{m+1}^{-1}\omega(\delta_{m+1}) \leq AR\omega\left(\frac{1}{R}\right).$$

On the other hand, we have

$$TP_j(g_j)(x) = \|g_j\|_c TP_j(\bar{g}_j)(x)$$
$$\leq AR\omega\left(\frac{1}{R}\right) TP_j(\bar{g}_j)(x).$$

Hence we can get

$$\left|S_R^{\frac{n-1}{2}}(f;x) - f(x)\right| \leq A\omega\left(\frac{1}{R}\right)\left\{S_*^{\frac{n-1}{2}}(\bar{h};x) + \sum_{j=1}^{n} TP_j(\bar{g}_j)(x) + 1\right\}. \tag{4.10.32}$$

Similarly, when $x \in Q_f$, we have

$$\left|\widetilde{S}_R^{\frac{n-1}{2}}(f;x) - f^*(x)\right| \leq A\omega\left(\frac{1}{R}\right) \widetilde{S}_*^{\frac{n-1}{2}}(\bar{h};x) + \left|\widetilde{S}_R^{\frac{n-1}{2}}(g;x) - g^*(x)\right|.$$

Due to Lemma 10.2, we have

$$\left|\widetilde{S}_R^{\frac{n-1}{2}}(g;x) - g^*(x)\right| \leq \frac{A}{R}\left\{\sum_{j=1}^{n}\sum_{l=0}^{\nu_k} TP_{j,k+1-2^l}(g_j)(x) + \widetilde{\theta}_R(g(x))\right\}$$
$$\leq A\frac{1}{R}m\left\{\sum_{j=1}^{n}\sum_{l=0}^{\nu_k} TP_{j,k+1-2^l}(\bar{g}_j)(x) + 1\right\},$$

where $M = \max\{\|g_j\|_\infty : j = 1, 2, \ldots, n\}$. It easily implies that
$$M \leq A\delta_{m+1}^{-1}\omega(\delta_{m+1}) \leq AR\omega\left(\frac{1}{R}\right).$$

Consequently, we have

$$\left|\widetilde{S}_R^{\frac{n-1}{2}}(f;x) - f^*(x)\right|$$

$$\leq A\omega\left(\frac{1}{R}\right)\left(\widetilde{S}_*^{\frac{n-1}{2}}(\overline{h};x) + \sum_{j=1}^{n}\sum_{l=0}^{\nu_k} TP_{j,k+1-2^l}(\overline{(g_j)})(x)) + 1\right). \quad (4.10.33)$$

Combining (4.10.32) with (4.10.33) and applying the estimate on the type of the operator TP, similar to Case 1, we have that the inequality (4.10.33) in Case 2 that still holds.

Thus we have
$$\left|\bigcup_{m=0}^{\infty}\Delta_m(\lambda)\right| \leq A\sum_{m=0}^{\infty}(m+2)^{-\frac{\lambda}{A}}.$$

If $\lambda \geq 3A$, by the estimate
$$\int_2^\infty x^{-\frac{\lambda}{A}}dx = \left(\frac{\lambda}{A}-1\right)^{-1} 2^{-\frac{\lambda}{A}+1} < e^{-\frac{\lambda}{A}\log 2},$$

we have
$$\sum_{m=0}^{\infty}(m+2)^{-\frac{\lambda}{A}} \leq e^{-\frac{\lambda}{A}},$$

which implies
$$\left|\bigcup_{m=0}^{\infty}\Delta_m(\lambda)\right| \leq Ae^{-\frac{\lambda}{A}} \quad (4.10.34)$$

for any $\lambda > 0$, where $A = A_n(P)$.

We define
$$C(x) = \sup_{R\geq 1}\rho_R(f;x)\left(\omega\left(\frac{1}{R}\right)\log\log\frac{9}{\Omega(1/R)}\right)^{-1}.$$

Then for any $\lambda > 0$, it is easy to have that
$$\{x \in Q_f : C(x) > \lambda\} \subset \bigcup_{m=0}^{\infty}\Delta_m(\lambda).$$

4.10 Approximation of continuous functions

Hence, we have

$$|\{x \in Q_f : C(x) > \lambda\}| = |\{x \in Q : C(x) > \lambda\}| \leq A e^{-\frac{\lambda}{A}}.$$

Thus we have obtained the conclusion (I).

Denote the constant $A = A_n(P)$ in (4.10.30) by A_0 and let $\lambda_0 = 2A_0$. We have

$$\left\{x \in Q_f : \limsup_{R \to \infty} \rho_R(f;x) \left\{\omega\left(\frac{1}{R}\right) \log\log \frac{9}{\Omega(1/R)}\right\} > \lambda_0\right\} \subset \limsup_{m \to \infty} \Delta_m(\lambda_0).$$

From the estimate

$$\left|\limsup_{m \to \infty} \Delta_m(\lambda_0)\right| = \lim_{m \to \infty} \left|\bigcup_{j=m}^{\infty} \Delta_m(\lambda_0)\right|$$

$$\leq \lim_{m \to \infty} \sum_{j=m}^{\infty} A_0(j+2)^{-2}$$

$$= 0,$$

it follows that

$$\limsup_{R \to \infty} \rho_R(f;x) \left\{\omega\left(\frac{1}{R}\right) \log\log \frac{9}{\Omega(1/R)}\right\}^{-1} \leq 2A_0$$

for a.e. $x \in Q^n$. Thus we get the conclusion (II). ∎

As a direct consequence of Theorem 4.10.3, we have the next theorem.

Theorem 4.10.4 *If $f \in Lip\alpha$, $0 < \alpha < 1$, that is,*

$$\omega(f;t) = O(t^\alpha),$$

then we have

$$\left|S_R^{\frac{n-1}{2}}(f;x) - f(x)\right| + \left|\widetilde{S}_R^{\frac{n-1}{2}}(f;x) - f^*(x)\right| \leq C(x) R^{-\alpha} \log\log R,$$

for $R \geq 9$, where $C(x)$ is finite for a.e. $x \in Q^n$ and satisfies

$$|\{x \in Q : C(x) > \lambda\}| < A e^{-\frac{\lambda}{A}}$$

for any $\lambda > 0$, and

$$\limsup_{R \to \infty} \rho_R(f;x) R^\alpha (\log\log R)^{-1} \leq A$$

for a.e. $x \in Q^n$.

By similar discussions, we easily get the following theorem which is described by the modulus of continuity of second order. Here we omit the proof.

Theorem 4.10.5 *Suppose that $f \in C(Q)$ satisfies*

$$\lim_{\delta \to 0+} \frac{\omega_2(f;\delta)}{\delta^2} = +\infty.$$

If we replace w in Theorem 4.10.3 by ω_2, then the conclusion is still true.

Bibliography

[ABKP1] Alvarez, J., Bagby, R. J., Kurtz, D. S., and Pérez, C. Weighted estimates for commutators of linear operators, *Studia Math*, **104** (1993), 195–209.

[Ba1] Bateman, H. *Higher Transcendental Function, Vol. 1*, McGraw-Hill Company, Inc., 1953.

[Ba2] Bateman, H. *Tables of Integral Transforms, Vol. 2*, McGraw-Hill Company, Inc., 1954.

[Bo1] Bochner, S. and Summation of multiple Fourier series by spherial means, *Trans. Amer. Math. Soc.*, **40** (1936), 175-207.

[BoC1] Bochner, S., and Chandransekharan, K. On the localization property for multiple Fourier series, *Ann. of Math.*, **49** (1948), 966-978.

[Bou1] Bourgain, J. Besicovitch type maximal operators and applications to Fourier analysis, *Geom. Funct. Anal.*, **1** (1991), 147-187.

[BG] Bourgain J, and Guth, L. Bounds on oscillatory integral operators based on multilinear estimates, *Geom. Funct. Anal.*, **21** (2011), 1239-1295.

[CZ1] Calderón, A. P., and Zygmund, A. On the existence of certain singular integrals, *Acta Math.*, **88** (1952), 85-139.

[CZ2] Calderón, A. P., and Zygmund, A. Singular integrals and periodic functions, *Studia Math.*, **14** (1954), 249-271.

[Ca1] Carbery, A. The boundedness of the maximal Bochner-Riesz oprerator on $L^4(\mathbb{R}^2)$, *Duke Math. J.*, **50** (1983), 409-416.

[CRV1]   Carbery, A., Rubio de Francia, J. L., and Vega, L. Almost everywhere summability of Fourier integrals, *J. London Math. Soc.*, **38** (1988), 513-524.

[CSo1]   Carbery, A., and Soria, F. Almost everywhere convergence of Fourier integrals for functions in Sobolev spaces and a localization principle, *Rev. Mat. Iberoamericana*, **4** (1988), 319-337.

[CS1]   Carleson, L., and Sjolin, P. Oscillatory integrals and multiplier problem for the disc, *Studia Math.*, **44** (1972), 287-299.

[CM1]   Chandrasekharan, K., and Minakshisundaram, S. *Typical Means*, Oxford University Press, 1952.

[Cha1]   Chang, C. P. On certain exponential sums arising in conjugate multiple Fourier series, Ph. D. dissertation, University of Chicago, Chicago, 1964.

[Cha2]   Chang, C. P. Bochner-Riesz summability almost everywhere of multiple Fourier series, *Studia Math.*, **26** (1965), 25-66.

[Chan1] Chang, Y. C. Two theorems on spherical convergence and spherical Riesz summation of multiple Fourier series (in Chinese), *J. Beijing Normal Univ.* (Natural Science), **3** (1979), 50-65.

[Che1]   Cheng, M. T. Uniqueness of multiple trigonometric series, *Ann. Math.*, **52** (1950), 403-416.

[CheC1] Cheng, M. T., and Chen, Y. H. The approximation of functions of several variables by trigonometric polynomials (in Chinese), *J. Peking Univ.* (Natural Science), **4** (1956), 411-425.

[Chr1]   Christ, M. On almost everywhere convergenee of Bochner-Riesz means in higher dimensions, *Proc. Amer. Math. Soc.*, **95** (1985), 16-20.

[CRW1]   Coifman, R. R. Rochberg, R., and Weiss, G., Fractorization theorems for Hardy spaces in several variables, *Ann. of Math.*, **103** (1976), 611-635.

[Co1]   Cordoba, A. The Kakeya maximal function and the spherical summation multiplier, *Amer. J. Math.*, **99** (1977), 1-22.

[Co2]   Cordoba, A. A note on Bochner-Riesz operators, *Duke Math. J.*, **46** (1979), 505-511.

[CoL1]   Cordoba, A. and Lopez-Melero, B. Spherial summation: A problem of E. M. Stein, *Ann. Inst. Fourier*, **31** (1982), 147-152.

[CLWY1] Cui X. N., Liu X., Wang R., and Yan D.Y., A characterization of multidimensional multi-knot piecewise linear spectral sequence and its applications, *Acta Mathematica Sinica*, DOI: 10.1007/s10114-013-2099-y, 2013.

[DC1]   Davis, K. M., and Chang, Y. C. *Lectures on Bochner-Riesz Means*, London Mathematical Society Lecture Notes Series, **114**, Cambridge Univ. Press, 1987.

[Fa1]   Fan, D. S. Salem type theorem of Fourier series on compact Lie groups (in Chinese), *Ann. of Math.* (China), **6A** (1985), 397-410.

[Fe1]   Fefferman, C. Inequalities for strongly singular convolution operators, *Acta Math.*, **124** (1970), 9-36.

[Fe2]   Fefferman, C. Pointwise convergence of Fourier series, *Ann. of Math.*, **98** (1973), 551-571.

[Fe3]   Fefferman, C. The multiplier problem for the ball, *Ann. of Math.*, **94** (1972), 330-336.

[Fe4]   Fefferman, C. A note on spherical summation mulipliers, *Israeli J. Math.*, **15** (1973) 44-52.

[Fef1] Fefferman, R. A. A theory of entropy in Fourier analysis, *Adv. in Math.*, **30** (1978), 171-201.

[GR1]   Garcia-Cueva, J., and Rubiao de Francia, J. L. *Weighted norm inequalities and related topics*, North-Holland, 1985.

[Go1]   Golubov, L. D. On the convergence of spherical Riesz means of multiple Fourier series (in Russian), *Mat. Sb.*, **96** (1975), 189-211.

[Go2]   Golubov, L. D. On the summability of spherical Riesz means of Fourier integrals (in Russian), *Mat. Sb.*, **104** (1977), 577-596.

[He1]   Herz, C. On the mean inversion of Fourier and Hankle transforms, *Pro. Nat. Acad. Sci.* USA, **40** (1954), 996-999.

[HL1]   Hu, G. E., and Lu, S. Z. The commutator of the Bochner-Riesz operator, *Tohoku Math. J.*, **48** (1996), 259-266.

[HL2]   Hu G. E., and Lu S. Z. A weighted L^2 estimate for the commutator of the Bochner-Riesz operator, *Proc. Amer. Math. Soc.*, **125** (1997) 2867–2873.

[Ig1]   Igari, S. Decomposition theorem and lacunary convergence of Riesz-Bocher means of the Fourier transforms of two variables, *Tohoku Math. J.*, **33** (1981), 413-419.

[J1]   Jiang, X. Some problems on approximation, convergence and saturation in Fourier analysis of several variables (in Chinese), *Master Dissertation*, Beijing Normal University, 1985.

[J2]   Jiang, X. On Approximations of periodic functions by spherical linear means of multiple Fourier series (in Chinese), *J. Beijing Normal Univ. (Natural Science)*, **21** (1985), 1-12.

[KS1]   Kaczmarz, S., and Steinhaus, H. *Orthogonal series theory* (in Russian), Moscow, 1958.

[Le1]   Lee, S. Improved bounds for Bochner-Riesz and maximal Bochner-Riesz operators, *Duke Math. J.*, **122** (2004), 205-232.

[Li1]   Lippman, G. E. Spherial summability of conjugate multiple Fourier series and integrals at the critical index, *SIAM J. Math. Anal.*, **4** (1975), 681-695.

[Lu1]   Lu, S. Z. Convergence of spherical Riesz means of multiple Fourier series (in Chinese), *Acta Math. Sinica*, **23** (1980), 385-397.

[Lu2]   Lu, S. Z. Spherical integrals and convergence of spherical Riesz means (in Chinese), *Acta Math. Sinica*, **23** (1980), 609-623.

[Lu3]   Lu, S. Z. Spherical Riesz means of conjugate multiple Fourier series, *Scientia Sinica*, **24** (1981), 595-605.

[Lu4]   Lu, S. Z. (C,1) Summability of multiple conjugate Fourier integrals, *Chinese Science Bulletin*, **26** (1981), 200-204.

[Lu5]   Lu, S. Z. Spherical summability of Fourier series of bounded variation functions in several variables, *Chinese Science Bulletin*, **26** (1981), 774-778.

[Lu6]   Lu, S. Z. A note on the almost everywhere convergence of Bochner-Riesz means of multiple conjugate Fourier series, *Lecture Notes in Math.*, **908** (1982), 319-325.

[Lu7]   Lu, S. Z. Strong summability of Bochner-Riesz spherical means, *Scientia Sinica (Ser.A)*, **28** (1985), 1-13.

[Lu8]   Lu, S. Z. A class of maximal functions on spherical Riesz means, *Scientia Sinica (Ser.A)*, **29** (1986), 113-124.

[LDY]   Lu, S. Z., Ding, Y., and Yan, D. Y. *Singular Integrals and Related Topics*, World Scientific Publishing Co. Pte. Ltd, 2007.

[LTW1]   Lu, S. Z., Taibleson, M. H., and Weiss, G. On the almost everywhere convergence of Bochner-Riesz means of muliple Fourier series, *Lecture Notes in Math.*, **908** (1982), 311-318.

[LTW2]   Lu, S. Z., Taibleson, M. H., and Weiss, G. *Spaces generated by blocks*, Beijing Normal University Press, 1989.

[LW1]   Lu, S. Z., and Wang, S. M. Spaces generated by smooth blocks, *Constructive Approx.*, **8** (1992), 331-341.

[LX1]   Lu, S. Z., and Xia, X. A note on commutators of Bochner-Riesz operator, *Front. Math. China*, **2** (2007), 439-446.

[MLL1]   Ma, B. L., Liu, H. P., and Lu, S. Z. Norm inequality with power weights for a class of maximal spherical summation operators (in Chinese), *J. Beijing Normal Univ. (Natural Science)*, **2** (1989), 1-4.

[MTW1]   Meyer, Y., Taibleson, M. H., Weiss, G. Some functional analytic properties of the spaces B_q generated by blocks, *Ind. Univ. Math. J.*, **34** (1985), 493-515.

[O1]   Oskolkov, K. E. Rate estimates of approximations of continuous functions and their conjugates by Fourier sum on set of total measure (in Russian), *Izv. Akad. Nauk USSR Math. Ser.*, **38** (1974), 1393-1407.

[P1]   Pan, W. J. On localization and convergence of spherical sum of multiple Fourier integrals (in Chinese), *Scientia Sinica (Ser.A)*, **11** (1981), 1310-1321.

[S1]   Saks, S. Remark on the differentiability of the Lebesgue indefinite integral, *Fund. Math.*, **22** (1934), 257-261.

[Sa1]   Sato, S. Entropy and Almost everywhere convergence of Fourier series, *Tohoku Math. J.*, **33** (1981), 593-597.

[Sh1] Shi, X. L. Function families ΛBMV and its applications to Fourier series (in Chinese), *Scientia Sinica (Ser. A)*, **9** (1984), 593-597.

[ShS1] Shi, X. L., and Sun, Q. Y. Weighted norm inequalities for Bochner-Riesz operators and singular integral operators. *Proc. Amer. Math. Soc.*, **116** (1992), 665-673.

[Sj1] Sjolin, P. Convergence almost everywhere of certain singular integrals and multiple Fourier series, *Ark. for Matematik*, **91** (1971), 65-90.

[S1] Sogge, C. D. *Fourier Integrals in Classical Analysis*, Cambridge Tracts in Math., No. 105, Cambridge Univ. Press, 1993.

[St1] Stein, E. M. Localization and summability of multiple Fourier series, *Acta Math.*, **100** (1958), 93-147.

[St2] Stein, E. M. On certain exponential sums arising in multiple Fourier series, *Ann. of Math.*, **73** (1961), 87-109.

[St3] Stein, E. M. On limits of sequences of operators, *Ann. of Math.*, **74** (1961), 140-170.

[St4] Stein, E. M. *Singular intergrals and differentiability properties of functions*, Princetion University Press, 1970.

[St5] Stein, E. M. *Harmonic analysis: real-variable methods, orthogonality and oscillatory integrals*, Princeton University Press, 1993.

[SW1] Stein, E. M., and Weiss, G. *Introduction to Fourier analysis on Euclidean spaces*, Princeton University Press, 1971.

[SWe1] Stein, E, M., and Weiss, N. J. On the convergence of Poisson integrtals, *Trans. A. M. S.*, **140** (1969), 35-54.

[TW1] Taibleson, M. H., and Weiss, G. Certain function spaces connected with almost everywhere convergence of Fourier series, *Proceeding of the conference on Harmonic analysis in Honor of A.Zygmund*, Wadsworth Publ. Co. Vol. I (1980), 95-113.

[Ta1] Tao, T. The Bochner-Riesz conjecture implies the restriction conjecture, *Duke Math.J.*, **96** (1999), 363-375.

[Ta2] Tao, T. On the maximal Bochner-Riesz conjecture in the plane, for $p < 2$, *Trans. Amer. Math. Soc.*, **354** (2002), 1947-1959.

[TVV1]   Tao, T., Vargas, A., and Vega, L. A bilinear approach to the restriction and Kakeya conjectures, *J. Amer. Math. Soc.*, **11** (1998), 967-1000.

[To1]   Tomas, P. Restriction theorems for the the Fourier transform, Harmonic Analysis in Euclidean space, *Proc. Sympos. Pure Math., Part I, Amer, Math. Soc.*, **35** (1979), 111-114.

[Tr1]   Trigub, R. M. Summability of multiple Fourier series. Growth of Lebesgue constants, *Anal. Math.*, **6** (1980), 255-267.

[Wa1]   Wang, K. Y. On estimates of square remainder for multiple de la Vallee Poussin summation (in Chinese), *Ann. Math. (China)*, **3** (1982), 789-802.

[Wa2]   Wang, K. Y. (C,1) summation of multiple conjugate Fourier series (in Chinese), *Bull. of Science (China)*, **24** (1984), 1473-1477.

[Wa3]   Wang, K. Y. (H,q) summation of M-type for double Fourier series and its conjugate series (in Chinese), *Acta Math. Sinica*, **27** (1984), 811-816.

[Wa4]   Wang, K. Y. A note on Almost everywhere convergence of Walsh-Fourier series (in Chinese), *J. Beijing Normal Univ. (Natural Science)*, **1** (1985), 11-15.

[Wa5]   Wang, K. Y., Generalized spherical Riesz means of multiple Fourier series, *Adv. in Math. (China)*, **1** (1985), 77-80.

[Wa6]   Wang, K, Y. Approximation for contimuous periodic function of several variables and its conjugate function by Riesz means on set of total measure, *Approximation theory and its applications*, **1** (1985), 19-56.

[Wa7]   Wang, K. Y. Strong summability of spherial Riesz means of multiple conjugate Fourier series (in Chinese), *Acta Math.Sinica*, **29** (1986), 156-175.

[Wan1]   Wang, S. L. Convergence of spherical Riesz means of multiple Fourier series (in Chinese), *Acta Math. Sinica*, **28** (1985), 41-52.

[Wat1]   Watson, G. N. *Theory of Bessel functions*, Cambridge, 1952.

[We1]   Welland, G. N. Norm convergence of Riesz-Bochner means for radial function, *Can. J. Math.*, **27** (1975), 176-185.

[Wo1]　　Wolff, T. An improved bound for Kakeya type maximal functions, *Rev.Mat. Iberoamericana*, **11** (1995), 651-674.

[Wu1]　　Wu, Z. J. Bochner-Riesz means of multiple conjugate Fourier integrals (in Chinese), *J. Hangzhou Univ.*, **10** (1983), 436-447.

[XL1]　　Xia, X., and Lu, S. Z. Commutators of Bochner-Riesz operators for a class of radial functions, *Acta Math. Scientia*, **31** (2011), 477-482.

[Zy1]　　Zygmund, A. *Trigonometric series*, Vol. I and II, Cambridge, 1959.

Index

B_q, 211, 213, 214, 216, 217, 224–226, 229–232, 332, 371
BMO, 129, 130, 136–140
$L \log L$, 177
L^1, 211, 261, 280
L^∞, 226, 250, 261, 263, 264, 331
L^q, 225, 226

a.e. approximation, 347
Abel-Poisson means, 38, 39, 263, 265, 266
almost everywhere convergence, viii, 105, 194, 368, 370–373

Banach space, 1, 225
Bessel function, 61, 98, 100, 101, 131, 136, 145, 192, 242, 245, 295, 301, 304, 373
Bessel potential space, 232
block, viii, 208, 211, 213, 214, 216, 218, 219, 223–232, 241, 243, 333, 334, 371
Bochner-Riesz, 48, 59, 61, 334, 368–372, 374
Bochner-Riesz conjecture, 96, 113, 372
Bochner-Riesz means, vii, viii, 3, 39–41, 45–47, 105, 113, 128, 141, 146, 152, 166, 177, 201, 208, 231, 259, 276, 289, 294, 303, 307, 316, 319, 324, 332, 368, 370, 371, 374

Calderón-Zygmund, viii, 294, 309, 310, 333
Carleson-Sjölin theorem, viii, 61, 65, 72, 73, 78, 85, 96, 97, 109
Cauchy sequence, 7, 213
commutators, viii, 129, 130, 367, 371
conjugate Fourier integral, 293, 294, 303, 307, 342, 370
conjugate Fourier series, 309, 310, 316, 332, 337, 370

conjugate function, 312, 315, 320, 347, 350, 356, 373
critical index, vii, viii, 40, 128, 130, 141, 146, 166, 177, 187, 194, 201, 208, 231, 244, 307, 309, 319, 324, 336, 370

Dirichlet kernel, 2, 229
disc conjecture, 51
duality, 85

entropy, 208–211
equivalence, 323
Euler number, 182
exponential sums, 368, 372

Fefferman, 15, 51, 60, 89, 96, 208–210, 369
Fefferman theorem, viii, 51
Fejér mean, 2
Fourier series, 373
Fourier transform, 237, 263, 267, 294, 310, 370, 373
fractional integral operator, 139

Hölder's inequality, 114, 220, 282, 338
Hardy-Littlewood maximal function, 61, 107, 116, 137, 234
Hardy-Littlewood maximal operator, 144, 145, 259
Hardy-Littlewood-Sobolev theorem, 139
Hausdorff-Young's inequality, 92, 292, 346
Hilbert transform, 78, 103, 293

interpolation, 73, 98
interpolation of operators, 280

Kakeya maximal function, 78, 79, 85, 88, 96, 368

kernel, 3, 37, 39, 41, 62, 108, 122, 166, 231, 258, 264, 293, 302, 310, 312, 316, 324, 325, 331, 333, 355
Kolmogorov, 152

Lebesgue, 9, 15, 35–37, 39, 44–46, 52, 115, 166, 167, 169, 197, 198, 200, 201, 225, 237, 248, 249, 309, 331, 346, 349, 350, 356, 371, 373
Lebesgue constant, 166, 167, 248, 249, 331
Lebesgue point, 15, 35–37, 39, 198, 200, 201, 225, 350
Leibniz formula, 329
linear operator, 48, 81, 105, 107, 128, 129, 183, 236, 241, 260, 266, 268, 367
localization, 42, 45, 146, 165, 166, 334, 336, 337, 367, 368, 371

maximal function, 78, 79, 85, 88, 96, 232, 368, 371, 374
Minkowski's inequality, 35, 61, 346
multiple Fourier integral, vii, 41, 371
multiple Fourier series, vii, viii, 2, 3, 10, 141, 231, 367–370, 372, 373
multiplier, viii, 48–50, 52, 86, 116, 120, 236, 260, 368, 369

orthogonality, 372
oscillatory integral, 60–62, 66, 367, 368, 372

Plancherel theorem, 49, 91, 119, 120, 127, 238
Poisson summation formula, 10, 11, 147, 264, 267, 269, 318

quasi-norm, 213, 225

radial function, 32, 90, 92, 94, 97, 122, 267, 295, 374
rectangular partial sum, 2, 15
restriction conjecture, 372
restriction theorem, 78, 89, 92, 93, 96
Riesz potential, 67

saturation, 259–261, 265, 266, 269, 272, 275–277, 370
Schwartz function, 114, 123
singular integrals, 129
Sobolev, 45, 118, 120, 368
spherical partial sum, 2
spherical Riesz means, 369–371, 373
Stein, vii, 67, 78, 89, 95, 98, 109, 110, 112, 129, 138, 145, 147, 152, 154, 165, 177, 183, 186, 194, 196, 198, 207, 214, 237, 241, 245, 280, 294, 322, 324, 369, 372
strong summability, 371, 373
strong summation, 280, 337, 342, 347
strong type, 106, 190, 333

the the Fourier transform, 32
Trigonometric series, 374

weak type, 106–108, 129, 241, 294, 315, 333
Weiss, viii, 98, 129, 145, 154, 183, 208, 210, 213, 214, 226, 229, 230, 368, 371, 372

Zygmund, viii, 155, 293, 294, 309, 315, 367, 372